X24.80/N 333

PHYSICS OF THE SOLAR CORONA

ASTROPHYSICS AND SPACE SCIENCE LIBRARY

A SERIES OF BOOKS ON THE RECENT DEVELOPMENTS
OF SPACE SCIENCE AND OF GENERAL GEOPHYSICS AND ASTROPHYSICS
PUBLISHED IN CONNECTION WITH THE JOURNAL
SPACE SCIENCE REVIEWS

Editorial Board

J. E. BLAMONT, *Laboratoire d'Aéronomie, Verrières, France*

R. L. F. BOYD, *University College, London, England*

L. GOLDBERG, *Harvard College Observatory, Cambridge, Mass., U.S.A.*

C. DE JAGER, *University of Utrecht, Holland*

Z. KOPAL, *University of Manchester, Manchester, England*

G. H. LUDWIG, *NASA, Goddard Space Flight Center, Greenbelt, Md., U.S.A.*

R. LÜST, *Institut für Extraterrestrische Physik, Garsching-München, Germany*

B. M. MCCORMAC, *Lockheed Palo Alto Research Laboratory, Palo Alto, Calif., U.S.A.*

H. E. NEWELL, *NASA, Washington, D.C., U.S.A.*

L. I. SEDOV, *Academy of Sciences of the U.S.S.R., Moscow, U.S.S.R.*

Z. ŠVESTKA, *Czechoslovak Academy of Sciences, Ondřejov, Czechoslovakia*

Secretary of the Editorial Board

W. DE GRAAFF, *Sterrewacht 'Sonnenborgh', University of Utrecht, Utrecht, Holland*

VOLUME 27

PHYSICS OF THE SOLAR CORONA

PROCEEDINGS OF NATO ADVANCED STUDY INSTITUTE
ON PHYSICS OF THE SOLAR CORONA
HELD AT CAVOURI-VOULIAGMENI, ATHENS, GREECE
6–17 SEPTEMBER 1970

Edited by

CONSTANTIN J. MACRIS

*Research Center for Astronomy and Applied Mathematics,
Academy of Athens, Athens, Greece*

D. REIDEL PUBLISHING COMPANY

DORDRECHT-HOLLAND

Library of Congress Catalog Card Number 76–154741

ISBN 90 277 0204 7

All Rights Reserved
Copyright © 1971 by D. Reidel Publishing Company, Dordrecht, Holland
No part of this book may be reproduced in any form, by print, photoprint, microfilm,
or any other means, without written permission from the publisher

Printed in The Netherlands by D. Reidel, Dordrecht

PREFACE

The Advanced Study Institute on the 'Physics of the Solar Corona' was held in Greece, September 6–17, 1970 under the auspices of the Scientific Committee of the NATO and the Astronomical Institute of the National Observatory of Athens.

The subject of the Summer School concerned one of the most exciting regions of the solar atmosphere, the solar Corona. This outer part of the atmosphere of the Sun with the small density, high temperature, great extention, internal motions and the variety of physical conditions that prevail within it, poses difficult problems to solar Physics.

The great interest that the study of the Corona exhibits has led me to propose to the NATO Scientific Committee the organization of a School in Greece on the 'Physics of the Solar Corona'.

After the approval of the proposal, I wrote to the most eminent solar Physicists and asked them to lecture in the organized Advanced Study Institute. Most of them have graciously replied to my call and I consider it my duty to express my deepest thanks to them for the acceptance of the invitation.

The courses were attended by 50 students from 12 different countries. The Summer School took place in Cavouri-Vouliagmeni, located on the Saronic gulf, 20 kilometers south-east of Athens.

During the inaugural ceremony welcoming addresses were given by Dr C. J. Macris, Director of the Advanced Study Institute, by Professor J. Xanthakis, President of the Greek National Committee for Astronomy and by Mr O. Angelopoulos, President of the Board of Directors of the National Observatory. His Excellency the Undersecretary of State for National Education, Mr S. Dimitrakos, opened the Institute. On behalf of the NATO Scientific Committee, Professor P. Bourgeois, talked on 'New Trends in the Activities of the NATO Science Committee'.

At the beginning of the first session the introductory lecture was presented by Dr J. W. Evans and at the last session Professor L. Goldberg gave a summary of the topics discussed during the meetings. The various speakers, with their advanced level lectures gave a very precise picture of the present state of coronal research and at the same time many questions and new problems arose, such as the ones posed by Professor L. Goldberg at the end of the course. The answers to the new problems will greatly increase our knowledge of Solar Coronal Physics.

This volume contains all the papers presented at the Institute and I think it gives us a great part of our recent knowledge of the Corona. The prompt publication of the book by D. Reidel Publishing Company must be considered as another contribution. This is because we all know how our viewpoints change in a particular field due to the rate at which our knowledge is being enriched in our time.

I want to address my deep gratitude to the Scientific Committee of the NATO for the grant that made it possible to organize this Summer School, as well as to the Greek Government, whose moral and material support helped greatly to the success of the School.

I wish to extend my personal thanks to my Secretary, Mr A. Tsobanoglou and to my collaborators, Mr C. Alissandrakis and Miss H. Dara, for their important assistance before and during the Advanced Study Institute.

CONSTANTIN J. MACRIS

Assistant Professor, University of Athens
Director of the NATO Advanced Study
Institute on Physics of the Solar Corona

OPENING ADDRESS

New Trends in the Activities of the NATO Science Committee

It is with the greatest pleasure that I shall take part, as representative of the NATO Science Committee in this second Summer Course on the Sun, dedicated this time to the Physics of the Solar Corona and that I address to you on behalf of our Committee our best wishes of success.

The first Course, five years ago was an important one and I am convinced that it will be the same this time too.

Born twelve years ago from the necessity for NATO to develop the scientific and technical potentialities of the member Nations of the Alliance, the Science Committee is still expanding its programmes and preparing new ways.

Due to the Science Fellowships Programme each two years about one thousand young scientists are sent abroad to improve their research training.

The links provided by this programme between Universities and Institutes of member countries enhance the scientific potential of the Alliance.

The Advanced Study Institutes programme perhaps the most outstandingly successful of the Science Committee's programmes, is developing in such a way that this year 46 Advanced Study Institutes are sponsored.

The Research grants' programme through which collaboration is established on a large scale between laboratories and institutes of member countries was maintained in twelve years about 400 projects selected by the Advisory Panel were realized.

Programmes implementing a much larger cooperation were developed or created in Oceanography, Metereology, Radiometereology, Operational Research and in the problems of human factors.

The recommandations of the Armand Study Group contained in its report under the title 'Increasing the Effectiveness of Western Science' are now being applied in several countries with important results; at the same time the conclusions of the Killian report on the establishment of an International Institute of Science and Technology are being studied for action by an international group of scientists.

The aim of the Science Committee since its formation was to help the member countries of the Alliance develop their own scientific and technical programmes on a national basis enhancing in this way their potentialities in new fields.

At the beginning the Committee's policy was to aid preferentially the developing countries of the Alliance and in fact, this policy gave satisfactory results. Since several years, on the contrary, it seemed more realistic to support larger programmes asking for collaboration among more than two countries. It is to this connection that problems referring to materials' research, stress under tension, computer science, sea pollution, and to the realization of an Atlantic oceanographic platform were successfully studied. However the case of the International Institute of Science and Technology mentioned above had already shown that it was impossible for the Committee to take in charge many expensive programmes.

Therefore it was decided that in the future the Committee should understate only the preparatory study of such programmes and leave the responsibility of their own budgets.

These preparatory studies are carried out by small groups of experts and discussed later on in international conference of specialists sponsored by the NATO Science Committee.

Several such conferences were already held with great success and their proceedings published.

In conclusion, I wish to congratulate the organizers of this Advanced Study Institute of Physics of the Solar Corona for their efforts and especially for the first class group of specialists they have gathered here as lecturers; as far as I am concerned, I thank them heartily for their kind reception.

Best wishes for good and fruitful work.

P. BOURGEOIS
Representative of the Science Committee

TABLE OF CONTENTS

Preface	
by C. J. Macris	V
Opening Address	
by P. Bourgeois	VII
Contributing Authors	XI
1. Introduction to Research on the Solar Corona	
by J. W. Evans	1
2. Atomic Processes in the Solar Corona	
by G. Noci	13
3. Magnetohydrodynamics and Plasma Physics of the Solar Corona	
by F. Meyer	29
4. The Chromosphere-Corona Transition Region	
by R. G. Athay	36
5. Coronal Magnetic Fields	
by G. Newkirk, Jr.	66
6. The Extended Coronal Magnetic Field	
by J. M. Wilcox	88
7. Investigations on Coronal Monochromatic Emissions in the Optical Range	
by A. Dollfus	97
8. Coronal Events Observed in 5303 Å	
by R. B. Dunn	114
9. The Solar Corona in the Eleven-Year Cycle	
by M. Waldmeier	130
10. Coronal Active Regions and Flare-Associated Events	
by J. B. Zirker	140
11. Observational Effects of Flare-Associated Waves	
by S. F. Smith and K. L. Harvey	156
12. The Surges	
by C. J. Macris	168
13. Relations between the Areas Index and Different Phenomena in the Chromosphere, the Corona and the Interplanetary Space	
by J. Xanthakis	179
14. Models of the Quiet and Active Solar Atmosphere from Harvard OSO Data	
by R. W. Noyes	192
15. The Determination of Chromospheric-Coronal Structure from Solar XUV Observations	
by C. Jordan and R. Wilson	219

16. X-Ray Spectroscopy of Solar Active Regions and Flares
    by W. M. Neupert — 237
17. A Note on a Recent Identification of the Solar Flare 1.9 Å Line Feature
    by K. J. H. Phillips — 254
18. Recent Investigations about Solar X-Rays Emitted by the Solar Corona by Means of SOLRAD Satellites
    by M. Landini and B. C. Monsignori Fossi — 257
19. Radio Emission of the Quiet Sun
    by M. Felli and G. Tofani — 267
20. Solar Bursts at Decameter and Hectometer Wavelengths
    by M. R. Kundu — 287
21. Models of the Solar Transition Region Chromosphere-Corona
    by G. Noci — 308
22. Studies of the Outer Corona through Space Radio Astronomy
    by M. D. Papagiannis — 317
23. Summary
    by L. Goldberg — 333
Index of Names — 342

CONTRIBUTING AUTHORS

1. R. G. Athay, High Altitude Observ., P.O. Box 1558, Boulder, Colo. 80302, U.S.A.
2. A. Dollfus, Observatoire de Meudon, Meudon-92 (Hauts de Seine), France.
3. R. B. Dunn, Sacramento Peak Observatory, Air Force Cambridge Research Laboratories, Sunspot, N. M. 88349, U.S.A.
4. J. W. Evans, Director, Sacramento Peak Observatory, Air Force Cambridge Research Laboratories, Sunspot, N. M. 88349, U.S.A.
5. M. Felli, Osservatorio Astrofisico di Arcetri, Largo Enrico Fermi 5, Arcetri, Firenze, Italy.
6. L. Goldberg, Harvard College Observatory, 60 Garden Street, Cambridge, Mass. 02138, U.S.A.
7. Carole Jordan, Spectroscopy Division, Culham Laboratory, Abingdon, Berkshire, England.
8. M. R. Kundu, University of Maryland, Department of Physics and Astronomy, College Park, Md 20742, U.S.A.
9. M. Landini, Osservatorio Astrofisico di Arcetri, Largo Enrico Fermi 5, Arcetri, Firenze, Italy.
10. C. J. Macris, Director, Research Center for Astronomy and Applied Mathematics, Academy of Athens, 14 Anagostopoulou Street, Athens (136), Greece.
11. F. Meyer, Max-Planck-Institut für Physik und Astrophysik, Föhringer Ring 6, 8 München 23, G.F.R.
12. B. C. Monsignori-Fossi, Osservatorio Astrofisico di Arcetri, Largo Enrico Fermi 5, Arcetri, Firenze, Italy.
13. W. Neupert, Goddard Space Flight Center, Greenbelt, Md., U.S.A.
14. G. Newkirk, Director, High Altitude Observatory, P.O. Box 1558, Boulder, Colo. 80302, U.S.A.
15. G. Noci, Osservatorio Astrofisico di Arcetri, Largo Enrico Fermi 5, Arcetri, Firenze, Italy.
16. R. W. Noyes, Smithsonian Astrophysical Observatory and Harvard College Observatory, 60 Garden Street, Cambridge, Mass. 02138, U.S.A.
17. M. Papagiannis, Department of Astronomy, Boston University, 725 Commonwealth Ave., Boston, Mass. 02215, U.S.A.
18. K. J. H. Phillips, Mullard Space Science Laboratory, Holmbury Street, Mary Dorking, Surrey, England.
19. Sara F. Smith, Lockheed Observatory, Rye Canyon, P.O. Box 551, Burbank, Calif. 91503, U.S.A.
20. G. Tofani, Osservatorio Astrofisico di Arcetri, Largo Enrico Fermi 5, Arcetri, Firenze, Italy.

21. M. Waldmeier, Director, Swiss Federal Observatory, Schmelzbergstrasse 25, Zürich 6, Switzerland.
22. J. Wilcox, Space Science Laboratory, University of California, Berkeley, Calif. U.S.A.
23. R. Wilson, Spectroscopy Division, Culham Laboratory, Abingdon, Berkshire, England.
24. J. Xanthakis, Research Center for Astronomy and Applied Mathematics, Academy of Athens, 14 Anagnostopoulou Street, Athens, Greece.
25. J. B. Zirker, University of Hawaii, Institute for Astronomy, 2525 Correa Road, Honolulu 96822, Hawaii, U.S.A.

1. INTRODUCTION TO RESEARCH ON THE SOLAR CORONA

JOHN W. EVANS

*Sacramento Peak Observatory of the Air Force Cambridge Research Laboratories,
Sunspot, N.M., U.S.A.*

1. Introduction

The purpose of this introduction is to provide a review of the past milestones in the development of coronal science, and to point out some of the immediate current problems. It is intended as an aid to the non-specialist in placing the following detailed discussions in their proper context. This published version of the introductory talk at the Advanced Study Institute has been modified in several places to correct errors pointed out by my colleagues at the time and to remove remarks that were meaningful only in the context of the time and place of the lecture.

It is surprising that the corona, one of the most striking spectracles of nature, appears to have gone practically unnoticed through all recorded history up to 1842. The eclipse of that year crossed southern Europe and was seen by many astronomers, who for some reason chose that year to become interested. Since then, most accessible total eclipses have been observed and the corona became the object of serious scientific research.

The observational description of the corona advanced rapidly with the application of photography, spectroscopy and polarimetry, but by the early 1930's the physical interpretation of the observations was in a state of coma. Given the entirely reasonable thermodynamic and spectroscopic premises of the time, it was clear that the corona was physically impossible. There it stood, a warm gas at no more than 6000°, without any visible means of support. It was extremely thin, since comets passed through it undamaged. Its spectrum consisted of a continuum broken only by an assembly of emission lines, not one of which could be identified. The otherwise undiscovered element coronium had gone down the drain because there was no slot for it in the periodic table. Coronal science was in a very fascinating state of confusion.

The break came in 1939 to 1942. Grotrian noticed that two coronal lines of similar behavior at 6374 and 7892 A matched the wavelengths found by Edlén for forbidden lines of Fe X and Fe XI. Edlén then proceeded to identify 17 other coronal lines, all due to forbidden transitions in highly ionized atoms, mostly iron, nickel and calcium. The presence of these ions unequivocally required a high temperature of several hundred thousand degrees or more. The only escape from this surprising conclusion would involve a flat denial of Edlén's identifications.

Acceptance of the high temperature of the corona took us over the divide and made the observed corona a believable physical entity. It could now stand on its own feet by hydrostatic pressure. The thorough ionization of hydrogen and helium provided an abundance of free electrons to scatter photospheric white light, and smear out the Fraunhofer lines in the process. The extraordinary widths of the coronal lines were simply explained as thermal Doppler broadening. Thus the major observed characteristics of the corona fell into place. If we closed our eyes to the questions of how it got there and why it is so hot, we could now in all good conscience believe that it exists and conforms to the physical laws of nature. This was the great step ahead in coronal science. But on opening our eyes we were confronted with the one overriding fundamental problem of hot material. Where does the material come from and how is it heated? This question was recognized in the 1940's as a formidable one, and it caused some short-lived foot dragging before acceptance of the high temperature became wholehearted and universal. However, the favorable arguments proved irresistible, and were considerably reinforced by the measurements of the meter wave solar radio flux, which confirmed the high temperature. By 1950 no serious doubts remained. The problem of material and energy, which is really the problem of how the corona can exist at all, became the ultimate target of coronal research, standing isolated above all other questions in its basic importance to our understanding of the corona. Partial solutions have been proposed, but none are fully satisfactory. If anything, the problem has become more difficult since 1950. We now know, as we did not then, that the corona loses material and energy at such a prodigious rate that the whole instantaneous coronal supply of both must be completely replaced every few hours.

Most modern research on the corona is not an explicit attempt to solve the hot material problem directly, but everything we can learn about the corona must surely contribute to the ultimate solution. We are busily engaged in elaborating a quantitative picture of the interplay of temperature, density, magnetic fields and motions, and the relations of these to the phenomena of the chromosphere and photosphere inside, and the whole planetary and cometary system outside. Some day this picture will reach a stage at which the sources of matter and energy will be evident. In the process, I would be very surprised if new fundamental problems did not make their appearance.

2. New Methods of Observation

We now know a great deal more about the corona than we did in the 1940's. Three new observational techniques have been very important in enlarging the kinds of data available, leading to revolutionary new concepts of the nature of the corona. They observe the radio, XUV, and particle radiations of the corona directly. Each tells us things about the corona that can be determined in no other way.

Solar radio astronomy began in the late 1940's with simple radiometers that observed integrated coronal noise in the low frequency bands. The early results unambiguously confirmed the radical new high temperature concept of the corona, and disclosed short-lived bursts of enormously enhanced radiation. From these early beginnings, radio observing equipment developed rapidly, with the introduction of interferometer arrays capable of locating the positions of radiating elements of the corona, sweep frequency or multiple frequency receivers to define the radio spectra of the bursts, and polarimeters which in some cases measured all of the Stokes parameters.

Additional radio data of importance have come from observations of the scatter of the radio radiation of cosmic sources, mainly the Crab Nebula, as they are occulted by the corona. The occultations show a detectable coronal effect normally out to about 60 $R_\odot$, but the outer limit is highly variable and reaches out to more than 100 $R_\odot$ in some cases. The observations provide a surprising amount of semi-quantitative information on the structure of the corona between 10 and 100 $R_\odot$, and some rough estimates of electron density, which appears to be no more than 200 or 300 cm^{-3} at 60 $R_\odot$.

Radar observations of the corona are difficult, but are potentially among the most exciting methods available to us. The problem is to concentrate the necessary power into a sufficiently narrow beam to show coronal structures. The main advantage of the radar technique is its use of a discrete frequency which is Doppler shifted on reflection from moving elements. The observations to date have detected an overall expansion of the corona of about 16 km/sec at a height near 1.3 $R_\odot$, an extremely important bit of information which by itself justifies the very considerable effort required. The spatial resolution is still insufficient to isolate coronal elements less than about 1 $R_\odot$ in diameter. Further refinement by the current technology is discouragingly difficult and expensive, but the promise of the radar method is tremendous, and we can only hope for a sufficient improvement in the technology to make high resolution observations feasible.

Identification of the coronal lines intensified the urge to explore the ultraviolet spectrum from above the atmosphere. Instead of a few forbidden lines of a few highly ionized atoms, we should find permitted lines from all stages of ionization and we could hope to see the corona against the disk of the sun. The Naval Research Laboratory group headed by Friedman and Tousey began the first experiments in 1948 with rocket borne detectors. The early observations were necessarily primitive by later standards, but showed that the solar ultraviolet spectrum was considerably brighter than expected and very rich in lines. Since then the field has expanded rapidly with dozens of rocket launches and the Solrad, Vela, and OSO satellites. The productivity of this very difficult form of solar observation has exceeded our wildest dreams. Spectrographs, spectroheliographs and X-ray telescopes have now completed the preliminary exploration. The Fraunhofer spectrum extends down to a rather abrupt end at about 1550 A. At shorter wavelengths we find an emission spectrum of some thousands of lines showing most stages of ionization for a large number of atoms and all stages for a few. The monochromatic images of the sun do indeed show the corona on the disk in great detail. Among other things, we find that the active regions in the corona are upward extensions of those observed in optical wavelengths, but surpass the background energy of the whole quiet sun by a large factor which tends to increase steadily toward shorter wavelengths.

In the following seminars you will learn the details of the radio and space observations of the corona and their interpretation. At this point, I stress their peculiar characteristics that are so important. They are attuned to the radiations emitted by high temperature and superthermal phenomena that are invisible at optical wavelengths. This is the basic advantage. It means, first of all, that we see the high temperature corona directly on the disk with little of the line of sight ambiguity inherent in limb observations. Equally important, we also see directly a variety of nonthermal particle phenomena that could only have been suspected without these techniques. This is a bonus that could not have been predicted with any conviction because we could not know that such phenomena were really present on the sun. We now know that they are not only present, but play the dominant role in the emission of the radiations that link the sun to the earth's upper atmosphere in solar terrestrial effects. The radio and space techniques provide the tools for studying and eventually understanding them.

At present, both the radio and space techniques are deficient in spatial resolving power, which is to say that they are inferior to optical telescopes in this respect. We know quite definitely how to improve the resolving power and are restrained in affecting the improvements only by economic factors. To me, this means that sooner or later the improvements will be made, although I suspect that we will have to contain our impatience for some time.

There is one other aspect of observation from above the atmosphere of great importance to coronal science. This is the continuous observation of the white electron corona. Tousey and Newkirk have refined the externally occulted coronagraph to reduce instrumental scattered light to the neighborhood of 10^{-9} of the photospheric brightness. They have demonstrated its ability to show the K corona in observations from balloons and rockets, and are preparing instruments for continuous observations from satellites over long periods of time. The expectation is that these instruments will greatly surpass the present ground based K coronameter in spatial resolution and ability to see the corona at large distances from the limb.

The third new approach in coronal observation is the direct sampling of coronal material and magnetic fields in space beyond the terrestrial magnetosphere. The neighborhood of the earth has been explored in some detail by a number of Explorer, Vela and Imp satellites, and in less detail in the region between the orbits of Venus and Mars by Mariners 2, 4 and 5. Different space vehicles have carried an assortment of detectors that record particle flux and velocity, the direction and strength of the magnetic field, and yield some estimate of the 'temperature' of the coronal material. The observations brilliantly confirmed the existence of the solar wind. The measured densities and velocities near the earth vary considerably around averages of 5 particles/cm^3 at 450 km/sec. Both quantities show rapid fluctuations and appear to be related to the sunspot cycle. Magnetic fields of a few gamma have a distinctive broad pattern directly related to the broad pattern of photospheric fields, superposed on finer structures.

The space probes yield a most satisfying kind of data, direct unambiguous measurements of the quantities of interest. It is the kind of data that most coronal astronomers are not accustomed to, but would certainly like to be. Nature has met us half way by providing celestial probes in the ionic comet tails, which are blown away from the nuclei by the solar wind. Their aberration angles are direct measurements of the solar wind velocities, and fine structures in the tails tell much about the structure of the solar wind. Hoffmeister was the first to study interplanetary particles by this method in 1943 before the words 'solar wind' had been breathed. The velocities calculated from his data agree quite accurately with the probe determinations. Biermann and his associates have greatly extended the work since 1950, and you will hear about it from Dr. Schmidt later on. The availability of the probe data will enhance the value of the comet observations by removing or reducing some of the uncertainties of interpretation. The comets are helpful in having orbits that traverse parts of the solar system that are beyond the reach of present day probes, particularly those regions well removed from the ecliptic plane where our information about the extended corona is very scanty.

3. Coronal Landmarks of the Last Twenty Years

Looking back over the past twenty years of coronal research, we see a few discoveries and theoretical developments that have especially influenced our thinking and the direction of research. It would be difficult to name all of them because some, like the whole concept of the role of magnetic fields, rather grew like Topsy and diffused through all the thinking without any identifiable impact time.

I think my favorite is the idea proposed by Sydney Chapman in 1957 of an extended corona filling the solar system out to one astronomical unit and probably much beyond. The concept of such a corona probably occurred to others, but when one considered any barometric extrapolation of the inner corona, the density at 1 AU (200 $R_\odot$) came out to considerably less than the interstellar background. Chapman's special contribution was to show that the thermal conductivity of coronal material is so high that the whole extended corona would be abundantly heated by conduction from the inner corona. We should then expect the magnitude of the negative density gradient to be enormously reduced, and the corona should extend much farther out. This was the concept that put the extended corona in business.

Chapman considered the case of a corona in static equilibrium, and showed that it should extend beyond one AU, and that a high temperature of hundreds of thousands of degrees should be expected near the earth. This model had a peculiarity, however. Unless one assumed an unreasonably low temperature for the inner corona, the pressure at infinity remained much higher than that of the interstellar background. In other words, the static corona could not be contained.

Parker reasoned in 1960 that an uncontained corona must necessarily be expanding, and the solar system should be permeated by a solar wind. He showed that for a given temperature and density at the inner boundary, there is a unique model meeting the requirements of zero velocity of expansion at the base of the corona and a vanishing pressure at infinity. The model showed a velocity of expansion continually increasing outward, powered by heat conducted from the inner corona. As the material passed a well defined critical distance, its velocity became supersonic. The temperature at 1 AU should be high, but somewhat lower than for the static model.

Parker's theory of a supersonic solar wind was not universally accepted without some hot debate. I will not take the time here to discuss the alternatives, because numerous observations from space probes confirm Parker's supersonic solar wind. These direct observations have inspired a number of investigators to elaborate the theory to include such elements as the observed magnetic field, and we are in a fair way to knowing more about the solar corona at 1 AU than at 1 $R_\odot$. Since the two are intimately related, we have every reason to hope that advances in knowledge of the solar wind will solve some of the problems of the inner corona, such as the temperature gradient. Some people consider this a slightly embarrassing prospect, but I think it not unnatural that the study of the corona might begin with the part we can get our hands on and grow toward the more distant sun, It is, after all, somewhat accidental that our means of observation were originally more sensitive to the inner corona than to the corona immediately surrounding us, which reversed the natural order of things.

Burgess made another important contribution in 1964 with his calculation of the effects of dielectric recombinations on the ionization balance in the corona. Earlier calculations assumed that the temperature of the corona must be such that for each ion species the collisional ionization rate would equal the radiative recombination rate. On this basis, the temperature of the quiet corona always came out less than a million degrees, while the line profiles, the hydrostatic density gradients and the more sophisticated radio observations, all persisted in yielding temperatures of 1.5 to 2.5×10^6 degrees. Burgess showed that dielectric recombination rates were very much higher than radiative rates. As a result, the calculated coronal temperature had to be approximately doubled to attain a collisional ionization rate high enough to balance these higher recombination rates. This discovery removed a long standing fundamental discrepancy that had plagued coronal astronomers since they had begun thinking in terms of a high temperature corona.

A third class of discoveries which has simultaneously enlarged our view of solar activity in the corona and posed an array of tough problems is due to the radio and XUV observers. They find clear evidence for a variety of superthermal phenomena characterized by streams of high energy particles with decidedly non Maxwellian velocity distributions. In some instances of flare associated phenomena, they have observed radiation brightnesses equivalent to those of Planck radiators at temperatures of 10^8 or 10^9 degrees, and disturbances moving through the corona at several tenths of the velocity of light. Somewhere in the complex of strong magnetic fields that constitute an active center, particle accelerators of tremendous power must exist. After about 15 years of intensive and enthusiastic effort, no one has yet been able to describe any of the solar particle accelerators satisfactorily. This is certainly one of the major problems of solar physics, and one of the most fascinating.

4. Consideration of Immediate Coronal Problems

Right now we have much unfinished business in the corona. It is a vast depot for energy and matter where the inflow of both commodities balances the outflow over day long time intervals, and the very considerable fluctuations of both are taken up by a surprisingly small inventory of stock on hand. The briskness of the traffic changes slowly with the solar cycle by a factor of around 5. The inner corona is a very inhomogeneous structure with large variations from point to point in temperature and density. We should not expect descriptions of the 'average corona' to be any more than the average of its characteristics over the sun. Regions of the corona that conform to the average are decidedly limited, and the deviations from the average are very much larger than in the lower regions of the solar atmosphere. While we have certainly learned a great deal about the corona since 1842, we can look forward to a goodly number of years of exciting research before the questions presently confronting us are answered. There will, of course, be others by then. Let us now look at some of the current questions. The most basic problem is that of the sources of matter and energy required to replace the steady outflow that we observe. We do not even know enough to say with any conviction whether or not the supplies of matter and energy are directly related, or represent two separate processes. At the moment, it is more convenient to consider them separately, since the mechanism of coronal heating has attracted rather more attention than the resupply of material.

Most discussions of coronal energy begin quite properly with an estimate of the losses that must be balanced. The known losses are those due to radiation and thermal conduction, both outward and inward. Since the advent of measurements from above the earth's atmosphere, it has been possible to estimate the outward radiation of the corona in the XUV spectrum. The average flux at one AU is something like 2 erg cm^{-2} sec^{-1}. At the base of the corona, this comes to 10^5 erg cm^{-2} sec^{-1}. If we assume that the inward flux equals the outward flux, the total radiative loss, then, is about 2×10^5 erg cm^{-2} sec^{-1} in the XUV wavelengths. The radiative losses at longer wavelengths are relatively insignificant.

The total conductive losses are not as clear. First, consider the conductive flux outward. Estimates of this based on the negative temperature gradient in the corona are uncertain because the gradient itself is very uncertain. I think we do better to base the estimate on the measured energy flux in the solar wind. If we accept the Parker mechanism, this flux is due entirely to the heat conducted into the outer corona from the coronal base. The averages of particle flux and velocity determined by detectors in space are about 2×10^8 protons cm^{-2} sec^{-1} travelling at a velocity of 400 km/sec. The sum of the kinetic and potential energies of these particles represent a flux of 5×10^4 erg/cm^{-2} sec^{-1} at the base of the corona. This is a minimum figure, since any significant sinks in interplanetary space would require additional energy. If we assume that this minimum is about right, we find that the total outward energy flow due to radiation and conduction is something like 1.5×10^5 erg/cm^{-2} sec^{-1} at 1 $R_\odot$ from the center of the sun.

The downward conductive flux from the corona into the chromosphere depends on the temperature profile between the two, which is quite unknown at present and probably varies widely over the solar surface. However, the observations show temperatures of 2×10^6 deg at heights below 3×10^4 km, a datum which permits the calculation of a lower limit to the flux. Assume that $T = 2 \times 10^6$ deg at a height Δh above the level where $T = 10^5$ deg, and that no energy is deposited in the corona between these two levels. Then, the downward conductive flux is $F_c = 3 \times 10^{15}/\Delta h$ erg cm^{-2} sec^{-1}. For the observed $\Delta h < 3 \times 10^9$ cm, we find $F_c > 10^6$ erg cm^{-2} sec^{-1}. Incidentally, the temperature gradient is $dT/dh = 10^6 \, F_c T^{-5/2}$. This gives 3×10^4 deg/km at the $T = 10^5$ level, practically a temperature discontinuity in the sense that one mean free path in the vertical direction traverses a wide range of temperature. In sum, we find that the downward conductive flux dominates the energy losses in the corona, which must exceed 10^6 erg cm^{-2} sec^{-1}, and may be very much larger. How is this energy transported into the corona and deposited there?

Most investigators of this problem assume that the ultimate source of the energy is in the mass motions of the hydrogen convective zone. We need to define a process for transporting the energy from the convective zone through the photosphere and chromosphere. Then we must find some mechanism for converting the energy into heat. These are really separate problems of transport and dissipation.

In 1948, Biermann and Schwarzschild independently proposed that the energy is transported by acoustic waves excited by convection just below the photosphere. This idea was considerably reinforced in 1960 when Leighton and his associates discovered the 5 min vertical oscillations in the photosphere and lower chromosphere. Subsequent observations by a number of workers have defined these waves in some detail, but there are a number of critical properties that have not yet been determined with sufficient precision to decide whether they are vehicles of sufficient energy to heat the corona. It is clear that these waves are not the simple acoustic waves that Schwarzschild and Biermann had in mind. Some of the uncertainties are the velocity of propagation (which appears to be several times the sonic velocity), the actual velocity amplitudes of the oscillations (i.e., the interpretation of the observed Doppler shifts), the effects of various damping mechanisms like energy losses from compression crests by conduction and radiation, and the coupling of the waves with magnetic fields. Musman has made the most recent estimate of the energy carried by the 5 minute oscillations. He finds that at most the energy flux is 2×10^6 erg cm^{-2} sec^{-1}, which he considers insufficient for heating the chromosphere and corona. He is the first to admit, however, that this finding is certainly not the last word.

If we assume, Musman notwithstanding, that the 5 minute oscillations are in fact the mechanism of energy transport up through the lower chromosphere, we must still face the problem of converting this energy into heat in the higher layers. If energy losses are negligible, the velocity amplitude of a wave as it moves up into a medium of decreasing density must necessarily increase as $\varrho^{-1/2}$ (ϱ = density). When ultimately the velocity amplitude exceeds the sonic velocity, the oscillation becomes a shock wave capable of dissipating its energy as heat. This principle has been the basis for a number of theories of coronal heating. In the light of what we now think we know about the velocity amplitudes and the structure of the chromosphere, none of these theories provide sufficient heat dissipation in the corona.

Analogous mechanisms involving hydromagnetic waves that develop into hydromagnetic shocks have been proposed. They meet much the same objections. The most promising of the hydromagnetic shock theories is one proposed by Uchida. He pictures energy brought through the photosphere and lower chromosphere mechanically by acoustic waves being converted into hydromagnetic waves which develop into shocks as they advance into the low density of the corona. The high conductivity of the corona enhances the dissipative efficiency, and the required energy is deposited. However, to meet the full energy requirement of the corona and transition region, Uchida assumes an input of acoustic energy approximately 10^2 times the observed energy of the 5 minute oscillations.

Billings has suggested an ingenious hydromagnetic process for heating the corona over active centers where the required energy flux is greater by several orders of magnitude. Much of the coronal structure in the active centers consists of filamentary loops. Billings considers what would happen if the filaments are magnetic tubes of force in which both the magnetic field and the material density are slowly varying lengthwise, and rapidly varying in a radial gaussian distribution with maxima at the central axis of the filament. A plane hydromagnetic wave entering the base of a filament would then be drawn out into a sort of sheath, since its velocity would be greatest along the central axis and decrease with increasing distance from the axis. As the wave advances along the axis no shock develops since the gradients of density and magnetic field strength are small. However, the sides of the sheath are hydromagnetic waves advancing at right angles to the axis into strong negative radial gradients, and strong hydromagnetic shocks could be expected. These shocks emanating from the sides of the filaments heat the coronal material. We should expect, therefore, that the axial regions of the filaments would be cool, and the temperatures should rise toward their outer boundaries, a configuration that fits comfortably with the observed instances of Hα loops which appear to coincide with more diffuse coronal loops. Billings points out that the Osterbrock theory forbids hydromagnetic waves from the convective layer below the photosphere ever reaching the upper chromosphere. However, Osterbrock assumed small magnetic fields of less than 20 gauss. In an active center, the fields could easily be several hundred gauss and the energy losses to the photosphere and chromosphere would then be enormously reduced. Thus, this qualitative picture appears to provide a possible mechanism for heating coronal active regions, which, in turn, might heat the whole isothermal region of the corona by conduction. The question of the amount of energy that can enter the rather constricted base of a filament is crucial. The whole process should be examined quantitatively.

The second part of our fundamental existence problem is the supply of coronal material. The mass losses are notably less well known than the energy losses. The measured flux of protons in the solar wind implies an average flux at the base of the corona of 2×10^{12} protons cm^{-2} sec^{-1}. Since the total number of protons resting on

1 cm^2 of the base of the corona is probably around 10^{19}, we conclude that this flux can be expressed as approximately 10^{-7} corona per sec, and the solar wind 'replacement time' is of the order of a hundred days. This is the only process for which we know the outward mass flux quantitatively. It is probably very much less than downward losses through the prominences.

A sampling of the hundreds of motion pictures of prominences shows a great predominance of downward over upward motion of prominence material. Jefferies and Orrall stated that one loop prominence they studied had a downward flux of material equivalent to about 10^{-4} corona sec^{-1}. However, the loops and surges are special cases in that they almost surely lie along strong magnetic fields which form a strong barrier against condensation. The quiescent prominences, on the other hand, are supported in magnetic fields more or less perpendicular to the prominence axes, according to the Kippenhahn-Schluter model. We can reasonably accept the theory that they draw their material from the corona by condensation. Hence the observed downward flow from these prominences is ultimately a loss of coronal material. I am not aware of any systematic estimates of the downward mass flux, but the steady flow from the several dozen prominences present on the sun at any one time looks comparable to the flux in a single loop. Hence, in a very qualitative way, we should expect the prominence losses to be of the order of a few coronas per day. This is probably the dominating mass drain in the corona.

Since their rediscovery by Roberts in 1948, the spicules have been regarded as likely candidates for the source of coronal material. If we think in terms of 10^5 spicules on the sun at any one time, with an average diameter of 1000 km, and density of 10^{11} hydrogen atoms per cm^3, moving upward with a mass velocity 20 km/sec, we calculate an average mass flux for the whole sun of 2.5×10^{15} hydrogen atoms cm^{-2} sec^{-1}, or or something like 4×10^{-3} corona/sec, or about 60 coronas /day. If our premises are correct, it appears, then, that spicule mass flux is ample to maintain the corona. This hypothesis is very attractive because we see the spicules clearly, and they look exactly like jets shooting material up into the corona. However, Billings and others point to some awkward facts that remain to be explained. The inhomogeneities of the corona are much more pronounced than the fluctuations in spicule distribution on the sun appear to be. In particular, the surface density of spicules under coronal condensations is not obviously enhanced. There does not seem to be any strong relation between spicules and the solar cycle, while the corona responds very handsomely. Billings suggests that effects of this sort might be explained in terms of the magnetic fields associated with high levels of solar activity. Properly tailored fields could impede the loss of coronal material and thus cause a buildup of density even though the input mass flux remained constant. On the whole, I believe the spicules are the most promising source for coronal material. Before they can be definitely accepted, however, we must learn more about them. We do not even know the mass flux in a spicule with any certainty. The diameters are uncertain, the apparent upward velocities measured at the limb are not necessarily the mass velocities, and the assumed density could easily be wrong by an order of magnitude.

The thought that spicules might replace the losses of energy as well as matter in the corona is a most appealing one which I am sure has occurred to many. The majority of spicules appear to terminate a little above the level of the transition layer, and thus might put their energy where it is required most. However, it could not be a simple matter of utilizing the spicule kinetic energy, which falls short of the requirement by a factor of 100. I suggest, however, that some variant of the mechanism proposed by Billings for feeding hydromagnetic energy into coronal condensations through their filamentary structures might work in the spicules. The objection that hydromagnetic energy cannot penetrate the photosphere if the magnetic fields are in the 1–20 G range does not seem insuperable. In measuring photospheric fields we determine the average flux over a resolution element usually several arc seconds in diameter. In every instance I know of, improvement in the spatial resolution of magnetic measurements invariably leads to the discovery of highly concentrated fields of hundreds of gauss. These occur in the bright coarse network boundaries where the spicules are most numerous. I do not know of any careful theoretical analysis of the Billings' process, and would like to encourage some industrious theoretical astrophysicist to look at it.

A second major coronal problem is the role of magnetic fields and their delineation. There can be no doubt of their existence. They are almost as essential for epxlaining coronal phenomena as is the high temperature. No one has proposed any alternative to magnetic constraint of the coronal plasma to shape the conspicuous polar plumes, coronal arches and loops, and the decidedly non spherical large structures of the sunspot minimum corona, particularly the helmet structures and the overlying streamers. Many of the radio phenomena are caused by synchrotron radiation, for which magnetic fields are a necessity. The evidence for the existence of these fields is overwhelming, but they have been remarkably successful in eluding direct observation and measurement, except for the extended corona near the earth. However, order of magnitude information on the fields of the inner corona is slowly accumulating.

Magnetometers on space probes show fields of a few gammas which vary in broad patterns that accurately reflect corresponding patterns in the photospheric fields. This can only mean that the lines of force must extend from the photosphere to the earth's orbit. If we assume that they are radial, the field strength near the sun must be two or three gauss. This is a minimum figure, however, because we see in the inner corona many closed structures that must contain closed field lines that do not reach out beyond a few solar radii.

The radio bursts in wavelength ranges from centimeters to 10 or 20 meters yield estimates of magnetic field strengths which vary fairly systematically with height, from thousands of gauss just above the chromosphere to the order of 10 G near 1 $R_\odot$. These figures are not characteristic of the general corona, however. The bursts are active region phenomena and their strong fields are probably of the closed type. Observations of the bursts at low frequencies with radioheliographs have defined the general character of the fields at high levels. They show for the first time the high arching fields that join widely separated photospheric fields of opposite sign, occasionally in opposite hemispheres.

Newkirk and Schatten have tried calculating the potential field in the corona from the measured photospheric longitudinal fields. The simplifying assumption here is that the fields are current free at all heights above the observed level. The calculated fields agree remarkably well with some of the observed high coronal structures, and certainly have the same character as the fields deduced from the radio observations. The evidence suggests that the potential fields may not deviate seriously from the actual fields, and this extrapolation of the photospheric observations will be very useful. The agreement found by Wilcox and his associates between the field pattern at one AU with the photospheric pattern is further encouragement.

Another promising approach is the study of the polarization of coronal emission lines. The direction of polarization in the absence of magnetic fields is quite regular and predictable. If fields are present, however, they change the polarization direction. At coronal levels where collisions are infrequent, these changes in direction are related to the field direction in a rather complicated manner. Charvin has succeeded in defining the high level magnetic arches in several instances by this technique, and is developing the method into a useful one.

Thus, it is apparent that there are ways in which we can learn about the coronal magnetic fields by observation, and we are slowly gathering the information we need to describe their role in coronal physics.

The problem of the temperature gradient in the inner corona out to a few solar radii has been an exceedingly stubborn one. The various methods for measuring coronal temperatures are in good agreement, and they show a considerable range of temperatures between quiet and active regions. There are no certain indications, however, of a systematic variation of temperature with height for the corona as a whole out to about 2 $R_\odot$. The best we can do right now is to consider the inner corona outside the active centers as radially 'isothermal', but still with appreciable irregular fluctuations from point to point. Since the coronal temperature at one AU is around 3×10^5, a temperature maximum must exist somewhere above the chromosphere. I suggest that the height of this maximum is probably as variable from point to point and in time as the other coronal features, and an average height for it may not be very meaningful. The distribution of densities is much better defined, and various investigators have derived average curves of density variation with height. Again I suspect that the fluctuations are more important than the average curves, which seem to represent the actual distribution of a rather small fraction of the corona at any one time.

In concluding this introduction to our coronal institute, I hope that the various speakers will give us a much more precise picture of the state of coronal research than this very general one. I also hope they will not find too many errors in fact or emphasis in what I have said.

2. ATOMIC PROCESSES IN THE SOLAR CORONA

GIANCARLO NOCI

Osservatorio Astrofisico di Arcetri, Firenze, Italy

1. Introductory Remarks

The quiet solar corona is characterized by high kinetic temperatures ($\sim 10^6$ K) and low densities ($\sim 10^9$–10^8 cm^{-3} in its inner and denser part). Incident upon it are the chromospheric and photospheric radiation fields with radiation temperatures of some thousand degrees.

These conditions imply that LTE does not exist. In fact, because of its low density, the transfer of energy between radiation and matter is negligible, except for the free states. With this exception, therefore, there is no interaction – apart from scattering, involving no energy transfer – between the coronal plasma, and the radiation crossing it, both the photospheric and chromospheric fields and that created in the corona itself: matter and radiation are uncoupled. Consequently both the radiative and the collisional transitions between two generical states do not balance, requiring the first ones, to balance, a different population ratio than the second ones: designating with N_2, N_1 the populations of the states 2, 1, with C_{21} and R_{21} the number of collisional and radiative transitions per second per ion from level 2 to level 1 and with C_{12}, R_{12} the analogous quantities in the converse process, the equalities

$$N_2 C_{21} = N_1 C_{12} \tag{1.1}$$

$$N_2 R_{21} = N_1 R_{12}, \tag{1.2}$$

expressing detailed balancing, cannot hold together, since $R_{21}/R_{12} \neq C_{21}/C_{12}$ due to the uncoupling of radiation and matter. Hence LTE does not exist.

However, in the case of the free states the collisional rates populating and depopulating a given state are so much larger than the radiative rates that the population ratios are controlled by the collisions. Since these latter are at random (quiet corona) thermal population ratios arise. (Equation (1.1) holds while Equation (1.2) does not, but it is unimportant.) We can therefore assume that the velocities of the free electrons in the solar corona are thermally distributed. Consequently a description in terms of LTE is tenable for the radiation due to the free-free transitions, as, for instance, the radio-emission. Furthermore, for low energy variations the radiation absorption probabilities are sufficiently large to allow the optical depths to reach high values, so that a true coupling between radiation and matter exists at long (metric) radio wavelengths in the solar corona. (Equation (1.2) holds, for transitions corresponding to long radio wavelengths, together with Equation (1.1)).

Going back to the general case, the LTE hypothesis must be substituted with that of the steady states, requiring constancy in time of the populations: instead of assuming that every particular process happens as many times as the reverse one, it is assumed that all the processes populating a particular state balance all the processes depopulating it. To solve the corresponding equations one must know the transition rates, which depend on atomic parameters (cross-sections, transitions probabilities) and on ambient conditions. The latter ones consist of the energy distribution of the colliding particles and of the photons.

Among the colliding particles the electrons are the most important: since they have a Maxwellian distribution of the velocities, their behaviour can be described by the parameter temperature only (T_e), besides the number density (N_e); as said before we can assume, in the solar corona: $T_e \approx 10^6$ K, $N_e \approx 3 \times 10^8$ cm^{-3}, in the inner layers. As far as the positive ions (essentially protons) are concerned, they become important in transitions involving very close levels, as in the hydrogen fine structure (Purcell, 1952), and may also have effects in the excitation of the forbidden lines (Section 2.4). The ions suffer a continuous redistribution of the velocities by collisions, as the electrons: they possess, therefore, a Maxwellian velocity distribution characterized by a kinetic temperature which is usually assumed to be equal to the electron temperature. This point, however, should be considered with care since the cross-sections for electron-ion momentum transfer are small owing to the large mass difference. Indeed a difference in the kinetic temperatures of ions and electrons can exist for the solar wind transfers ions of some kinetic temperature through regions of different kinetic temperatures. This can influence – besides other phenomena – the ionization equilibrium (Delache, 1965).

Taking into account the photon frequency distribution is a less simple task since the radiation spectrum is not thermally shaped and therefore not one parameter only, but a number of them is needed. Useful parameters are the radiation temperatures: designating with index 2 the upper level, with index 1 the lower one, and with v_{12} the central frequency in the transition, an effective radiation temperature ϑ_{12} can be defined by the equation:

$$B_{v_{12}}(\vartheta_{12}) = \frac{1}{w} \int_0^\infty J_v \phi_{12}(v)\, dv, \qquad (1.3)$$

where $B_{v_{12}}$ is the Planck function at the frequency v_{12}, J_v the radiation mean intensity, $\phi_{12}(v)$ the absorption profile in the transition 1–2, and w a dilution factor.

The part at radio wavelengths of the radiation field crossing the solar corona originates from populations thermally distributed (free–free transitions); the values of the ϑ's match therefore the values of the temperature of the layers where the radiation originates ($J_v = B_v$ in Equation (1.3)). They go from $\sim 10^6$ K at metric wavelengths to 6000–5000 K in the millimetric region. In the infrared the assumption of a thermal distribution of the emitting populations is still tenable (transitions involving free and upper bound states, see Section 2.3): since the opacity of the solar atmosphere decreases with the frequency in this region of the spectrum, one gets ϑ's which decrease, reaching the value of the temperature minimum (~ 4700 K). In the visible the radiation comes from the photosphere: LTE exists there and the ϑ's increase up to ~ 6500 K at 4200 Å. Afterwards, because of the Balmer discontinuity, there is a decrease in the radiation temperature : $\sim 6200\,\mathrm{K}$ at 3600 Å (Holweger, 1970). At shorter wavelengths the opacity of the solar atmosphere increasing again, the radiation originates in higher photospheric layers and ϑ decreases once more to the temperature minimum value at ~ 1750 Å (Tousey, 1963). Finally, below 1750 Å the photospheric radiation, corresponding to the temperature minimum, is much less than the chromospheric radiation which consists of a non LTE emission having the maximum intensity in the Lyman series and continuum (Ly-α: $\vartheta_{12} = 7680\,\mathrm{K}$, Ly-$\beta$: $\vartheta_{12} = 6850\,\mathrm{K}$, Ly-cont: $\vartheta = 6620\,\mathrm{K}$, according to the observations of Hall *et al.* (1963)), and in few other lines.

Excepting the Lyman series, the quoted values refer to the continuum spectrum: they must therefore be decreased at the wavelengths corresponding to strong absorption lines.

2. Bound–Bound Transitions

2.1. Radiative transitions

The transition rates (number of transitions per second per atom) due to radiative processes can be written, using the Definition (1.3) and the well known relations among the Einstein coefficients:

$$R_{21} = A_{21}\left(1 + \frac{w}{\exp[h\nu_{12}/K\vartheta_{12}] - 1}\right) \qquad (2.1)$$

$$R_{12} = A_{21}\frac{g_2}{g_1}\frac{w}{\exp[h\nu_{12}/K\vartheta_{12}] - 1}. \qquad (2.2)$$

In Equation (2.1) the first term (spontaneous transitions) is dominant if $h\nu_{12}/K\vartheta_{12} \gg \ln(1+w)$, or, assuming $w = 0.5$:

$$\lambda_{12}\vartheta_{12} \ll 3.55 \quad \text{(cgs)}, \qquad (2.3)$$

being λ_{12} the wavelength corresponding to the transition 2–1. In other words, remembering the values of ϑ_{12} (Section 1), the stimulated emissions begin to be important – compared with the spontaneous ones – in the infrared, i.e. they can be neglected when the lower levels are involved.

2.2. Permitted collisional transitions

Since a collisional transition is due to the transient electric field produced by the colliding charged particle, the collisional cross-sections can be expressed by the oscillator strengths or Einstein coefficients of the corresponding transition.

With some exceptions (see Sections 1 and 2.4) the electrons are by far the most efficient particles. For transitions induced by electron collisions the most simple and correct general expression for the collisional transition rates (C_{21}, C_{12}) has been given by Van Regemorter (1962): putting in the Bethe approximation formula for the de-excitation cross-section σ_{21} instead of the Gaunt factor a function of the ratio $\varepsilon_2/h\nu_{12}$ (ε_2 = kinetic energy of the colliding electron), such that σ_{21} agrees with the best available observations and computations from threshold to high energies, and using a Maxwell distribution of the electron velocities, Van Regemorter finds the formula:

$$C_{21} = 20.60 \lambda_{21}^3 \frac{N_e}{\sqrt{T_e}} A_{21} P\left(\frac{hv_{12}}{KT_e}\right) \quad \text{(cgs)}. \tag{2.4}$$

The function $P(x)$ is given by Van Regemorter: it is of the order unity for $x = 10^{-3}$ both for neutral atoms and posititive ions, while it decreases to 0.2 for $x \geqslant 1$ for the second ones and to $\sim 10^{-2}$ at $x = 10^2$ for the first ones.

Since the colliding electrons have a thermal distribution of the velocities, one gets, for the excitation transition rates:

$$C_{12} = C_{21} \frac{g_2}{g_1} e^{-hv_{21}/KT_e}. \tag{2.5}$$

For some transitions reliable Gaunt factors are known. They permit a better precision than attainable with Van Regemorter's $P(x)$.

2.3. Total transition rates

The collisional and radiative de-excitation rates calculated with the above formulae for the physical conditions in the inner corona referred to in Section 1 (the data of Labs and Neckel (1968) have also been used), are compared in Figure 1 (below $\lambda_{12} = 10^{-5}$ the full curve refers to positive ions only; above $\lambda_{12} = 10^{-5}$ to neutral atoms and positive ions). It results that in the solar corona radiative de-excitations are dominant unless the levels are very close, i.e. at λ_{21} of the order of 10^{-1} millimeters. Using the hydrogenic approximation one gets the result that the de-excitations towards the nearest level become collisionally dominated for values of the principal quantum number $n \approx 13Z^{2/3}$. Still considering transitions to the nearest level, for lower values of n radiatively induced de-excitations are the important processes until, at $n \approx 5Z^{2/3}$ in hydrogenic ions, spontaneous decays become more efficient ($Z - 1 =$ ionization degree).

Analogously to Figure 1 collisional and radiative excitation rates are compared in Figure 2. It results that radiative excitations are dominant from ~ 1600 Å to the same infrared limit as the de-excitations. We must remark, however, that the dashed curve has been calculated by using the radiation temperatures corresponding to the continuum spectrum; on the contrary, the vertical lines have been obtained using the values of the radiation temperatures at the flat top of a few lines.

It is interesting to consider the behaviour of the curves in Figures 1 and 2 at different heights h above the solar surface.

Since the dilution factor decreases slowly with the increasing of h (less than a factor 4 at $h = 0.5 R_\odot$), the curves referring to radiative transitions lower slightly from the transition region up to large heights (Equations (2.1) and (2.2)). On the contrary the curves referring to collisional transitions suffer much stronger variations: moving from the considered point ($T_e = 10^6$, $N_e = 3 \times 10^8$) down in the transition region, the curve referring to collisional de-excitations essentially shifts upwards, keeping parallel,

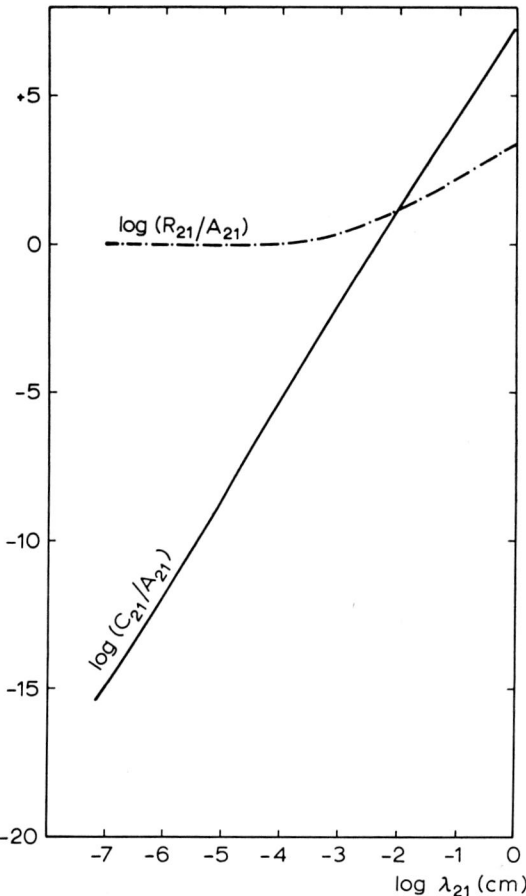

Fig. 1. Collisional and radiative de-excitations in the solar corona. $T_e = 10^6$ K, $N_e = 3 \times 10^8$ cm^{-3}, radiation temperatures as described in the text.

of quantities equal to the variation of $\log N_e - \frac{1}{2} \log T_e$ (Equation (2.4)); the one referring to collisional excitations suffers the same variation at long wavelengths, while at short ones it falls down rapidly because of the decreasing of the exponential factor in Equation (2.5) with T_e. At larger heights than those considered in Figures 1 and 2, both the excitation and de-excitation curves shift downwards, keeping parallel, of quantities equal to the variation of $\log N_e$ ($T_e \approx$ constant in corona). Accordingly a different behaviour with the height for lines originating from levels populated by radiation or by collisions results. Furthermore levels which have populations controlled by collisions in the lower corona will have populations controlled by radiation at greater heights (Woolley and Allen, 1948).

Consider for instance resonance lines. The processes populating the upper level are: transitions from the ground level, transitions from higher levels and recombinations. In spite of the much greater value for the ratio $(C+R)/A$ for the second ones with

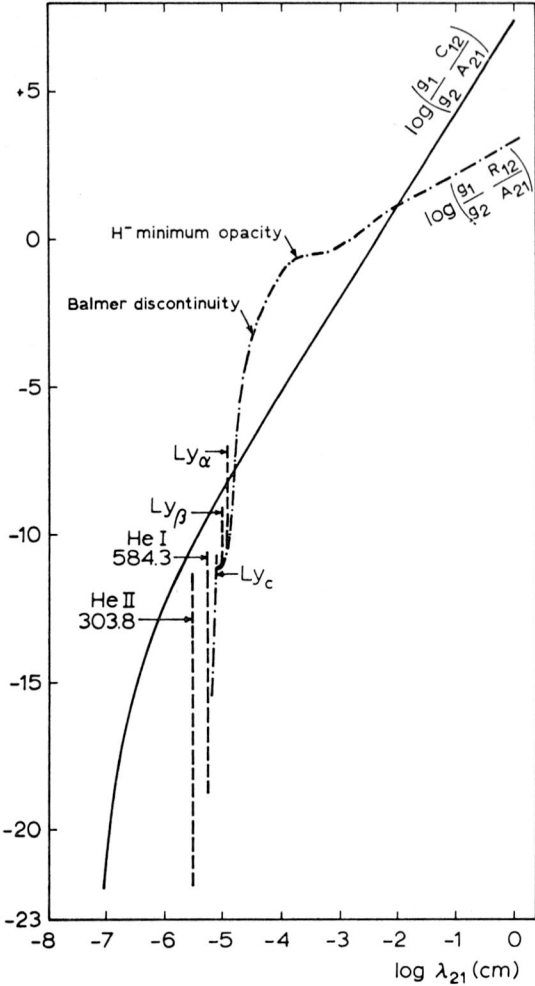

Fig. 2. Collisional and radiative excitations in the solar corona. $T_e = 10^6$ K, $N_e = 3 \times 10^8$ cm^{-3}, radiation temperatures as described in the text.

respect to the first ones (Figures 1 and 2: the energy differences involved correspond, for coronal ions, to wavelengths in the ultraviolet or in the visible), the first ones are more important because of the predominant population of the ground level and of the fact that the A_{21} of the resonance line is among the greatest in the ion. As far as the recombinations are concerned, we shall see in Section 3.6 that $N_1 T_{1c} > N_c T_{c1}$ and in Section 3.3 that $T_{c1} > T_{c2}$ (c stands for continuum, 1 and 2 designate here the first and second level respectively, N_c is the concentration of the ionization stage next to that of N_1, and $T = C + R$); since A_{k1} decreases very rapidly when the level k approaches the ionization limit, Figure 2 suggests, and detailed calculations prove, that $T_{12} > T_{1c}$. From this it follows: $N_1 T_{12} > N_c T_{c2}$; still more, as a consequence of a chain of three sensible inequalities: $N_1 T_{12} \gg N_c T_{c2}$. Hence, direct recombinations can also be neglected as processes populating the upper level of the resonance line. The processes depopulating it are certainly spontaneous decays (Figure 1). If therefore J_ν^p represents the mean intensity due to the photospheric and chromospheric radiation, the total intensity of a resonance line, off the solar limb, where I_ν is the intensity towards the observer, is given by

$$I = \text{const} \int_x^\infty \left[N_2 A_{21} - N_1 B_{12} \int_0^\infty I_\nu \phi_{12}(\nu) \, d\nu \right] dx =$$

$$= \text{const} \int_x^\infty N_1 \left[B_{12} \int_0^\infty J_\nu^p \phi_{12}(\nu) \, d\nu + C_{12} \right] dx, \qquad (2.6)$$

where the steady state hypothesis has been used, and x is a coordinate along the line of sight. If $N_1 \approx f(T_e) N$, being N the total concentration of all the ionization stages of the considered element (Section 3.6), since in the solar corona, where hydrogen is almost completely ionized, $N/N_e = \text{constant}$, Equation (2.6) can be written:

$$I = \begin{cases} \text{const}' \int_x^\infty N_e f(T_e) R_{12}(\vartheta_{12}, w) \, dx \\ \\ \text{const}' \int_x^\infty N_e^2 f(T_e) q_{12}(T_e) \, dx \end{cases} \qquad (2.7)$$

according to the mechanism populating the upper level. Here $q_{12}(T)$ is defined by: $C_{12} = N_e q_{12}$ and const' contains the abundance of the element. The function $f(T)$ is sensibly different from zero only in some temperature intervals so that many lines arise in thin layers in the transition region, given the large temperature gradient in it. In the proper corona, since T_e is nearly constant and w varies slowly, the variation of the intensity of a line is essentially due to the variation of N_e. In other words, the total concentration of the ion is proportional to the coronal density (N_e) and the excitation rate either is constant (first equation) or is also proportional to N_e.

The first equation (2.7) holds for Ly-α, the second one for the other resonance lines (Figure 2). The arguments leading to Equations (2.7) can be true for some non-resonance lines, like Ly-β, for which the second equation holds (fig. 2). The behaviour of Ly-α and Ly-β in the coronal spectra obtained by Speer *et al.* (1970) in the last eclipse illustrates beautifully the Equations (2.7): while the intensity of the Ly-β corona drops off rapidly with height, the Ly-α corona has a much larger extension, similar to the K corona, i.e. to the continuous component of the coronal emission, whose intensity is proportional to N_e since it is due to the photospheric radiation scattered by the electrons.

2.4. Forbidden transitions

A brief hint only will be given here to forbidden transitions, since they will be discussed later by Goldberg.

For forbidden transitions Equation (2.4) is no longer valid because the Bethe approximation, on which Equation (2.4) relies, assumes short range interactions between the atomic electron and the perturbing electron to be negligible, which is only true for dipole transitions (allowed transitions). Instead of expressing the cross-sections by spontaneous transition coefficients and Gaunt factors, one can use the collision strengths Ω_{21} introduced by Hebb and Menzel (1940):

$$\sigma_{21}(\varepsilon_2) = \frac{I_H}{\varepsilon_2} \frac{\Omega_{21}(\varepsilon_2)}{g_2} \pi a_0^2, \qquad (2.8)$$

where I_H is the hydrogen ionization energy, ε_2 the kinetic energy of the colliding electron and a_0 the radius of the first Bohr orbit. Using an average value $\bar{\Omega}_{21}$ for $\Omega_{21}(\varepsilon_2)$ and integrating over a Maxwellian velocity distribution gives

$$C_{21} = 8.63 \times 10^{-6} \frac{\bar{\Omega}_{21}}{g_2} \frac{N_e}{\sqrt{T_e}} \quad \text{(cgs)}. \qquad (2.9)$$

For many forbidden transitions in positive ions the energy variations of the Ω's are sufficiently small to be neglected. $\bar{\Omega}_{21}$ can then be replaced with Ω_{21} in Equation (2.9). Values of Ω_{21} have been calculated for individual forbidden transitions.

For levels connected to the ground level by a forbidden transition cascade population can be important. To give an example, let us consider the level $3p\ ^2P_{3/2}$ in the ground configuration of Fe XIV (excitation energy corresponding to $\lambda_{21} = 5303$ Å). For this level the populating processes of radiative excitation, direct collisional excitation and collisional excitation to higher levels followed by cascade are in the ratios 8:1:2.5, respectively, for $N_e = 2 \times 10^8$ cm^{-3} and $T = 10^6$ K. A further contribution of 2.5% – which increases strongly with temperature – is due to proton impact (Seaton, 1964). The importance of proton collisions is, however, not yet too clear (see Zirker, 1969).

3. Ionization and Recombination

3.1. The thermal limit

Figures 1 and 2 show that in the solar corona the transitions among the high levels are collisionally induced. Furthermore it is a general result that not only the ratios C/A, but also the C_{21} themselves, i.e. the collisional transition rates between consecutive levels, increase with increasing principal quantum number, since A_{21} decreases less than $\lambda_{21}^{-3} P^{-1}(h\nu_{21}/kT_e)$. In other words, both because the matrix elements increase rapidly and because the energy difference decreases. More generally, for almost all levels the cross-sections for excitation are much larger than those for de-excitation. Therefore, at a sufficiently high value of $n(n_t)$ the collisional transition probability from one level to all the higher levels and to the continuum becomes equal to the transition probability towards all the lower lying levels, consisting essentially of spontaneous decays.

From this phenomenon the following picture emerges: for $n<n_t$ the levels are mainly connected with the lower lying levels; for $n>n_t$ the levels are mainly connected with the higher levels and with the continuum. Since the free electrons have a Maxwellian velocity distribution, the populations of these levels are given by the Boltzmann-Saha formula referred to the ions of the following ionization stage.

This picture, introducing the concept of a thermal limit corresponding to n_t, is based on works of Bates and Kingston (1961), D'Angelo (1961), McWhirter (1961), Ivanov-Kholodnii and Nikol'skii (1961). It has been developed by Wilson (1962), who gives for the principal quantum number n_t (non-integral) the equation:

$$n_t^7 = 1.4 \times 10^{15} (z+1)^6 T_e^{1/2} N_e^{-1} \quad \text{(cgs)}, \tag{3.1}$$

where z is the ionization degree. This result assumes the radiation field to be negligible and n_t to be large (as it is in the solar corona).

A slightly different expression for n_t, and for C_{1t}^z, R_t^z (see below) as well, is due to Griem (1964).

Equation (3.1) has been obtained with hydrogenic approximations; it is correct for any ion, with said assumptions, if dielectronic recombination is negligible: the effect of dielectronic recombination in hindering the establishing of thermal equilibrium in the high levels by populating them via spontaneous decays will be discussed later.

With this picture the problem of evaluating the level populations is solved by using Saha equation for the levels above the thermal limit and the steady state equations for the levels below it. For these it is often sufficient to consider the transitions towards the ground level (spontaneous decays, see Figure 1) and those from it (collisional excitations for permitted transitions, given the high energies involved for coronal ions, see Figure 2). This assumes the total population of the ion or, better, the total population of the ion below the thermal limit, to be known. This requires the solution of equations which express the balance of all the ionization processes with all the recombination ones. Such equations do not need the inclusion of transitions from the continuum to levels above the thermal limit and vice versa, since they balance exactly.

3.2. Bound–free transitions

It is usually assumed that the ionization processes originate essentially from the ground level, since the ground level is by far the most populated.

Since the ionization potentials of the ions existing in the solar corona are very high, corresponding to ultraviolet frequencies, one can deduce from Figure 2, by reasons of continuity, that collisional ionizations are much more important than radiative ionizations.

The ionization by electron collision of a z times ionized atom in the ground level can be described by the formula:

$$X_1^z + e \rightarrow X^{z+1} + e + e, \tag{3.2}$$

where X^{z+1} includes all possible levels. The corresponding rate is given by:

$$C_{1c}^z = N_e q_{1c}^z = N_e \times 2.0 \times 10^{-8} T_e^{1/2} \sum_l \frac{\zeta_z(n,l)}{I_z^2(n,l)} 10^{-5040\,I_z(n,l)/T_e}, \qquad (3.3)$$

according to Seaton (1964; see also Burgess and Seaton, 1964). Here n is the principal quantum number of the outer electrons of X_1^z, $\zeta_z(n, l)$ is the number of electrons with quantum numbers n, l, and $I_z(n, l)$ the corresponding ionization energy. $I_z(n, l)$ is here in eV and the other quantities are in cgs units. Equation (3.3) has been obtained using an expression for the cross-section (whose form is suggested by approximate classical theories) which agrees with the best experiments and calculations for coronal conditions. According to Seaton this formula should be correct within a factor of two.

Another ionization process must be considered, namely the collisional ionization via bound levels. It consists in collisional excitations from the levels below the thermal limit to those above it, which result in a decreasing of the total population of the levels below the thermal limit and also of the total ion population: in fact immediate collisional transitions to the continuum compensate the increase in population in the thermal levels.

The corresponding hydrogenic rate coefficient is given by Wilson (1962, 1967)

$$C_{1t}^z = N_e q_{1t}^z = N_e \times 4.8 \times 10^{-5} \frac{e^{-(I_z - I_t^z)/KT_e}}{n_t^2 (z+1)^2 \sqrt{T_e}} \quad \text{(cgs)}, \qquad (3.4)$$

where I_z is the ionization potential of the ion and I_t^z the ionization potential from the energy level corresponding to $n_t: I_t^z = (z+1)^2 I_H/n_t^2$. However it has been shown (see for instance Jordan, 1969) that this process can be neglected both in the transition region and in the solar corona.

3.3. Free-bound transitions

The two processes of recombination to consider now, the radiative and the collisional, are photorecombination and three-body recombination. In their comparison we may be guided again by Figure 1. Actually detailed calculations confirm what Figure 1 suggests, namely that for the low levels photorecombinations are by far more important than three-body recombinations, given the energy differences involved.

The same argument used to find the relation between the Einstein coefficients permits to infer easily the radiative recombination cross-section from the bound-free absorption coefficient. Using the well-known expression for hydrogen-like ions, after integration over a Maxwellian distribution of the electron velocities, one gets for the process:

$$X_I^{z+1} + e \rightarrow X_n^z + h\nu \qquad (3.5)$$

the rate coefficient:

$$R_{\text{rad},cn}^z = N_e \alpha_{\text{rad},cn}^z = N_e \times 2.065 \times 10^{-11} \frac{(z+1)^2}{\sqrt{T_e}} \frac{f_n((z+1)^2/T_e)}{n} \quad \text{(cgs)},$$
$$(3.6)$$

where the function f_n, including the Gaunt factor, is tabulated by Seaton (1959). Stimulated recombinations are neglected in this formula.

The function f_n is such that the ratio $R^z_{\text{rad},c1}/R^z_{\text{rad},c2}$ is in the range 2–3 in coronal conditions (Seaton, 1959). Since dielectronic recombinations into the second level are negligible, this justifies the assumption $T_{c1} > T_{c2}$ made in Section 2.3.

With a suitable approximated expression for f_n valid at high temperatures, one gets, after summation over all n:

$$R^z_{\text{rad}} = N_e \alpha^z_{\text{rad}} = N_e \times 1.3 \times 10^{-9} (z+1)^2 \frac{\sqrt{I_z}}{T_e} \tag{3.7}$$

as the total recombination rate coefficient (Burgess and Seaton, 1964). The ionization potential I_z is here in electron volts and the other quantities in cgs units. This hydrogenic formula can be considered correct also for non hydrogen-like ions given the high ionization degree of atoms in the solar corona.

For higher values of $(z+1)^2/T_e$, namely in the transition region, the treatment of the correction factors of Elwert (1952) is preferable. This gives:

$$R^z_{\text{rad}} = N_e \alpha^z_{\text{rad}} = N_e \times 0.97 \times 10^{-12} I_z \frac{n_1}{\sqrt{T_e}} f, \tag{3.8}$$

with the same units as in Equation (3.7). n_1 is the ground state principal quantum number and f a correction factor near unity incorporating the Gaunt factor.

One has to consider, at this point, the radiative decays from the levels above the thermal limit: they increase the total population below the thermal limit and also the total population of the ion since the places left free in the upper levels by these radiative decays are immediately filled by collisional transitions from the continuum. Such a process is not negligible in the transition region (Jordan, 1969); the corresponding hydrogenic rate coefficient has been given by Wilson (1962, 1967):

$$R^z_t = N_e \alpha^z_t = N_e \times 1.2 \times 10^{-6} \frac{(z+1)^4 e^{I^{t+1}_z/KT_e}}{n_t T_e^{3/2}} \quad \text{(cgs)}. \tag{3.9}$$

3.4. Dielectronic Recombination

Another process exists which permits communications between the free and the bound levels, namely resonance between continuum and doubly excited levels above the first ionization limit. This can be described as (Figure 3):

$$X_i^{z+1} + e \rightleftarrows X_p^z \tag{3.10}$$

If now, after the electron capture, a spontaneous decay of the doubly excited ion to a level not subject to autoionization takes place, the right to left version of this process can not happen and the electron is definitely captured. This kind of recombination is the well known dielectronic recombination. The second part of the process can be described as:

$$X_p^z \to X_q^z + h\nu_{pq}. \tag{3.11}$$

The recombination rate into level q through level p, if X^{z+1} is initially in level i, is given by

$$N_e \alpha_{di,ipq}^z = N_{pi}^z A_{pq}/N_i^{z+1}, \tag{3.12}$$

where N_{pi}^z is the fraction of the population of the doubly excited level p coming from recombinations of X_i^{z+1}. N_{pi}^z is obtained balancing depopulating and populating processes:

$$N_{pi}^z \left(A_{pc} + \sum_q A_{pq} \right) = T_{ip} N_i^{z+1}, \tag{3.13}$$

where A_{pc} represents the total number of autoionizations per second per ion from level p: $A_{pc} = \sum_i A_{pi}$, where A_{pi} refers to the final state i of X^{z+1}. (Transitions out of level p other than autoionizations and stabilizing decays are neglected in Formula (3.13).) As the velocity distribution of the electrons is Maxwellian, the populating

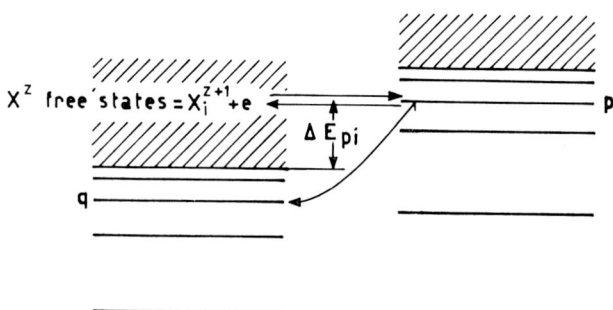

Fig. 3. Dielectronic recombination.

transitions $T_{ip} N_i^{z+1}$ are the same as in thermal equilibrium; since in this case detailed balancing holds, one can put $T_{ip} N_i^{z+1}$ equal to the frequency per cubic centimeter of the inverse transitions in thermal conditions, namely to $A_{pi} N_p^{z*}$, being N_p^{z*} related by the Boltzmann-Saha formula to N_i^{z+1} and N_e. Equations (3.12) and (3.13) then give:

$$N_e \alpha_{di,ipq}^z = \frac{A_{pi} A_{pq}}{A_{pc} + \sum_q A_{pq}} N_e \frac{g_p}{2g_i^+} \frac{h^3}{(2\pi m KT)^{3/2}} e^{-\Delta E_{pi}/KT}, \tag{3.14}$$

where g_i^+ is the statistical weight of the level where is the recombining ion, and ΔE_{pi} the energy difference between level p and the ionization limit of the recombined ion X^z, which corresponds to X^{z+1} in the level i (Figure 3).

We will now consider, among the initial states of X^{z+1}, the ground state only, which, because of its predominant population, gives the essential contribution to the recombinations. Moreover, we can assume that one $p \to q$ transition only is important, which consists in the return to the initial (ground) state of the inner excited electron. We can also assume that, for the levels p into which the free electron of the continuum corresponding to X_1^{z+1} can be captured, it is: $A_{p1} \simeq A_{pc}$. We exclude now from the summation the p levels for which the selection rules give $A_{p1} = 0$, and, since $A_{p1} \gg A_{pq}$ for the important p levels, get, for the total dielectronic recombination rate:

$$R_{di}^z = \alpha_{di}^z N_e = N_e \sum_p \alpha_{di,\,1pq}^z = N_e \frac{h^3}{(2\pi m KT)^{3/2}} \sum_p A_{pq} \frac{g_p}{2g_1^+} e^{-\Delta E_{p1}/KT}. \quad (3.15)$$

This quantity was considered for years much less than the direct photorecombination rate, until in 1964 Burgess realized the full effect of the statistical weight g_p. In fact g_p is proportional to the square of the principal quantum number of the upper electron, n_u, while when n_u tends to infinity ΔE_{p1} is bound. Then the expression just written diverges. However a limit for n_u arises by the fact that, actually, the autoionization probabilities A_{pi} decrease with the quantum number n_u, so that for a sufficiently high value of n_u, instead of being $A_{p1} \gg A_{pq}$ the opposite inequality is true and a finite value of R_{di}^z then results. Burgess (1965a) has developed a simplified general expression for α_{di}^z, in which the summations over the states of the upper electron have been accomplished, but not those over the states of the inner electron. Calculations must therefore be made separately for every ion. This formula should give results to an accuracy of about 20 per cent if the effect of collisions on the high levels is neglected.

The calculations show that dielectronic recombination is generally much more important than photorecombination in coronal conditions: for He II → He I Burgess (1964) found a value $\sim 10^2$ of the ratio R_{di}^z/R_{rad}^z above $T_e \approx 4 \times 10^5$. Therefore, for a given temperature, recombination is strongly favoured with respect to the case when photorecombination only is present. As a consequence at that temperature lower ionization stages are found. Alternatively, the same ionization stage as without dielectronic recombination is found at a different temperature, more exactly at higher temperatures, since both the recombining rates decrease with temperature in coronal conditions, while the ionization rate increases.

Let us now discuss briefly the effect of collisions on dielectronic recombination.

Burgess (1965b) argues that the states above the thermal limit should not be included in the summation for the total dielectronic rate. He makes calculations for the case Ca II → Ca I to establish the value n_d of the principal quantum number n_u such that for $n_u > n_d$ the terms in the summation, due to the decreasing of the coefficients A_{p1}, give negligible contribution to the total dielectronic rate $N_e \alpha_{di}^z$. In the examined ion 80% of the contribution arises at $n_u < 150$, so that we can put $n_d \approx 150$. If $n_t > n_d$ the reduction of the dielectronic recombination by subsequent collisional ionization is negligible. This point has been studied by Dupree (1968) who – although she does not use explicitly the concept of thermal limit – essentially calculates n_t for O v ($T_e = 2.6 \times 10^5$ K) and Fe xv ($T_e = 2.75 \times 10^6$ K) for both values of the electron density $N_e = 10^8$

cm^{-3} and $N_e = 10^9$ cm^{-3}; comparing it with the quoted calculation of Burgess she concludes that $n_d < n_t$ in the solar corona, but not in the transition region, where therefore reduction in α_{di}^z by collisions has to be considered. The reduction by photo-ionization is negligible, according to Dupree (see also the beginning of Section 3.1).

When $n_d > n_t$ it is clear that the cut off in the summation for α_{di}^z is better determined by the collisions and that the definition of thermal limit has to be modified.

In fact this definition makes reference to the depopulating transitions and the ratio between upward and downward rates is not modified by dielectronic recombination. However in hydrogen-like ions, where this process is absent, the levels above the thermal limit are not only depopulated but also populated by collisions with upper levels and with the continuum, whence their thermal population. In non-hydrogen-like ions, on the contrary, dielectronic recombinations overpopulate the high levels, whose occupation numbers become greatly in excess with respect to their thermal value (Goldberg and Dupree, 1967). If now one includes in the definition of the thermal limit the condition that above it also the populating mechanism must consist essentially of collisional transitions, the occupation numbers of the levels above the thermal limit are still given by the Boltzmann-Saha formula.

The picture made in Section 3.1 is therefore changed at high temperatures, i.e. when dielectronic recombination is important. If n_t is the principal quantum number of the newly defined thermal limit, calling n_c the old n_t, we have that for $n > n_c$ the levels are still essentially connected with the continuum – whence the validity of Equation (3.4) – but the populations are not necessarily thermal – whence the necessity of correcting the expression (3.9) which assumes a thermal population for $n > n_c$, while this last property holds for $n > n_t$. We have also that $n_t \geq n_c$, the equality being true when $n_d < n_c$, and conversely. To make the picture complete we recall that for $n < n_d, n_t$ there is a group of greatly populated levels filled by dielectronic recombination and depleted by spontaneous decays. Still below are the low levels essentially connected with the ground state.

3.5. Autoionization

To conclude, we must consider the autoionization process, whose importance in determining the ionization equilibrium was first discovered by Goldberg *et al.* (1965). This phenomenon consists in the right to left version of process (3.10), following the excitation of an inner shell electron. As seen in Section 2.3, excitations consist essentially in collisional transitions from the ground level, in the solar corona. Therefore, if $N_e q_{1p}^z$ is the excitation rate from the ground level to level p, of the $N_e N_1 q_{1p}^z$ excited ions, the fraction $A_{pc}/(A_{pc} + \sum_q A_{pq})$ will autoionize, giving the rate coefficient:

$$C_{1,\text{auto}}^z = N_e q_{1,\text{auto}}^z = N_e \sum_p q_{1p}^z \frac{A_{pc}}{A_{pc} + \sum_q A_{pq}} \approx N_e \sum_p q_{1p}^z, \qquad (3.16)$$

since $A_{pc} \gg \sum_q A_{pq}$ for the important levels. The coefficients $q_{1p}^z (= C_{1p}^z/N_e)$ are discussed in Section 2.

The effect of the summation in Equation (3.16) has resulted to be more importatu than supposed at the beginning, when the contribution of more inner electron shells and that arising from forbidden transitions were taken into account (Bely, 1967).

3.6. Ionization equilibrium

According to Sections 3.1–3.5, the steady state hypothesis, with regard to ionization, gives:

$$N^z N_e (q_{1c}^z + q_{1t}^z + q_{1,\text{auto}}^z) = N^{z+1} N_e (\alpha_{\text{rad}}^z + \alpha_t^z + \alpha_{di}^z), \tag{3.17}$$

where N^z, N^{z+1} are the concentrations of the ions in the z and $z+1$ ionization stages, respectively. From Equation (3.17):

$$\frac{N^{z+1}}{N^z} = \frac{q_{1c}^z + q_{1t}^z + q_{1,\text{auto}}^z}{\alpha_{\text{rad}}^z + \alpha_t^z + \alpha_{di}^z}. \tag{3.18}$$

In this equation q_{1t}^z, α_t^z and α_{di}^z are density dependent. However in the solar corona, to a good approximation, the first two can be neglected and the third is density independent $(n_t = n_c > n_d)$; this gives the well known result that the ratio

$$\frac{N^{z+1}}{N^z} = \frac{q_{1c}^z + q_{1,\text{auto}}^z}{\alpha_{\text{rad}}^z + \alpha_{di}^z}. \tag{3.19}$$

does not depend on N_e. This is not true, however, in the transition region, where an intermediate situation between the coronal equilibrium (3.19) and the photospheric local thermodynamic equilibrium exists.

Equation (3.17) shows that recombinations upon all levels are essentially balanced by ionizations from the ground level; accordingly: $N_1 T_{1c} > N_c T_{c1}$, an unequality; used in Section 2.3.

To conclude, we note that the increase of temperature due to dielectronic recombination, discussed in Section 3.4, is indeed reduced by a density dependent term (only effective in the low corona) and, above all, by the inclusion of autoionization in the ionization balance. However, the temperatures derived are still up to a factor of two greater than those obtained prior to the discovering of the importance of doubly excited levels. This removes the well known discrepancy in the coronal temperatures deduced from line broadening and from ionization theory, existing previously.

References

Bates, D. R. and Kingston, A. E.: 1961, *Nature* **189**, 652.
Bely, O.: 1967, *Ann. Astrophys.* **30**, 953.
Burgess, A.: 1964, *Astrophys. J.* **139**, 776.
Burgess, A.: 1965a, *Astrophys. J.* **141**, 1588.
Burgess, A.: 1965b, *Proceedings of the Second Harvard-Smithsonian Conference on Stellar Atmospheres*, p. 47.
Burgess, A. and Seaton, M. J.: 1964, *Monthly Notices Roy. Astron. Soc.* **127**, 355.
D'Angelo N.: 1961, *Phys. Rev.* **121**, 505.
Delache, P.: 1965, *Compt. Rend. Acad. Sci.* **261**, 643.
Dupree, A. K.: 1968, *Astrophys. Letters* **1**, 125.

Elwert, G.: 1952, *Z. Naturf.* **7a**, 703.
Goldberg, L. and Dupree, A. K.: 1967, *Nature* **215**, 41.
Goldberg, L., Dupree, A. K., and Allen, J. W.: 1965, *Ann. Astrophys.* **28**, 589.
Griem, H. R.: 1964, *Plasma Spectroscopy*, McGraw-Hill, Chap. 6.
Hall, L. A., Damon, K. R., and Hinteregger, H. E.: 1963, *Space Res.* **3**, 745.
Hebb, M. H. and Menzel, D. H.: 1940, *Astrophys. J.* **92**, 408.
Holweger, H.: 1970, *Astron. Astrophys.* **4**, 11.
Ivanov-Kholodnii, G. S. and Nikol'skii, G. M.: 1961, *Soviet Astron. – AJ* **5**, 31.
Jordan, C.: 1969, *Monthly Notices Roy. Astron. Soc.* **142**, 499.
Labs, D., and Neckel, H.: 1968, *Z. Astrophys.* **69**, 67.
McWhirter, R. W. P.: 1961, *Nature* **190**, 902.
Purcell, E. M.: 1952, *Astrophys. J.* **116**, 457.
Seaton, M. J.: 1959, *Monthly Notices Roy. Astron. Soc.* **119**, 81.
Seaton, M. J.: 1964, *Planet. Space Sci.* **12**, 55.
Speer, R. J., Garton, W. R. S., Morgan, J. F., Nicholls, R. W., Goldberg, L., Parkinson, W. H., Reeves, E. M., Jones, T. J. L., Paxton, H. J. B., Shenton, D. B., and Wilson, R.: 1970, *Nature* **226**, 249.
Tousey, R.: 1963: *Space Sci. Rev.* **2**, 3.
Van Regemorter, H.: 1962, *Astrophys. J.* **136**, 906.
Wilson, R.: 1962, *J. Quant. Spectrosc. Radiat. Transfer* **2**, 470.
Wilson, R.: 1967, 'Plasmas in Space and in the Laboratory', ESRO-SP-20, p. 373.
Woolley, R. v. d. R. and Allen, C. W.: 1948, *Monthly Notices Roy. Astron. Soc.* **108**, 292.
Zirker, J. B.: 1969, *Solar Phys.* **11**, 68.

The following papers on ionization equilibrium, dielectronic recombination and autoionization are also recommended:
Cox, D. P. and Tucker, W. H.: 1969, *Astrophys. J.* **157**, 1157.
Shore, B. W.: 1969, *Astrophys. J.* **158**, 1205.
Temkin, A. (ed.): 1966, *Autoionization*, Mono Book Corp., Baltimore.

3. MAGNETOHYDRODYNAMICS AND PLASMA PHYSICS OF THE SOLAR CORONA

FRIEDRICH MEYER

Max-Planck-Institut für Physik und Astrophysik, Munich, Germany

1. Macroscopic and Microscopic Description

At its temperature T of about 1.6×10^6 K the solar corona is a fully ionized plasma. The self collision frequency γ (Spitzer, 1962) $\gamma = n \ln \Lambda / (11.4 \, A^{1/2} \, T^{3/2})$ at this temperature and a density of $n = 3 \times 10^8 / \text{cm}^3$ is for electrons $\gamma_e = 11.5$/sec and for protons $\gamma_p = 0.27$/sec, with a mean free path for both species of $\lambda = 0.75 \times 10^8$ cm. The collision frequency for energy exchange between protons and electrons is $\gamma_{ep} = 1.2 \times 10^{-2}$/sec. The frequency of electrostatic oscillations of the electrons relative to the proton center of mass, i.e. the plasma frequency, is $\omega_p = (4\pi n e^2 / m_e)^{1/2} = 10^9$/sec. The corresponding length scale is the Debye distance $D = (kT/4\pi n e^2)^{1/2} = 0.5$ cm. In a coronal magnetic field B of 4 G the gyrofrequency $\omega_g = eB/mc$ is for electrons $\omega_{ge} = 0.7 \times 10^8$/sec and for protons $\omega_{gp} = 0.4 \times 10^5$/sec with gyroradii R_g of $R_{ge} = 12$ cm and $R_{gp} = 5.2 \times 10^3$ cm respectively. (A in Spitzer's formula is the ratio m/m_p of the particle mass m to the proton mass m_p. The term $\ln \Lambda$ results from the shielding of the otherwise diverging electrostatic interaction between particles by a polarization of the plasma around a charged particle, and has a value of about 20 in the corona.)

These frequencies and length scales characteristic for the species composing the coronal plasma divide the coronal structure and behaviour distinctly into macroscopic and microscopic. The macroscopic state of the corona is adequately described by the local averages mass density ϱ, bulk velocity v, gas pressure p, temperature T, magnetic field $\mathbf{B}$, electric current density $\mathbf{j}$, plus other macroscopic means, all taken over spatial regions and times large compared to the microscopic scales listed above. The microscopic state enters into this description only via the transport coefficients, which can be determined as functions of the macroscopic variables if the plasma is close to its local thermodynamic equilibrium (Chapman and Cowling, 1960).

The microscopic state is characterized by the particle distribution functions in velocity and space for each species, and their interaction. It is the only feasable description when time and space scales considered are of the above listed order of magnitude and when large deviations from the local thermodynamic equilibrium occur.

2. Frozen-in Magnetic Fields

A characteristic feature of the solar corona is its high electrical conductivity which is of the order $ne^2/m_e \gamma_e$ for a fully ionized plasma and is given by

$$\sigma = 1.4 \times 10^{17} \, T^{3/2} / \ln \Lambda$$

(Spitzer, l.c.; for strong magnetic fields $\omega_{ge} \gg \gamma_e$ the conductivity perpendicular to the magnetic field is reduced to about half this value). An estimate for the time of decay τ of any electric current system in the corona can be obtained by deviding the energy content $B^{*2}/8\pi$ of the non-potential part B^* of the magnetic field with the rate of Ohmic dissipation j^2/σ and relating $j = c/4\pi |\nabla \times \mathbf{B}| \approx cB/4\pi L$, where L is a length of the order of the dimension of the system. This gives the order of magnitude (cgs-units)

$$\tau = (4\pi\sigma L^2)/c^2 .$$

For $L = 10^8$ cm τ becomes 3×10^5 yr. This is extremely long compared to any characteristic dynamical time scale in the corona. The corona is therefore an ideal conductor. Ohm's law in its simple macroscopic form reduces to

$$0 = \mathbf{j}/\sigma = \mathbf{E} + (\mathbf{v}/c) \times \mathbf{B}.$$

The electric field in a system of reference moving with the fluid vanishes. Using Maxwell's equation $\nabla \times \mathbf{E} = -(1/c)\,\partial \mathbf{B}/\partial t$ one derives

$$\partial \mathbf{B}/\partial t = \nabla \times (\mathbf{v} \times \mathbf{B}).$$

This equation means the conservation of magnetic flux Φ through any area S bounded

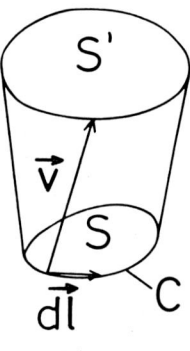

Fig. 1.

by a materially moving circumference C. The rate of change of

$$\Phi = \int_S \mathbf{B} \cdot d\mathbf{S}$$

consists of two parts, one by transport of S to a new position S' in space, the other by a change with time of $\mathbf{B}$. The latter is

$$\int_S \partial \mathbf{B}/\partial t \cdot d\mathbf{S},$$

the former is

$$-\int_C \mathbf{B} \cdot (d\mathbf{l} \times \mathbf{v})$$

which represents the flux passing through S but leaving through the mantle surface between S and S' instead of through S'. Using Stokes theorem one has

$$\frac{d\Phi}{dt} = \int_S \left[\frac{\partial \mathbf{B}}{\partial t} - \nabla \times (\mathbf{v} \times \mathbf{B}) \right] \cdot d\mathbf{S} = 0.$$

In particular the flux through the walls of any magnetic tube of force remains always zero. This allows to materially define field lines, since the material inside an infinitesimal flux tube always stays inside. The magnetic field is 'frozen-in', its Maxwellian stresses are directly coupled to the gas.

In the corona the magnetic field tends to be the dominant force, since the ratio β of gas pressure to 'magnetic pressure'

$$\beta = \frac{nkT}{B^2/8\pi}$$

is $\beta \approx 0.2$ for $B = 4$ G, and decreases upwards. The coronal gas is rather stringently confined in the 'corset' of the magnetic field.

3. Force-Free Magnetic Fields

The magnetic configuration approximates that of a 'force-free' field, in which no material force is available to compensate the Lorentz-force, which therefore has to vanish,

$$\frac{1}{c} \mathbf{j} \times \mathbf{B} = 0.$$

In contrast to the vacuum magnetic field the corona can carry electric currents parallel to the field. The difference is important. Since field lines are materially identifiable, flux tubes can be twisted by the convective motion of the foot points in the photosphere. The twisting and distortion and any magnetic connection that differs topologically from the vacuum magnetic field brings energy into the coronal magnetic field that could in principle be a source for release in dynamic processes in the corona and chromosphere (e.g. in flares). The difficulty lies in the high electrical conductivity that allows significant ohmic dissipation only for much too long time spans. Various suggestions have been made that take into account deviations of the microscopic state from its local equilibrium for small regions of space in which then strong dissipation of magnetic energy and reconnection of magnetic field lines become possible. It is conceivable that such local singularities come about automatically to a degree that is required by the macroscopic boundary condition, if the MHD-configuration is appropriate (Petschek, 1964).

The lateral stresses in a twisted magnetic field try to widen the magnetic flux tube and counteract the tension along the flux tube that keep it tied down. The result is a buoyancy effect that raise the twisted magnetic loop to a new equilibrium position relative to the surrounding field. Magnetic buoyancy probably plays a role in eruptive prominences rising in the corona.

Hydromagnetic instabilities develop when the twist of a magnetic flux tube becomes sufficiently strong. In the instability the flux tube tries to unwind and 'pays' for it by increasing its length ('knot instability' of a twisted rubber band). The strong shears resulting could be connected with increased dissipation.

4. Hydrostatic Equilibrium and Flows Within Flux Tubes

The equilibrium of the coronal gas within each magnetic flux tube is hydrostatic. The pressure from flux tube to flux tube, however, will vary, the difference being compensated by slight deformations of the magnetic field: From the magnetohydrostatic equation

$$0 = -\nabla p + \varrho \mathbf{g} + \frac{1}{c} \mathbf{j} \times \mathbf{B}$$

($\mathbf{g}$ = gravitational acceleration) one obtains by multiplication with a line element ds parallel to $\mathbf{B}$ (with the corresponding height difference dh, $\mathbf{g} \cdot d\mathbf{s} = -g\,dh$, and $dp = \nabla p \cdot d\mathbf{s}$)

$$0 = -dp - \varrho g\,dh.$$

With $\varrho = P/kT$ the usual 'barometric law' for an isothermal corona follows

$$p = p_0\,e^{-h/H}, \quad \varrho = \varrho_0\,e^{-h/H}.$$

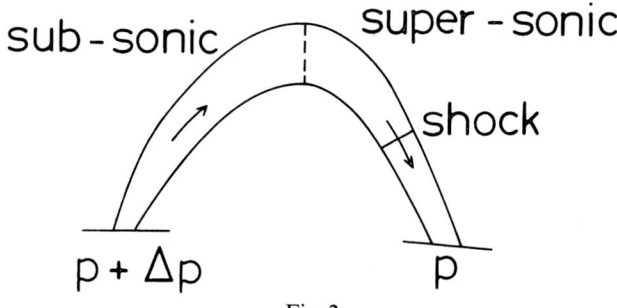

Fig. 2.

The scale height $H = RT/g$ is 0.9×10^{10} cm (pure hydrogen corona). p_0 varies with the footpoints of the magnetic flux tubes. A flux tube with both ends entering the photosphere can only be in hydrostatic equilibrium if the pressure at the base of the corona at both ends balances exactly. Any unbalance kept up over a sound travel time along the flux tube leads to a flow of gas along the field. This flow will reach supersonic velocities and pass through shock fronts on its descending branch (Meyer and Schmidt, 1968). This is due to the small viscosity $\mu = 2.2 \times 10^{-15}\,T^{-5/2}/\ln \Lambda = 0.35$ gr/cm sec of the coronal gas (Spitzer, l.c.). The Reynolds number $\mathrm{Re} = \varrho v l/\mu$ for $v \approx$ velocity of sound $V_s = 160$ km/sec and $l \approx 10^5$ cm is $\mathrm{Re} = 220$. Viscosity can only keep velocities below the speed of sound when the ratio

$$\frac{v}{V_s} \approx \mathrm{Re}\,\frac{\Delta p}{p}$$

is smaller than one. For any relative pressure difference $\Delta p/p > 1/\mathrm{Re} \approx 0.5\%$ shock fronts will occur. Since larger pressure differences could be expected between footpoints of coronal flux tubes it is surprising that velocities of the order of the sound velocity are not frequently observed in the corona where loop structures seems to be common.

5. Magnetohydrodynamic Waves

A. LONGITUDINAL WAVES

Consider a cube of gas of height h and cross sectional area S vertically penetrated by a magnetic field B with the lateral 'magnetic pressure' $p_m = B^2/8\pi$. In lateral compression ($h = \mathrm{const}$) the magnetic flux $\phi = B \cdot S$ and the mass $M = \varrho h S$ remain constant, that is, B is proportional to ϱ.

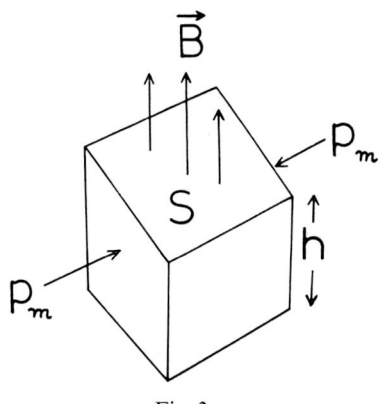

Fig. 3.

The increase of magnetic pressure p_m in compression leads to the same restoring action and wave motion as would a gas pressure p. Corresponding to the sound velocity $V_s = (dp/d\varrho)^{1/2}$ one has the magnetic wave velocity

$$V_A = (dp_m/d\varrho)^{1/2} = \left(\frac{B^2}{8\pi\varrho}\frac{d\ln B^2}{d\ln\varrho}\right)^{1/2} = B/(4\pi\varrho)^{1/2},$$

the Alfvén speed. If the gas pressure acts at the same time, the propagation speed of this longitudinal, compressional wave is

$$V = [d(p + p_m)/d\varrho]^{1/2} = (V_A^2 + V_s^2)^{1/2}.$$

This is the fast hydromagnetic mode for propagation at right angle to the magnetic field.

B. TRANSVERSAL WAVES

Another mode of wave motion made possible by the coupling of the magnetic field to the coronal gas is the compression-free, transversal 'pure' Alfvén wave. In it the field lines are laterally displaced from their straight equilibrium position, like the strings of a string instrument. They oscillate around their equilibrium position due to the inertia of the material tied to them. Alfvén waves travel along the magnetic field with the Alfvén speed.

For other directions of propagation the fast mode is a mixture of compressional and transversal motion. For propagation along the field it degenerates into the pure Alfvén mode. In the corona, where the magnetic field dominates, these two modes are the waves with which the field reattains equilibrium following a disturbance. The Alfvén speed for a field of 4 G and $n = 3 \times 10^8/\text{cm}^3$ is $V_A = 520$ km/sec.

C. SOUND WAVES

For strong magnetic fields the third MHD-wave mode becomes a sound wave traveling along the magnetic flux tubes confining it. The group velocity is the speed of sound, $V_s \approx 160$ km/sec. This is a simple consequence of the fact that the gas can freely move only along the field without bringing strong magnetic forces into play.

D. EFFECTS OF DENSITY LAYERING

All hydromagnetic modes in the corona are affected by the density layering. In magnetic field dominated regions of the corona the dispersion relation for the sound mode becomes

$$\omega^2 = V_s^2 (k_\parallel^2 + \cos^2\theta/4H^2),$$

where $k_\parallel$ is the wave number parallel to the magnetic field and θ is the angle of inclination between field and vertical direction. The cut-off frequency for $k_\parallel = 0$ is of the order of 10^{-3} sec^{-1}, but depends on $\cos\theta$ and is therefore not uniform through the corona. The corona does not present one unique resonance oscillation as do the chromosphere and photosphere in weak field regions. – From conservation of the energy flux density for linearized sound wave motion, $\varrho v^2 V_{gr} = $const, one obtains for the velocity amplitude $v = v_0 (\varrho/\varrho_0)^{-1/2} = v_0 e^{h/2H}$ in the isothermal corona: Sound waves steepen up into shock waves while traveling upwards, finally dissipating part of their energy and contributing to the heating of the corona. The sound mode is also responsible for shock fronts traveling into interplanetary space following strong flares.

Alfvén and fast hydromagnetic waves are strongly affected by the density dependence of their propagation speed, $V_A = B/(4\pi\varrho)^{1/2}$ proportional to $e^{h/2H}$. The refractive index $\sim 1/V_A$ decreases upwards and leads to a downward refraction of the wave energy, which therefore cannot escape upwards. A further reflection at the sudden density jump of the corona-chromosphere interface can lead to a wave guiding of the fast mode along the solar surface through the lower regions of the corona. This type of propagation is the cause for fast traveling disturbances frequently observed in the light of Hα following solar flares.

6. Importance of Microscopic Plasma State

The microscopic state does not seem to play an important separate role in the gross dynamical structure of the solar corona. However, a large number of events in the solar corona occur in which a non equilibrium microscopic structure of the plasma seems to be important. This is indicated by particle acceleration in solar flares and by radio and X-ray observations of non thermal coronal radiation. Obviously the corona can strongly deviate from local kinetic equilibrium, possibly through rapid instabilities. It is conceivable that this will also affect the macroscopic transport coefficients, especially the electrical conductivity of the corona. There is evidence for an 'anomalous', lower, electric conductivity in the coronal magnetic connection between active regions of different hemispheres, as visible in X-ray photographs of the sun (such connections could not come about if the ideal conductivity would hold), and in the surprisingly good agreement of currentfree magnetic configurations, computed from photospheric flux measurements as suggested by Schmidt (1964, 1968) with the observed coronal field structure (Newkirk and Altschuler, 1970).

7. Suggested Literature

For a concise text on nearly all aspects of a plasma bearing on the solar corona the reader is referred to L. Spitzer's book *Physics of Fully Ionized Gases*. Further text books of interest for coronal magnetohydrodynamics are those by Dungey, by Ferraro and Plumpton, and by Alfvén and Fälthammar. Attention is also called to articles by Kulsrud, Lüst and Sturrock in the Proceedings of the XXXIX Varenna Course on Plasma Astrophysics.

Note added in proof: The author wishes to express his apologies to Dr. Macris and the other authors of this volume for a long delay of this manuscript.

References

Alfvén, H. and Fälthammar, C.-G.: 1963, *Cosmical Electrodynamics*, 2nd ed., Clarendon Press, Oxford.
Chapman, S. and Cowling, T. G.: 1960, *The Mathematical Theory of Non-Uniform Gases*, University Press, Cambridge.
Dungey, J. W.: 1958, *Cosmic Electrodynamics*, Cambridge University Press.
Ferraro, V. C. A. and Plumpton, C.: 1961, *Magneto-Fluid Mechanics*, Oxford University Press.
Kulsrud, R. M.: 1967, 'Plasma Instabilities', in Proceedings of the International School of Physics 'Enrico Fermi', Course 39 *Plasma Astrophysics*, p. 46, P. A. Sturrock (ed.), Academic Press, New York, London.
Lüst, R.: 1967, 'Introduction to Plasma Physics', in Proceedings of the International School of Physics 'Enrico Fermi', Course 39 *Plasma Astrophysics*, p. 1, P. A. Sturrock (ed.), Academic Press, New York, London.
Meyer, F. and Schmidt, H. U.: 1968, *Z. Angew. Math. Mechanik* **48**, T218, and *Mitt. Astron. Ges.* **25**, 194.
Newkirk, G., Jr., and Altschuler, M. D.: 1970, *Solar Phys.* **13**, 131.
Petschek, H. E.: 1964, *AAS-NASA Symposium on Solar Flares*, p. 425, NASA, Washington.
Schmidt, H. U.: 1964, *AAS-NASA Symposium on Solar Flares*, p. 107, NASA, Washington; 1966, Atti del Convegno sui Campi Magnetici Solari, Firenze.
Spitzer, L., Jr.: 1962, *Physics of Fully Ionized Gases, Interscience Tracts on Physics and Astronomy*, 2nd ed., Interscience Publishers, New York, London.
Sturrock, P. A.: 1967, 'Waves in Plasmas', in Proceedings of the International School of Physics 'Enrico Fermi', Course 39 *Plasma Astrophysics*, p. 24, P. A. Sturrock (ed.), Academic Press, New York, London.

4. THE CHROMOSPHERE-CORONA TRANSITION REGION

R. GRANT ATHAY

High Altitude Observatory, National Center for Atmospheric Research*, Boulder, Colorado, U.S.A.

1. Introduction

Over the past two decades solar physicists have become increasingly aware that the chromosphere-corona transition region is a phenomenon of fundamental importance and of extremely unusual characteristics. This recent awareness was first mildly stimulated by solar radio observations at cm wavelengths and by improved observations of the optical continuum spectrum of the upper chromosphere and low corona and forbidden coronal line emission at the 1952 total solar eclipse. However, interest in the transition region remained at a relatively low ebb so long as these were the only types of observations available.

Intensities observed in the radio and optical continuum are characteristically slow functions of temperature. As a result, they do not sharply delineate the transition region. At no wavelength in the continuum does the observed radiation originate exclusively within the transition region. On the other hand, observations of coronal forbidden lines could, in principle, provide data that are sharply concentrated in the transition region. A given stage of ionization of a heavy element, such as Fe, exists in large abundance in a restricted temperature range only. It is expected, therefore, that spectral lines (both forbidden and permitted) of certain ionic species will originate entirely within the transition region. However, because the transition region is relatively thin in geometrical extent the forbidden lines originating there will be far weaker than those originating within the corona. Thus far, we have not succeeded in observing coronal lines as faint as is to be expected for forbidden lines arising within the transition region.

The real stimulus behind current interest in the transition region has come from observations of the permitted lines in the extreme ultraviolet. Under coronal conditions the permitted lines are not necessarily a great deal stronger than forbidden lines. Their uniqueness lies in the fact that they occur in the extreme ultraviolet where the continuum is orders of magnitude weaker than in the visual spectrum. Thus, an emission line with a given photon flux may be observed with a favorable signal to noise ratio if it occurs at $\lambda 1000$ whereas the signal to noise ratio may be very unfavorable if the same line photon flux were observed in the visual. At the center of the solar disk, for example, the photon flux at the solar surface in one angstrom of continuum is of the order of 10^{18} cm^{-2} sec^{-1} at $\lambda 5000$ and of the order of 10^{13} cm^{-2} sec^{-1} at $\lambda 1000$. The brightness of the inner corona just beyond the edge of the solar disk is about 10^{-5} the brightness of the disk. Hence, the signal to noise problem for

---

* The National Center for Atmospheric Research is sponsored by the National Science Foundation.

observing permitted coronal lines near $\lambda 1000$ against the solar disk is comparable to that for observing coronal forbidden lines in the visual spectrum above the solar limb at total eclipse.

Among the identified emission lines observed in the extreme ultraviolet are many that originate in the transition region. Their parent ions exist only at temperatures intermediate to those found in the chromosphere and those found in the corona. It is by virtue of such lines that we are now able to discuss the nature of the transition region and to draw attention to some of its remarkable properties. Radio and optical continuum data provide valuable supplemental data when used in conjunction with the permitted spectral lines. Nevertheless, it is the lines that provide the essential elements of the analysis. It is not too surprising therefore that our interest in the transition region should parallel closely our advances in space astronomy.

2. Historical Development

A. HOW THICK IS THE TRANSITION REGION?

The first suggestion that the chromosphere-corona transition region may exhibit unusually large temperature gradients came from Shklovskii (1944). His argument was based upon the decreased radiating efficiency of gases at low density together with the recognition that the gas density in the corona was at most a few percent of that in the upper chromosphere, and that the loss of radiating efficiency in the corona due to the lower density was at least partially compensated by the higher coronal temperature. He suggested, on this basis, that the temperature would rise quickly to its coronal value once it left the chromosphere regime.

Alfvén suggested in 1941 that thermal conduction from the corona may play an important role in removing energy from the corona and in determining the temperature structure of the adjacent regions. Giovanelli (1949) and Woolley and Allen (1950) computed the temperature structure of the transition region assuming that the flux of thermal conduction was constant through the transition region. They ignored all other sources and sinks of energy. Their models were characterized by high temperature gradients, hence a thin transition region.

In the purely conductive model, of course, the temperature gradient depends explicitly upon the adopted value for the conductive flux. Giovanelli (1949) adopted a flux of 6×10^5 erg cm^{-2} sec^{-1} as compared to 1.7×10^4 erg cm^{-2} sec^{-1} adopted by Woolley and Allen (1950). Thus, the temperature gradients in Giovanelli's model exceed by a factor of 35 those in the Woolley-Allen model, for a given temperature. For the flux picked by Giovanelli, $dT/dh \approx 10^{-1}$ at $T = 10^5$.

At about the same time that Giovanelli (1949) proposed the first conductive model of the transition region, Schatzman (1949) proposed a model in which energy was supplied by dissipation of shock waves and removed by radiation. Conduction was not explicitly taken into account. Interestingly, this model again led to a high temperature gradient; in this case, because of the refraction phenomenon of shock waves in atmospheres with steep gradients. The temperature gradient found by Schatzman was comparable to that found by Woolley and Allen and about an order of magnitude lower than that found by Giovanelli.

Schirmer (1950), Weymann (1960), Dubov (1960) and Osterbrock (1961) each considered the balance of shock wave dissipation and radiation in the upper chromosphere but they did not discuss the nature of the transition region per se.

In 1956, Athay and Thomas introduced a further argument favoring a sharp transition region. They reasoned from thermal stability arguments that only those temperatures will be found in the solar atmosphere for which the radiant energy loss is an increasing function of temperature. For any one element in a given stage of ionization the radiant energy loss per gram first increases with temperature then decreases after passing through a maximum. Thus, as the temperature increases through the solar atmosphere the dominant energy loss by radiation will shift progressively from one ionic specie to another representative of a higher excitation stage. Athay and Thomas (1956) identified the chromosphere with the dominant radiating species of H I and He II.

Beyond He there are no elements in the sun abundant enough to radiate as efficiently as hydrogen if they are in low states of ionization. However, in a rather crude sense the radiating efficiency of an ion increases as Z^4, where Z is the unbalanced nuclear charge. Thus, an element whose abundance relative to hydrogen is 10^{-4} would begin to have the same radiating efficiency as hydrogen if ionized ten times. It was argued by Athay and Thomas, therefore, that the temperature regime between that where He II was an efficient radiator ($\sim 40000°$) and that where the more abundant heavy elements were approximately ten times ionized ($\sim 10^{6\,\circ}$) would be essentially missing in the solar atmosphere. This meant that T_e would attempt to jump discontinuously from approximately $40000°$ at the top of the chromosphere to approximately $10^{6\,\circ}$ at the base of the corona; the intervening region being smoothed only by thermal conduction. The description of the transition region emerging from this picture does not differ essentially from that proposed by Giovanelli (1949). However, there was now a physical argument supporting the conductive model.

In 1961, de Jager and Kuperus included the simultaneous effects of shock dissipation, thermal conduction and radiation. They found temperature gradients in the transition region smaller than those found in the purely conductive models proposed by Giovanelli (1949) but much larger than those found by Schatzman (1949), who neglected conduction. Kuperus (1965) later extended and refined the initial computations of de Jager and Kuperus (1961) with similar results in the transition region.

Uchida (1963) added the effects of horizontal magnetic fields to those of thermal conduction, shock dissipation and radiation. He concluded that the effect of the horizontal magnetic field was to steepen the temperature gradient (because of the reduced conductivity) in the transition region and to raise the coronal temperature. Earlier, Osterbrock (1961) had shown that vertical magnetic fields increased the number of possible modes of wave propagation and could be effective in increasing the amount of wave energy available for heating the higher layers of the atmosphere, including the transition region. It is clear therefore that magnetic fields play an important role in the structure of the transition region.

Ulmschneider (1967) has considered the problem of hydrodynamic expansion together with shock dissipation, conduction and radiation. He again finds a temperature gradient in the transition region comparable to that found by Giovanelli (1949) and de Jager and Kuperus (1961). Kopp (1968) has considered a similar problem, but not with explicit reference to the transition region.

Models of the transition region discussed up to this point are essentially based on either theory or speculation and suffer the common difficulty of not relating directly to a critical observational quantity. Parallel attempts were made to construct models from observational data. Most of these models differed markedly from the theoretical models. However, the data used to construct the models usually either were of poor quality or they did not lead to a unique model. In the following, we shall comment on a few of the models representative of the different types of data.

Piddington (1954) constructed one of the early models of the transition region using a rather comprehensive set of radio and optical data. This model gave a very broad transition region in sharp contrast to the theoretical models. Near $T=10^5$ K the temperature gradient in the Piddington model is approximately 10^{-3} K cm^{-1} as opposed to 10^{-1} in the Giovanelli model. A similar analysis by Shklovskii and Kononovich (1958) lead to a similar result. However, Athay (1959a) was able to closely reproduce the optical and radio data with a stepped temperature distribution corresponding to an infinite temperature gradient in the transition region. Such a model is, of course, unphysical, but it demonstrated, nevertheless, that the type of model constructed by Piddington was in no way unique.

A model of the transition region based upon optical data and the assumption of hydrostatic equilibrium was proposed by Pottasch (1960). The optical data were used to obtain the mean electron density as a function of height. The temperature was then derived from the hydrostatic conditions. This model gave an even flatter transition region with $dT/dh \approx 10^{-4}$ K cm^{-1} near $T=10^5$ K. An obvious difficulty with this type of model is again the nonuniqueness of the interpretation of the data. Widely different electron density distributions will fit the observational data equally well.

By far the most useful data for constructing models of the transition region are the XUV spectral line fluxes observed on the solar disk. These data are both very sensitive to the nature of the transition region and subject to only a relatively small latitude of interpretation. Their usefulness in specifying the nature of the transition region was pointed out by de Jager (1959), and the technique of analysis was established by Ivanov-Kholodnyi and Nikol'skii (1961). The latter authors constructed a model of the transition region from XUV data that agreed, broadly speaking, with the radio models. However, their data were incomplete and preliminary. Later models using essentially the same analytical techniques gave much steeper temperature gradients.

Zirin and Dietz (1963) attempted to combine radio, optical, XUV and infrared data to construct a model of the transition region. However, their XUV data were again not complete enough to give a definitive model. Although these authors described the transition region as being very sharp, they did not give numerical values of dT/dh.

Pottasch (1964) was the first to analyze a large body of XUV data covering the transition region. His analysis yielded a temperature gradient of approximately 10^{-2} K cm^{-1} at $T=10^5$ K, which is near the value proposed by Schatzman (1949) and Woolley and Allen (1950). A similar analysis was carried out by Jordan (1965).

A reformulation of the analytical method used with the XUV data enabled Athay (1966a) to obtain a more direct evaluation of the temperature gradient. His analysis, as well as a subsequent one by Dupree and Goldberg (1967), gave $dT/dh \approx 10^{-1}$ at $T=10^5$ K. This is the highest value given by any of the models based on observational data and agrees with the value initially proposed by Giovanelli (1949). Because dT/dh is obtained from the observational data in a more direct manner in these latter analyses than in earlier analyses it is probably a more reliable result. The derivation of dT/dh by this method will be illustrated in a subsequent section of this paper.

B. WHERE IS THE TRANSITION REGION?

It was believed for some time that the high temperature of the corona was not reached until a height above the photosphere of some appreciable fraction of the solar radius. The first real data brought to bear on this concept was coronagraphic observations of forbidden coronal lines and eclipse observations of the extent of Hα in the chromosphere. The Hα chromosphere is irregular in extent because of prominences and spicules. At modest angular resolution, which is typical of most eclipse data, Hα emission can be traced to about 10000 km above the solar limb. Early coronagraphic observations suggested that the maximum intensity in the forbidden lines occurred at heights of the order of one minute of arc above the limb. This observation, coupled with the 10000 km extent of the Hα chromosphere led people to believe that the base of the corona was at least some 20000 km above the solar limb. It was later realized, however, that both sets of observations were misleading and essentially reflected only the properties of the observing conditions and apparatus.

Following the 1952 eclipse Athay and Roberts (1955) showed that the Fe XI forbidden line at $\lambda 7892$ reached maximum intensity definitely less than 7000 km above the limb and probably less than 3500 km above the limb. Also, Dunn's careful observations of the Hα chromosphere made in the mid-1950's showed that above 3000 km the chromosphere consisted only of spicules covering a small fraction of the solar surface. Still further evidence that the base of the corona was relatively close to the solar limb came from analysis of continuum data from the 1952 eclipse by Athay *et al.* (1955). These data indicated that a sharp rise in temperature set in at a height of about 3000 km above the limb.

Little has been done within recent years to refine our knowledge of just where the transition region is located. It seems very probable at this point in time, however, that the transition region is within about five arc seconds (3500 km) from the solar limb. In order to refine the observations much further we need either direct resolution of one or two arc seconds or an effective resolution of this order obtained from large statistical samples of data. At the time of this writing, analyses of OSO IV data in the XUV are being carried out for this purpose. The preliminary results confirm the suggestion that the transition region begins within about 5 arc sec above the limb (George Simon, private communication).

The traditional observational methods for locating the transition region involve limb observations, which are confused by a number of factors. These include obscuration of the transition region by spicules, integration along the solar tangent and irregular structure of the transition region itself. The former difficulty can be avoided by selecting coronal lines on the long wavelength side of the Lyman series limit. Satellite data for the limb darkening of XUV lines show clear evidence that spicules are opaque in the $\lambda 900$ to $\lambda 500$ region (Withbroe, 1970) but transparent at all $\lambda > 900$ except for a few of the stronger spectral lines of such ions at neutral hydrogen and ionized calcium.

The integration along the solar tangent, which is endemic to all limb observations, inevitably diffuses and partially obscures the appearance of the transition region. Also, since both the upper chromosphere and low corona have highly irregular geometries, it seems highly probable that the transition region itself will occur at quite different heights from point to point on the solar surface. Both of these effects tend to discourage the prospect for an accurate, definitive observation of the 'height' of the transition layer above the limb. Nevertheless the limb data provide the most promising observational evidence from which the height of the transition region can be studied and the existing data of this type can certainly be improved upon.

In the vicinity of active regions where individual bright emission features can be identified, it is possible to derive the height of the transition region from disk data in the XUV. If the emitting feature is displaced from the center of the disk, its apparent position on the disk will depend upon the height of the emitting region. Thus, if the position is known the relative heights of origin of different lines can be derived. This is the method used by Simon with OSO IV data.

Indirect, nevertheless strong, evidence that the transition region is not more than a few arc seconds above the limb is provided by model solar chromospheres. It is now fairly well established that the chromosphere below a height of about 1000 km above the limb has a mean temperature of the order of 6000 to 7000° and that hydrogen is predominantly neutral in these layers. The mean molecular weight of these layers of the chromosphere is therefore approximately 1.4, assuming that the relative abundance of helium is 0.1. For a mean molecular weight, μ, of 1.4 and a temperature of 6000°, the hydrostatic density scale height is 130 km. Hence, 1000 km above the limb the density in a hydrostatic chromosphere would have decreased by a factor $e^{7.7} = 2200$. At the limb, the hydrogen number density is 2×10^{16} cm^{-3}. Thus, at 1000 km we expect a density of about 10^{13} cm^{-3}. The gas pressure at this point will be of the order of $nkT = 8$ dyne cm^{-2}. The gas pressure in the upper photosphere, by comparison, is approximately 12000 dyne cm^{-2} and in the low corona it is approximately 0.2 dyne cm^{-2}.

If we make the physically plausible assumption that the gas pressure decreases steadily outwards in the solar atmosphere, in the mean, then the transition region must begin at a gas pressure no smaller than 0.2 dyne cm^{-2}. The preceding estimate placed the gas pressure at 1000 km above the limb as 8 dyne cm^{-2}. (More detailed models (cf. Athay, 1969) actually give a somewhat lower value.) It is possible that the low chromosphere is supported somewhat by random mass motions of the so-called 'microturbulence' class. However, the widths of the O I lines at $\lambda 1302$ and $\lambda 1304$, whose centers are formed well above 1000 km in the chromosphere show broadening velocities of only about 8 km/sec (Athay and Canfield, 1970), which is less than the thermal velocity of hydrogen atoms. Similar results are obtained for the K_3 profile of the Ca II H and K lines which are formed above 500 km in the chromosphere (cf. Athay and Skumanich, 1968; Dumont, 1967). Even if the microturbulent velocity equalled the mean thermal velocity the root mean square velocity would exceed the thermal velocity by only a factor 1.4, and the scale height would increase by 1.4. This would move the height of the surface of 8 dyne cm^{-2} gas pressure to 1400 km rather than 1000 km. It seems clear therefore that a gas pressure of the order of 8 dyne cm^{-2} is reached near 1000 km.

Above 1000 km the chromospheric model is still highly uncertain. Helium emission becomes strong at about this height and the chromosphere departs strongly from spherical symmetry (cf. Athay and Menzel, 1955). However, the gas pressure can drop only by a factor of 40 (less than e^4) between 1000 km and the transition region. The most pessimistic view, i.e., the one that makes the chromosphere as thick as possible, would be to assume that hydrogen and helium are both ionized so that $\mu = 0.7$ and that the temperature has some high value, say, 30000°. This gives a density scale height of 1300 km and reduces the gas pressure from 8 dyne cm^{-2} at 1000 km to 0.2 dyne cm^{-2} at a height of 6000 km, i.e., at about 8 arc sec above the limb.

We have, of course, been overly pessimistic in the upper chromosphere since we have allowed the temperature to jump from 6000° to 30000° without a reduction of pressure and we must expect the gas pressure to reach 0.2 dyne cm^{-2}, much before 6000 km. In the Athay-Menzel-Pecker-Thomas model of the chromosphere (1955), for example, the gas pressure is only 0.3 dyne cm^{-1} at 2000 km; in the Bilderberg model a pressure of 0.2 dyne cm^{-2} is reached at 1300 km above the limb; and in a model computed by Athay (1969), which is extended somewhat by microturbulent pressure, a pressure of 0.2 dyne cm^{-2} is reached at 2100 km. Each of these models is highly uncertain above 1000 km. Nevertheless, the clear suggestion is that the chromospheric gas pressure reaches a near coronal value in considerably less than 8 arc sec above the limb.

3. Derivation of the Conductive Flux and Temperature Model

A. DERIVATION OF THE CONDUCTIVE FLUX FROM XUV DATA

Spectral lines from permitted transitions in some multiply ionized metals arise entirely within the chromosphere-corona transition region. Observations of these lines against the solar disk permit a derivation of the geometrical thickness of the radiating region, assuming that the pressure and temperature are known in the radiating layer. The temperature and thickness combined give both the temperature gradient and an estimate of the energy flux carried by thermal conduction. The conductive flux, F_c, is given by

$$F_c = 1.1 \times 10^{-6} T_e^{5/2} \, dT_e/dh \qquad (1)$$

for a solar mixture of gases at a pressure of 10^{-1} dyne cm^{-2} and free of magnetic fields (Ulmschneider, 1969).

In the low corona the current estimates of n_e and T_e are 4×10^8 cm^{-3} and 1.5×10^6 K, respectively. In the upper chromosphere where Ly-α and the helium lines are formed $n_e \approx 2 \times 10^{10}$ and $T_e \approx 30000°$ (Athay, 1965). For both regions the product of n_e and T_e is 6×10^{14}, which corresponds to a gas pressure of 0.17 dyne cm^{-2}. Thus, we shall assume in the following that $n_e T_e = 6 \times 10^{14}$ throughout the transition region. This figure could be in error by a factor as large as two, but is not likely to be in error by a much larger factor. For convenience, in the following equations we denote the value of $n_e T_e$ by C_1.

The quantity $n_e T_e$ is useful also for estimating the opacities to be expected in the transition region. At $T_e = 10^5$, for example, we expect $n_e = 6 \times 10^9$. An element that has an abundance relative to hydrogen of 3×10^{-5}, will have a number density at $T_e = 10^5$ of 2×10^5. Assuming that all of the atoms of the element are in one stage of ionization, we obtain for the dominant ion stage a density 2×10^5 cm^{-3}. If this ion has a mean random velocity of, say, 20 km/sec and has a resonance line at $\lambda 1000$ with an oscillator strength of unity, its absorption coefficient at line center will be approximately 10^{-13} cm^{-2}. Thus, an optical thickness of unity will be reached in a distance given by $(2 \times 10^5 \times 10^{-13})^{-1} = 5 \times 10^7$ cm.

A photon generated within the transition region will escape provided the thickness of the transition region is less than a degradation length. For resonance lines formed at the densities considered here, the degradation lengths are typically greater than about 10^4 as measured in τ_0, the line center opacity (cf. Jefferies, 1960). Thus, we would not expect photons to be trapped within the transition region unless the thickness of the region exceeded $\tau_0 = 10^4$, or exceeded 5×10^{11} cm. The transition region, in fact, is more nearly 10^7 cm in thickness, which gives $\tau_0 < 1$. Thus, we may safely treat the transition region as being effectively thin.

In the effectively thin approximation the total energy emitted through an area 1 cm^2 at the top of the corona in a spectral line formed by a transition from level 2 to level 1 is

$$F = \tfrac{1}{2} \int_h^\infty h\nu A_{21} n_2 \, dh \qquad (2)$$

where A_{21} is the transition probability, n_2 is the population density of level 2 and dh is an element of geometrical height. The factor $\tfrac{1}{2}$ in Equation (2) accounts for the fact that half the radiation escapes to the photosphere where it is absorbed. We assume that electrons arrive in level 2 via electron collisions with atoms in the $n=1$ level and return to level 1 via spontaneous transitions. Thus, we write

$$n_1 C_{12} = n_2 A_{21}, \tag{3}$$

so that equation (2) becomes

$$F = \tfrac{1}{2} \int_h^\infty h\nu C_{12} n_1 \, dh. \tag{4}$$

Note that Equation (3) implies that transitions from level 1 with equal values of C_{12} will have equal values of $A_{21} n_2$ regardless of the value of A_{21}. It is for this reason that the coronal forbidden lines are of the same approximate intensity as the permitted lines, since the collision rates C_{12} are approximately the same for forbidden lines as for permitted lines.

Following Pottasch (1964), we write for C_{12}

$$C_{12} = 1.7 \times 10^{-3} n_e T_e^{-1/2} \chi^{-1} fg \exp(-11\,600\chi/T_e), \tag{5}$$

where χ is the excitation energy in electron volts, f is the oscillator strength and g is a Gaunt factor. For n_1, we write

$$n_1 = A n_e \frac{n_i}{n_{el}} = A n_e a_i. \tag{6}$$

A is the abundance of the element relative to hydrogen, and the quantity n_i/n_{el} ($=a_i$) is the fraction of the element that is in the ith state of ionization. This latter quantity may be computed independently as a function of T_e. Dielectronic recombination plays an important role in some ionization stages and should be taken into account in a_i. From Equations (4), (5) and (6) we obtain

$$F = 0.55 \times 10^{-15} gfA \int_h^\infty n_e^2 a_i T_e^{-1/2} \exp(-11\,600\chi/T_e) \, dh \tag{7}$$

or,

$$F = 0.55 \times 10^{-15} gfA \int_h^\infty n_e^2 g(T_e) \, dh. \tag{8}$$

The quantity $g(T_e)$ is a known function of T_e. It is a relatively narrow function with a maximum at T_{e1} and with a width given by the approximate limits $\tfrac{2}{3} T_{e1}$ and $\tfrac{4}{3} T_{e1}$. Pottasch (1964) used this result to rewrite Equation (8) as

$$F = 0.55 \times 10^{-15} gfAg(T_e) \int_{h(2/3\,T_{e1})}^{h(4/3\,T_{e1})} n_e^2 \, dh. \tag{9}$$

There are two unknowns in Equation (9): the abundance A and the emission measure $\int_{h(2/3T_{e1})}^{h(4/3T_{e1})} n_e^2 \, dh$. For a given element, A is assumed to be constant with depth and the different stages of ionization for the element define a curve specifying the emission measure as a function of T_{e1}. Since there must be a unique curve for the emission measure, the abundance of one element relative to another can be obtained by shifting the curves for each element, keeping T_{e1} constant, until they superpose. In this way Pottasch (1964) derived both the relative abundances and the emission measure.

Delache (1967) has drawn attention to the fact that thermal diffusion of ions in a sharp transition region will result in A being a function of depth. Mixing due to vertical motions through the transition region will tend to nullify this effect. Quantitative estimates of the height dependence (or constancy) of A, are difficult because of the unknown effects of mixing. We shall continue to treat A as constant, but we note that A may have some depth dependence.

The emission measure implicitly contains the temperature gradient, since it gives the value of $\int n_e^2 \, dh$ between two well defined temperature limits. As noted in the beginning of this section, the quantity $n_e T_e$ must be nearly constant through the transition region. Let us introduce this explicitly, therefore, and derive the temperature gradient directly as was done earlier by the author (Athay, 1966a). We find

$$F = 0.55 \times 10^{-15} gf A C_1^2 \int_{h(2/3\,T_{e1})}^{h(4/3\,T_{e1})} T_e^{-2} g(T_e) \, dh, \qquad (10)$$

or replacing dh with $dT_e (dT_e/dh)^{-1}$, we find

$$F = 0.55 \times 10^{-15} gf A C_1^2 \int_{2/3\,T_{e1}}^{4/3\,T_{e1}} T_e^{-2} g(T_e) \left(\frac{dT_e}{dh}\right)^{-1} dT_e. \qquad (11)$$

Finally, we assume that dT_e/dh is constant between $\tfrac{4}{3} T_{e1}$ and $\tfrac{2}{3} T_{e1}$ and write

$$F = 0.55 \times 10^{-15} gf C_1^2 G(T_e) \langle (dT_e/dh)^{-1} \rangle A, \qquad (12)$$

where $G(T_e)$ is given by the integral after $(dT_e/dh)^{-1}$ is removed. The unknowns now are $\langle (dT_e/dh)^{-1} \rangle$ and A, both of which can be derived following the method outlined by Pottasch (1964). Figure 1 shows the result obtained by Athay (1966a) in which dielectronic recombinations were not included and in which relative abundances of Goldberg et al. (1960) were adjusted slightly to improve the scatter in the data. Figure 2 shows the results obtained by Dupree and Goldberg (1967) for O and Si (the three points for iron do not really help in defining the curve) including the effects of dielectronic recombinations.

In both Figures 1 and 2 the straight portion of the dashed curve is the curve for $T_e^{5/2} dT_e/dh = \text{const}$, i.e., $F_c = \text{const}$. Between $T_e = 10^5$ and 10^6 K it is clear from the data that F_c is, in fact, nearly constant. The value of $T_e^{5/2} dT_e/dh$ for a Si abundance

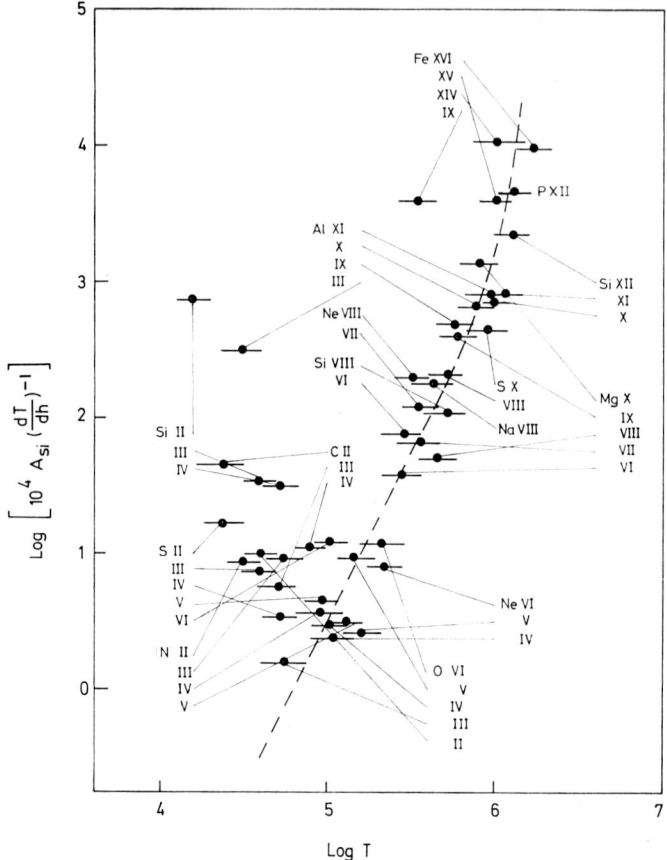

Figure 1. A plot of the relationship between T and $(dT/dh)^{-1}$ derived from XUV lines. Dielectronic recombinations have been ignored. The silicon abundance, A_{Si}, is 3×10^{-5}. Courtesy *Astrophys. J.* (Univ. of Chicago Press), Athay (1966a), copyright 1966 by the Univ. of Chicago Press.

of 3×10^{-5} is 3×10^{11} from Figure 1 and 10^{12} from Figure 2. Equation (1) gives for the corresponding heat fluxes 3.3×10^5 erg cm^{-2} sec^{-1} and 1.1×10^6 erg cm^{-2} sec^{-1}.

Two points concerning the above estimates of F_c are worth noting: (1) the constancy of F_c with T_e between $T_e = 10^5$ and 10^6 K is independent of whether dielectronic recombinations are included or not, only the constant value of F_c changes, and (2) F_c is very large. By comparison, for example, the Ly-α flux from the upper chromosphere, which is the major energy loss, is 2.7×10^5 erg cm^{-2} sec^{-1}. Also, the entire radiation loss from the corona and the transition region combined is of the same order as the Ly-α flux (Hinteregger *et al.*, 1965). Coronal energy losses due to the solar wind and conduction into space are small by comparison.

It is acceptable, perhaps, to have F_c as large as the radiation losses from the upper chromosphere and corona, but it would be strange, indeed, if F_c were greater than the radiation losses. Several factors could contribute to errors in our evaluation of F_c.

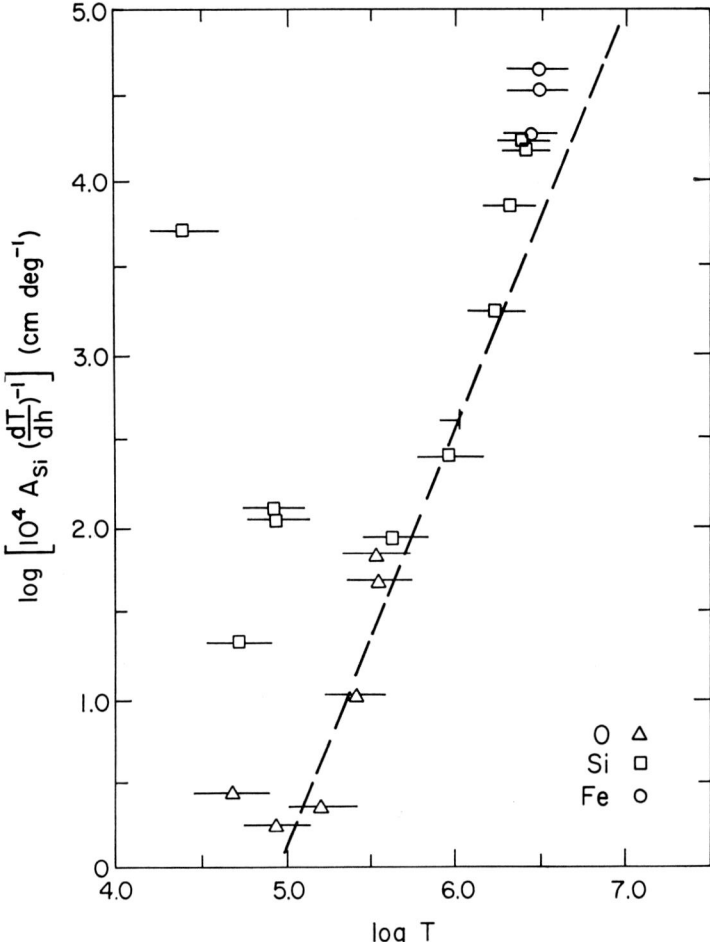

Fig. 2. Same as Figure 1 with dielectronic recombinations included (Dupree and Goldberg, 1967). Courtesy *Solar Phys.*

As already mentioned, C_1 could be in error by perhaps a factor of two. Since F_c is proportional to C_1^2, this would lead to a factor of four error in F_c. Values of C_{12} and the silicon abundance could easily be in error by factors of two. Also, for observed values of F we have used data in which the solar disk is unresolved. Thus, F represents an average taken over the solar disk with no allowance for spatial structure. This introduces additional errors. Furthermore, we have evaluated F_c assuming that the magnetic fields in the transition region are either vertical or non-existent. Horizontal magnetic field components will markedly reduce the conductivity and, consequently, the inferred value of F_c. We cannot conclude, therefore, that F_c really exceeds the coronal and upper chromospheric radiation.

It is clear, however, that F_c is large compared to the radiation loss from the transition region itself, and this perhaps explains the near constancy of F_c. The mere fact that F_c is large compared to the radiation losses within the transition region (presumably, to the dissipation of mechanical energy flux also) means that the divergence of F_c must be small compared to F_c itself, otherwise the thermal conduction would add or subtract too much energy to the transition region to be balanced by the other energy processes. If, for example, F_c changed by a factor of two between $T_e = 10^5$ K and $T_e = 10^6$ K we would have to add or subtract an amount of energy equal to F_c by some other process and we know of no process capable of doing this. Also, the slopes of the line in Figures 1 and 2 would change by only about 10 percent, which is well within the tolerance of the data. Thus, we should not conclude that F_c is strictly constant, only that it is approximately constant. It seems very possible, therefore, that some small gradient is present in F_c and that the divergence of F_c may be an important factor in the energy balance within the transition region.

B. DERIVATION OF THE CONDUCTIVE FLUX FROM RADIO DATA

Pottasch (1964) used disk values of the brightness temperature, T_b, at radiowavelengths to fix the abundance of silicon relative to hydrogen and an absolute calibration for the emission measures. A similar procedure was followed by Dupree and Goldberg (1967). Their results show that the radio brightness temperatures are well represented by the temperature model, that C_1 is indeed approximately constant at 6×10^{14} and that A_{Si} is within acceptable limits of the photosphere value. It is apparent from these results that the radio data verify the general conclusions but do not add any new results.

It is possible to use the radio data in an alternate way that is perhaps more meaningful, viz., to derive F_c in a manner somewhat analogous to the procedure followed for the XUV lines.

The brightness temperature of the sun at centimeter wavelength is given to good approximation by

$$T_b = \int_0^\infty T_e e^{-\tau_\lambda/\mu} \, d\tau_\lambda/\mu, \qquad (13)$$

where $\mu = \cos\theta$ and (Martyn, 1948)

$$d\tau_\lambda = -2.5 \times 10^{-23} (\lambda^2 n_e^2/\eta T_e^{3/2})(3.23 + \log T_e - \tfrac{1}{3} \log n_e). \qquad (14)$$

The latter term in parenthesis is the Gaunt factor and may be taken as a constant equal to 5 for our purposes. Also, we take the index of refraction, η, as unity and obtain

$$d\tau_\lambda = -1.3 \times 10^{-22} \lambda^2 n_e^2 T_e^{-3/2} \, dh. \qquad (15)$$

We next introduce the quantities C_1 and F_c to obtain

$$d\tau_\lambda = -1.4 \times 10^{-28} \lambda^2 \frac{C_1^2}{F_c} T_e^{-1} \, dT_e. \qquad (16)$$

By integration of Equation (16) with C_1 and F_c constant, we obtain

$$\tau_\lambda = 1.4 \times 10^{-28} \lambda^2 \frac{C_1^2}{F_c} (\ln T_{\text{cor}} - \ln T_e), \tag{17}$$

where we have abbreviated the coronal temperature as T_{cor}. Substituting Equations (16) and (17) into equation (15) with a_λ defined as

$$a_\lambda = 1.4 \times 10^{-28} \lambda^2 \frac{C_1^2}{F_c}, \tag{18}$$

we obtain the interesting result that

$$T_b = a_\lambda T_{\text{cor}}^{a_\lambda/\mu} \int_{T_{\text{chr}}}^{T_{\text{cor}}} T_e^{a_\lambda/\mu} \, dT_e/\mu \tag{19}$$

$$\approx \frac{a_\lambda}{a_\lambda + \mu} T_{\text{cor}}. \tag{20}$$

For brightness temperatures much below T_{cor} we must have $a_\lambda \ll 1$ so that $a_\lambda \ll \mu$ over most of the disk. At $\mu = 1$, then, Equations (18) and (20) give

$$F_c = 1.4 \times 10^{-28} \lambda^2 C_1^2 \frac{T_{\text{cor}}}{T_b}, \tag{21}$$

or, for $C_1 = 6 \times 10^{14}$

$$F_c = 50\lambda^2 \frac{T_{\text{cor}}}{T_b}. \tag{22}$$

From brightness temperatures given by Kundu (1964), we obtain $T_b = 1 \times 10^5$ at $\lambda = 25$ cm and $T_b = 3 \times 10^5$ at $\lambda = 40$ cm. These pairs of values give $T_b/\lambda^2 = 160$ and 190 respectively, consistent with the assumptions that C_1 and F_c are approximately constant.

With $T_b/\lambda^2 = 180$, we obtain from equation (22)

$$F_c = 0.3 \, T_{\text{cor}}. \tag{23}$$

Thus, with $T_{\text{cor}} = 1.5 \times 10^6$ K we obtain $F_c = 4.5 \times 10^5$ erg cm^{-2} sec^{-1} in good agreement with the results from the XUV data. Both results depend upon C_1^2. However, in the latter case F_c is proportional to T_{cor} and T_b/λ^2 whereas in the former it was proportional to the abundance of silicon and the observed line fluxes. The two determinations of F_c are quite independent, therefore.

The apparent agreement between the values of F_c obtained from radio and XUV data is open to scrutiny. In computing T_b, for example, we have ignored any contribution from the corona above the transition region, which, in fact, is not entirely permissable. The maximum brightness temperature of the sun occurs near 100 cm and is near 1.5×10^6 K. Part of this brightness temperature is contributed from the corona and part from the transition region. At a constant value of $T_b/\lambda^2 = 180$, we would expect $T_b = 1.8 \times 10^6$ at $\lambda = 100$ cm, which is reasonably consistent with observations. However, at these longer wavelengths the transfer of radio signals is complicated by refraction and this simple extrapolation is not valid. Also, if we assume that all of the brightness at $\lambda = 100$ cm is due to the corona above the transition region and that the brightness at $\lambda = 40$ cm due to the corona should again follow the law $T_b/\lambda^2 = $ constant we would conclude that nearly all of the brightness at 40 cm was due to the corona. The truth lies in between, of course, and it seems likely that by considering the observed values of T_b at 25 cm and 50 cm to originate entirely in the transition region we have perhaps somewhat underestimated T_b/λ^2 for this region and thereby somewhat underestimated F_c.

Finally, we note that a recent analysis of radio data for an active region by Chiuderi et al. (1970) yields a steep transition region with a conductive flux in good agreement with the XUV results. These new results are obtained using an improved analytical technique and do not depend upon the assumption of constant F_c.

C. DERIVATION OF $T_e(h)$

Having shown that F_c is approximately constant at a value near 5×10^5 erg cm^{-2} sec^{-1}, we can immediately derive the relationship between T_e and height. From Equation (1), we obtain

$$4.5 \times 10^{11} \, dh = T_e^{5/2} \, dT_e,$$

or

$$h - h_1 = 2(T_e^{7/2} - T_{e1}^{7/2}), \qquad (24)$$

where in the latter equation h and h_1 are in kilometers and T_e and T_{e1} are in units of 10^5 K.

Values of T_e and n_e are given as a function of $h - h_1$ in Table I for $T_{e1} = 10^5$ K and $n_e T_e = 6 \times 10^{14}$. As noted in the preceding discussion, h_1 is unknown but is believed to lie below 3500 km.

We note from Table I that T_e increases from 1×10^5 to 2×10^5 K in a distance of only 20 km. Zirin (1968) has argued that because of this extreme narrowness of the

TABLE I

T_e and n_e in the Transition Region

$h - h_1$ (km)	−1.8	0	21	92	260	560
T_e	5×10^4	1×10^5	3×10^5	3×10^5	4×10^5	5×10^5
n_e	1.2×10^{10}	6×10^9	3×10^9	2×10^9	1.5×10^9	1.2×10^9
$h - h_1$ (km)	1060	1810	2900	4400	6300	
T_e	6×10^5	7×10^5	8×10^5	9×10^5	1×10^6	
n_e	1×10^9	8.6×10^8	7.5×10^8	6.7×10^8	6×10^8	

transition region ions of a given specie will not remain in the transition region long enough to reach equilibrium and, hence, that the normal ionization equilibrium will not apply. This is an interesting argument that is neither easily dismissed nor easily accepted. Consider, for example, an ion of Ov in the atmosphere at a temperature of 2×10^5 K and at an electron and proton density of 3×10^9 cm^{-3}. The mean-free-path of an Ov ion in such an atmosphere is of the order of 0.1 km (Spitzer, 1962) and the mean collision time is of the order of 10^{-2} sec. The time for a collisional ionization from Ov to Ovi is of the order of 10^2 sec (House, 1964). Thus, during the time required for collisional ionizations the Ov ion will undergo approximately 10^4 encounters. Random walk theory predicts that in 10^4 encounters the ion will travel approximately 10^2 mean-free-path lengths, or approximately 10 km, which is of the same order as the thickness of the transition region.

In the higher transition region the thickness increases much faster than the mean velocity of the ions and the transition region becomes thicker in terms of the mean distance travelled between collisional ionization. It seems reasonable to assume, therefore, that for T_e above about 5×10^5 K the ionization equilibrium is established locally within the transition region. Near $T_e = 1 \times 10^5$ K, however, this may not be true and it may be necessary to reconsider the ionization equilibrium.

Figures 1 and 2 clearly suggest that the region of strong thermal conduction ends abruptly near $T_e = 1 \times 10^5$ K. In fact, this may not be the case, however. If the strong conductive flux were continued to 5×10^4 K, we note from Table I that the transition region would extend an additional 2 km only. It seems clear that in these regions Zirin's arguments are valid and that the ionization equilibrium will not be 'locally' established in this case. Thus, the assumptions underlying the plots in Figures 1 and 2 are no longer valid, and this, in itself, may account for the apparent failure of $F_c =$ const. Other reasons to suggest that the plots in Figures 1 and 2 fail below $T_e = 1 \times 10^5$ K have been given by Athay (1967), and by Kopp and Kuperus (1968).

4. Refined Models

A. CENTER-LIMB DATA

One disadvantage of the models of the transition region discussed in the preceding is the use of globally averaged fluxes for the XUV lines. The average includes active regions as well as center-limb variations. On the other hand, the radio data are supposedly representative of the center of the solar disk under quiet conditions. Since these latter data give essentially the same conductive flux as the XUV data, there is good reason to suppose that the globally averaged model obtained from the XUV lines is not greatly in error.

The globally averaged model can be defended from another point of view. Equations (2) and (4), which underlie the analysis of the XUV data, are based upon the assumption of 'effective thinness' rather than 'optical thinness'. The latter condition requires $\tau_0 < 1$ whereas the former requires only that $\tau_0 < \tau_{deg}$ where we may have $\tau_{deg} \gg 1$.

In the previous section we estimated that $\tau_0 = 1$ in a standard reference line would be reached in a path length of 500 km. A spherical shell of thickness Δh has a maximum path length at the solar limb of $2(2\pi R_0 \Delta h)^{1/2}$ and for a path length of 500 km at the limb we need only have $\Delta h = 0.05$ km. Thus, the transition region, even though very narrow, is wide enough to expect $\tau_0 \gg 1$ for many resonance lines at the limb. At $\Delta h = 100$ km, for example, the path length at the limb is 25 000 km. For the reference line used in estimating τ_0 the optical thickness of the layer at the center of the disk would be approximately 0.2 and at the limb it would be 50. Under such conditions part of the photons originally emitted in a direction tangential to the solar surface will be reabsorbed and scattered until their direction of travel has a substantial component in the direction vertical to the solar surface. Thus, the high value of τ_0 at the limb both reduces the limb intensity and increases the disk intensity. Under such circumstances Equations (2) and (4) are properly used only with globally averaged data. Preferably, the average should exclude active regions. For the transition region lines, however, present evidence suggests that active regions do not contribute a majority of the radiation and there is no reason to suppose that the global average fluxes for the quiet sun will differ by large factors from the present values, which include active regions.

There is, of course, more information in data giving center-limb variations of intensity than in globally averaged data and it is possible, in principle, to extract more information about the transition region from such data than has been extracted from the globally averaged data. Analyses of center-limb data have been attempted by Burton *et al.* (1969) and by Withbroe (1970). It is possible, using such data to derive the temperature gradient in the transition region independently of the absolute line fluxes and independently of the collisional excitation cross-section. However, the height of the emitting layer must be derived from the limb data independently for each ion and in the sharp transition region the height difference between two ions becomes a second order effect in the data. Also, the opacities must be known and this information can be obtained only from the data. Additional complications arise from the presence of atmospheric inhomogeneities, hence an unknown geometry. The assumption that the radiation is confined to a spherical shell may give meaningful results for globally averaged data whereas it may give completely meaningless results for comparisons of center-limb data.

Both Burton *et al.* (1969) and Withbroe (1970) found evidence that $\tau_0 \gg 1$ at the limb in many of the XUV lines. Also, they found evidence of strong departures from spherical symmetry. Withbroe (1970) attributed a systematic difference in the center-limb behavior of lines at $\lambda < 912$ from lines at $\lambda > 912$ to the large opacity of spicules in the Lyman continuum. Spicules are observed at the limb superposed on the transition region. Even though spicules cover less than 1 percent of the solar surface they merge at the limb to form essentially a complete shield of the limb at heights below about 4000 km (Athay, 1959b; also Thomas and Athay, 1961). Above 6000 km only a small fraction of the limb is obscured by spicules. Thus, Withbroe's (1970) results confirm both that spicules are opaque in the Lyman continuum and that the transition region occurs below 6000 km.

Center-limb data on the intensities of XUV lines have not been fully exploited. A model of the transition region derived by Burton *et al.* (1969) is not consistent with the conductive fluxes derived in the preceding section of this paper. However, the model was based upon spherical symmetry and upon the assumption that $\tau_0 < 1$ at the limb. The authors acknowledge that both restrictions are invalid and therefore do not consider the discrepancy between the models as meaningful.

It seems at this point that perhaps the greatest value of the center-limb data will be in determining the opacities and the nature of the geometry, which is the approach adopted by Withbroe (1970). Center-limb data together with the globally averaged data should then provide improved models of the transition region.

B. MAGNETIC FIELDS EFFECTS

The presence of magnetic fields in the solar atmosphere adds several complications to the discussion of the transition regions. Because of the variations in magnetic field strength and orientation from point to point on the sun we expect point to point variations in density, temperature and the vertical thermal conductivity, each of which may strongly influence the resultant model of the transition region.

Presumably the observations always sample the same general range in T_e so that changes in temperature structure from point to point influence mainly just the height of the emitting layer. However, changes in T_e are associated with changes in n_e and because the XUV line fluxes are proportional to n_e^2 they sample preferentially the regions of high n_e.

If we imagine that the Sun is covered fifty percent by randomly distributed, unresolved regions of very low density and fifty percent by regions of approximately $2\langle n_e \rangle$, almost all of the observed flux will come from the high density regions. To produce the same observed average flux from the sun and with $\langle n_e T_e \rangle = 6 \times 10^{14}$, the temperature gradients in the transition region would need to be doubled.

The influence of magnetic fields on the thermal conductivity produces even more severe results. In the presence of a magnetic field of field strength greater than about 0.1 G, which is a common condition in the solar atmosphere, the thermal conductivity remains appreciably large only in the direction of the field. This has the consequence in the transition region of making the thermally conducted flux large in regions of vertical magnetic fields and nil in regions of horizontal magnetic fields provided the temperature gradients are the same as in the zero magnetic field case. Also, in regions where the magnetic field lines are vertical just below the transition region and diverging through the transition region there will be a funneling of the heat flux due to thermal conduction into a localized area of the upper chromosphere. This effect is illustrated schematically in Figure 3.

Kopp and Kuperus (1968) have considered a specific model in which they identify regions of vertical magnetic field with the boundaries and centers of supergranules and regions of horizontal magnetic field with the remainder of the supergranule cells. They assume in their model that the flux of thermal conduction is constant along a given magnetic flux tube, i.e., they ignore radiative and convective losses and mechanical dissipation of energy within the flux tube, at least in the transition region. This is a highly questionable assumption, particularly in view of the fact that spicules occur preferentially at the borders of supergranules where Kopp and Kuperus concentrate the heat flux. Nevertheless the model is interesting because it leads to some detailed conclusions.

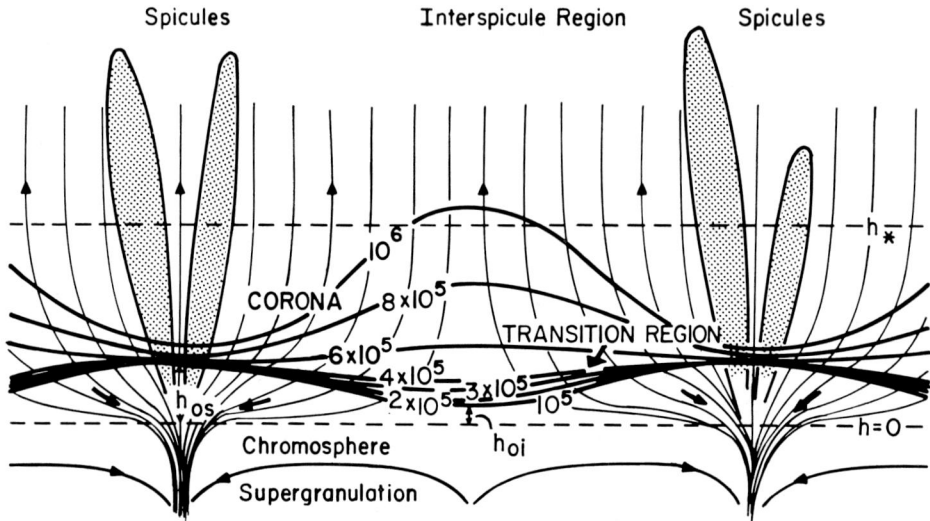

Fig. 3. Schematic cross-section of the upper chromosphere and low corona over a supergranule showing magnetic field lines, isotherms, fluid flow lines in the photosphere and the locations of spicules (Kopp and Kuperus, 1968). Courtesy *Solar Phys.*

Following Kopp and Kuperus (1968), we write

$$F_c(l) = -\kappa T^{5/2} \frac{\partial T}{\partial l}(l) \frac{B(l)}{|B(l)|} = \text{const} \tag{25}$$

and

$$\frac{\partial}{\partial l}\left(\kappa T^{5/2} \frac{\partial T}{\partial l}\right) = \kappa T^{5/2} \frac{\partial T}{\partial l} \frac{1}{|B|} \frac{\partial |B|}{\partial l}, \tag{26}$$

where l is measured in the direction of the field. For the monopole magnetic field model illustrated in Figure 3, which is the one considered by Kopp and Kuperus, the field lines are vertical at the supergranule border and at the supergranule centers. Thus, l may be replaced by the vertical coordinate h.

There is a sharp distinction in the Kopp-Kuperus model between the two vertical magnetic field regions. Just above the reference height $h=0$, which Kopp and Kuperus place in the upper chromosphere, the field strength at the border regions where spicules occur decreases as

$$B_s = ah^{-2} \tag{27}$$

whereas at the center regions it increases as

$$B_i = bh. \tag{28}$$

Both a and b are constant. Thus, the gradient of B is reversed in the two cases. The effect of this reversal is to steepen the transition region in the border (spicule) regions and flatten it in the center (interspicule) regions, as illustrated schematically in Figure 3.

Specifically, Kopp and Kuperus take $B_s = B_i = \text{const.}$ at $h = h_*$. This gives $a/b = h_*^3$ and by integration of Equation (26) with equations (27) and (28)

$$T_s^{7/2} = \frac{7}{2} \frac{F_{cs}^* h_*^2}{\kappa} (h_{0s}^{-1} - h_s^{-1}) \tag{29}$$

and

$$T_i^{7/2} = \frac{7}{4} \frac{F_{ci}^*}{\kappa h_*} (h_i^2 - h_{0i}^2). \tag{30}$$

The heights h_0 are the hypothetical locations of $T = 0$ if F_c were to remain constant. From equations (29) and (30) we obtain at $T_i = T_s$

$$\frac{(dT/dh)_s}{(dT/dh)_i} = \frac{F_{cs}^*}{F_{ci}^*} \frac{h_*^3}{h_s^2 h_i}, \tag{31}$$

and from Equations (27) and (28) we obtain

$$\frac{(dT/dh)_s}{(dT/dh)_i} = \frac{F_{cs}^*}{F_{ci}^*} \frac{B_s}{B_i}, \tag{32}$$

also at $T_i = T_s$.

The ratio of line fluxes from the two regions will be given by

$$\frac{F_i}{F_s} = \frac{P_i^2 (dT/dh)_i^{-1}}{P_s^2 (dT/dh)_s^{-1}} = \frac{P_i^2}{P_s^2} \frac{F_{cs}^*}{F_{ci}^*} \frac{B_s}{B_i}. \tag{33}$$

Kopp and Kuperus argue that $P_i^2 F_{cs}^* / P_s^2 F_{ci}^*$ is of order unity, hence that $F_i / F_s = (0) B_s / B_i$. Since B_s / B_i is expected to be much larger than unity, they conclude that most of the line flux comes from the interspicule regions. (Note that in this context interspicule means specifically the central regions of supergranule cells where the field is assumed to be nearly vertical. Thus, the fraction of the sun classified as interspicule by Kopp and Kuperus is less than 50 percent.)

Kopp and Kuperus further argue that in the interspicule region the conductive flux will be considerably lower than the value deduced in Section 3 of this paper because of the decreased temperature gradient and because of the funneling of flux away from these regions. They conclude that it may be possible to radiate away this reduced conductive flux in the interspicule regions. Thus, they set $\nabla \cdot F_c = \nabla \cdot F_{\text{rad}}$ in this region and write

$$\nabla \cdot F_{\text{rad}} = 2.3 \times 10^{-23} n_e^2 = 2.3 \times 10^{-23} P^2 T^{-2}, \tag{34}$$

following Orrall and Zirker (1961). They then obtain

$$\frac{d}{dh}\left(\kappa T_c^{5/2}\frac{dT_i}{dh}\right) = \frac{1}{h}\kappa T_i^{5/2}\frac{dT_i}{dh} + 2\times 10^{-23}P_i^2 T_i^{-2}, \tag{35}$$

which integrates to give the solid curve of $\log(dT/dh)^{-1}$ versus $\log T$ in Figure 4. The dashed lines in Figure 4 represent the solutions with no radiative losses for (a) the field free case and for (b) B given by Equation (28).

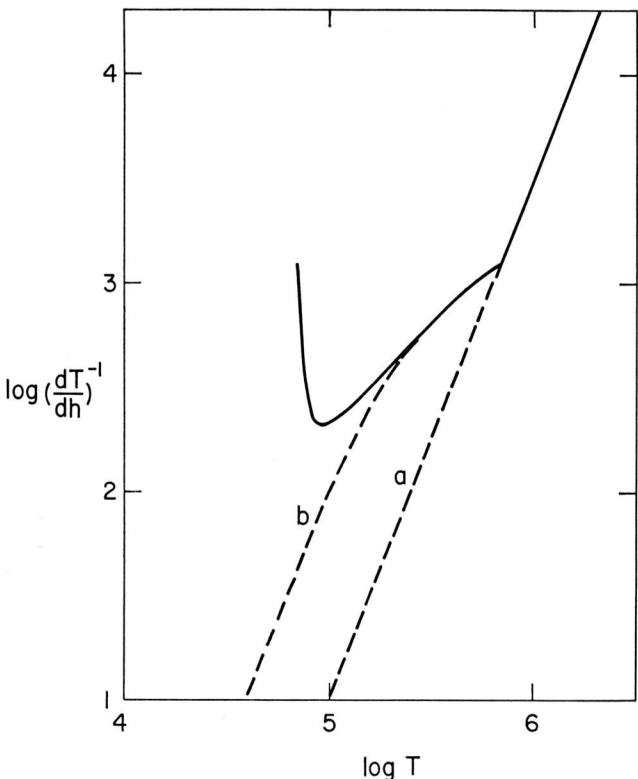

Fig. 4. Theoretical plots of $\log(dT/dz)^{-1}$ vs. $\log T$ in interspicule regions. The dashed lines represent the temperature distribution neglecting radiative losses (a) in the absence of a magnetic field, and (b) for the magnetic-field configuration described in the text. The solid line includes the effects of both magnetic channeling and radiative losses. Courtesy *Solar Phys.*

It is of interest to note that in the presence of a radiative loss a minimum occurs in $(dT/dh)^{-1}$ somewhat like the ones shown in Figures 1 and 2. The minimum in the empirical case is $(dT/dh)^{-1} \approx 10$ cm deg^{-1}, whereas the minimum in Figure 4 is $(dT/dh)^{-1} \approx 200$ cm deg^{-1}. Kopp and Kuperus attribute this to the fact that the interspicule regions cover only a fraction a_i of the sun and, in addition, to the supposition that P_i^2 is much less than P_s^2. For comparison with the minimum in Figure 4, the values of $(dT/dh)^{-1}$ in Figures 1 and 2 should be multiplied by a factor $a_i^{-1} P_e^2/P_i^2$, where $P_e = 6 \times 10^{14}$. If $a_i = 0.5$ and $P_i = \frac{1}{3} P_e$, we have $a_i^{-1} P_e^2/P_i^2 = 18$, which is enough to account for the differences between Figure 4 and Figures 1 and 2.

Although the Kopp-Kuperus model may be criticized in many details, it provides a useful illustration of the importance of magnetic fields in determining dT/dh in the transition region. There is no compelling reason for adopting a model for the magnetic field that has oppositely directed gradients at supergranule borders and supergranule centers. The flattened temperature gradient in the interspicule model of Kopp and Kuperus depends entirely upon the positive gradient of B, which is not a very likely configuration over much of the sun.

Both the radio and XUV data suggest strongly that the average value of dT/dh is near 10^{-1} deg cm^{-1} for $T \approx 1 \times 10^5$ K and that $T^{5/2} dT/dh$ is nearly constant between 10^5 K and 10^6 K. The inferred relationship between T and dT/dh near the lower limit of this range may be strongly affected by magnetic fields as well as by the breakdown of the assumptions about the ionization equilibrium and possible height gradients in the abundances of elements. Without more detailed data on both the spatial variations of the radiation intensity in the XUV lines and the magnetic field configuration it is difficult to arrive at any unique conclusion about the relative magnitudes of the temperature gradients in different solar regions. The conclusion of Kopp and Kuperus that dT/dh is larger in the spicule regions than in interspicule regions is entirely dependent upon the assumed geometry of B and upon the assumption that F_c is comparable in the two regions. It may well be that the main effect of the geometry of B is to determine the local values of F_c without having a strong effect upon dT/dh. On the other hand, one could even argue with some plausibility that if the structure of the transition region is controlled by thermal conduction and the thermal conduction, in turn, is limited by F_{mech}, and, further, that F_{mech} is greatest where the magnetic field is vertical, we need only have

$$\frac{F_c(B_\perp)}{F_c(B_\parallel)} \propto \frac{F_{\text{mech}}(B_\perp)}{F_{\text{mech}}(B_\parallel)} > 1.$$

Then, if

$$\frac{\kappa(B_\perp)}{\kappa(B_\parallel)} > \frac{F_{\text{mech}}(B_\perp)}{F_{\text{mech}}(B_\parallel)},$$

which is not too unlikely since $\kappa(B_\perp) \gg \kappa(B_\parallel)$, we will have

$$\frac{dT}{dh}(B_\perp) < \frac{dT}{dh}(B_\parallel).$$

This would identify regions of strong emission (smaller dT/dh) with the regions of vertical magnetic field, including those at the supergranule borders in contradiction to the Kopp-Kuperus model.

5. Energy Considerations

A. CONDUCTION RADIATION AND CONVECTION

The flux of energy due to thermal conduction was established in Section 3 to be of the order of 5×10^5 erg cm^{-2} sec^{-1}, assuming no magnetic field (or vertical magnetic fields). As already noted, this is a relatively large energy flux. By comparison, the radiation flux from the entire corona, including the transition region, is of the order of 3×10^5 erg cm^{-2} sec^{-1} (cf. Hinteregger et al., 1965; Athay, 1966b) and the radiation flux from the upper chromosphere (mostly Ly-α) is of this same order (cf. Athay, 1966b). Solar wind losses from the corona are nearly an order of magnitude smaller and, in any case, occur much higher in the corona. Balmer series radiation losses from the chromosphere are of the order of 1×10^6 erg cm^{-2} sec^{-1} (Athay, 1966b), but, again, these occur mainly in the low and middle chromosphere. Thus, the principle transport of energy to space from the transition region and adjacent chromospheric and coronal layers is by radiation and amounts to a total of approximately 6×10^5 erg cm^{-2} sec^{-1}.

When one considers that the energy removed from the transition region, the upper chromosphere and the low corona must be replaced by dissipation of mechanical energy but that we need replace only 6×10^5 erg cm^{-2} sec^{-1}, it is somewhat surprising that as much as 5×10^5 erg cm^{-2} sec^{-1} could be involved in thermal conduction through the transition region. This suggests that other energy transport processes are of importance within the upper chromosphere and corona. One need not search far for such a process.

Chromospheric spicules are observed at just the heights where we expect the transition region to occur. Furthermore the matter density in spicules (cf. Beckers, 1968) is typical of the upper chromosphere, which argues in favor of their origin in these layers. Spicule data still contain many uncertainties at heights below 4000 km where we expect the transition region to occur. Nevertheless it is possible to make crude estimates of their kinetic energy transport.

The kinetic energy flux through a fixed height surface in the sun is given by

$$F_{K.E.} = \tfrac{1}{2} n_H m_H v^3 a \tag{36}$$

where n_H and m_H are the number density and mass of hydrogen atoms within spicules, v is the spicule velocity and a is the fraction of the sun covered by spicules. We adopt for the spicule parameters in the transition region $n_H = 3 \times 10^{11}$ cm^{-3}, $v = 3 \times 10^6$ cm sec^{-1} and $a = 2 \times 10^{-2}$ (cf. Beckers, 1968). This gives $F_{K.E.} = 1.4 \times 10^5$ erg cm^{-2} sec^{-1}. As noted above, each of the spicule parameters are uncertain by factors of two or three and these could lead to more than order of magnitude errors in $F_{K.E.}$. Our point, however, is to illustrate that $F_{K.E.}$ may well be comparable to F_{rad} and F_c. Furthermore, we must, in all practicality, admit that we know virtually nothing about the structures and motions in the chromosphere at heights of, say, 3000–5000 km. Features other than spicules may be present and they may have sizable velocities. It has been recently suggested, in fact, that the violet asymmetry in the K_2 emission features may result from upper chromospheric structures moving downward with velocities of 10–20 km/sec (Athay, 1970). In any case, we cannot rule out the possibility that transport of energy by mass motion may be a vital part of the energy balance in the upper chromosphere, the transition region and the low corona. The spicules point clearly in this direction.

B. ENERGY BALANCE

It seems clear that energy is ultimately removed from the upper chromosphere and low corona by radiation and that the radiation flux represents all of the mechanical energy dissipated in these layers. Thus, we would write $\nabla \cdot F_{rad} + \nabla \cdot F_{mech} = 0$ for the atmosphere considered as a whole. How is it possible, then, that two additional energy fluxes (conduction and kinetic energy) are evident within these same layers with a magnitude comparable to the radiation flux. Two answers are possible: (1) the conduction and kinetic energy fluxes have been seriously overestimated; or (2) the conduction and kinetic energy fluxes represent an internal system.

The work by Kopp and Kuperus (1968) clearly indicates that magnetic fields may conspire to reduce F_c to a much lower value than has been inferred in Section 3. Also, spicule statistics clearly permit an order of magnitude reduction in $F_{K.E.}$. Thus, we must simply adopt this as one possibility, in which case the problem of explaining F_c and $F_{K.E.}$ is not particularly interesting.

The suggestion that F_c and $F_{K.E.}$ may represent an internal system in the transition region was made by Kuperus and Athay (1967). It presents a number of interesting possibilities and is supported by a certain amount of logic. In the following, we repeat the main points of the Kuperus-Athay argument. For further details, the reader is referred to their original paper.

To understand the reason for the Kuperus-Athay suggestion that the spicules may be driven by thermal conduction, consider the conditions at the base of the transition region. We suppose that the chromosphere and corona are heated by dissipation of mechanical energy generated in the convection zone. We then ask why it is that a large temperature jump occurs between the chromosphere and corona. One way to answer this question is with the postulate that the temperature rise occurs because the solar atmosphere is unable to dispose of the dissipated mechanical energy in a lower temperature configuration. Athay and Thomas (1956) (also, Thomas and Athay, 1961) showed that a solar mixture of gases has distinct maxima in radiating efficiency with increasing temperature. Two or more maxima were identified at temperatures below 1×10^5 K and a cluster of maxima were identified at temperatures near $1-2 \times 10^6$ K. The lower temperature maxima were associated with hydrogen and helium in the chromosphere and the higher temperature maxima were associated with the multiply ionized heavy elements in the corona. Following the Athay-Thomas line of reasoning, we would argue that the base of the transition region occurs when T has increased to the point where the last chromospheric maximum in F_{rad} occurs. This is a double edged argument. It argues on the one hand that the upper chromosphere is radiating at maximum efficiency and on the other that the radiation is, in effect, removing only the mechanical energy, i.e., that $\nabla \cdot F_{rad} + \nabla \cdot F_{mech} = 0$. If this latter conclusion were not true, it is clearly possible that the coronal temperature rise would not occur at all. Suppose, for example, that the upper chromosphere were, in fact, radiating away both the dissipated mechanical energy and the thermally conducted energy from the corona. This would imply that without the corona the chromosphere would be capable of radiating much more energy than is being supplied by mechanical energy, i.e., that the chromosphere alone could radiate away all of the mechanical energy dissipated in the chromosphere *and the corona*. Since the mechanical energy is believed to come from the photosphere this further implies that the mechanical energy passes through the upper chromosphere with little or no dissipation but is subsequently dissipated in the corona. Most of the chromospheric energy loss comes in the lower layers where the hydrogen Balmer series radiation is strong and there appears to be no currently acceptable way of carrying the conducted energy flux from the transition region into these lower chromospheric layers. We are therefore led to a picture in which the mechanical energy dissipation is strong in the low chromosphere and in the corona but is comparatively weak in the upper chromosphere. Such a picture seems much less plausible than one in which the mechanical energy dissipation is more or less monotonic with height. Thus, we shall assume that the corona exists only because the chromosphere is incapable of radiating away all of the mechanical energy carried up from below. Since the conductive energy flux through the transition region appears to be comparable to the radiation flux from the corona, it follows that the chromosphere is incapable of radiating away the mechanical energy dissipated in the chromosphere plus the energy carried back to the upper chromosphere by thermal conduction.

The above arguments present us with an apparent dilemma at the top of the chromosphere. This dilemma may be stated as follows: The coronal temperature jump occurs because the upper chromosphere is radiating at maximum efficiency, but, because of the temperature jump, the upper chromosphere receives a large new supply of energy in the form of F_c, which it cannot dispose of by radiation. The dilemma, of course, is non-physical and reflects either an error in logic or the omission of an important energy process. In light of what is known about spicules and $F_{K.E.}$, we consider the latter alternative.

Within the upper chromosphere-low corona system the full energy balance must include mechanical dissipation, radiation, conduction and convection. Thus, we must have

$$\nabla \cdot F_{mech} + \nabla \cdot F_{rad} + \nabla \cdot F_c + \nabla \cdot F_{K.E.} = 0, \tag{37}$$

where it is understood that $\nabla \cdot F$ must be negative for some terms. The apparent dilemma stated in the preceding paragraph resulted when we ignored one of these terms, viz., $F_{K.E.}$. Evidently this term cannot be ignored in the full energy balance. We have also argued that to a close approximation we should have

$$\nabla \cdot F_{mech} + \nabla \cdot F_{rad} = 0, \tag{38}$$

which implies that

$$\nabla \cdot F_c + \nabla \cdot F_{K.E.} = 0 \tag{39}$$

is also a good approximation.

Since there is no known way of removing energy from the upper chromosphere-low corona region other than by radiation and no known supply of energy to these layers other than by dissipation of mechanical energy, we expect Equation (38) to hold for the region as a whole. On the other hand, we cannot insist that it hold locally at each point within the region unless $\nabla \cdot F_c$ is large. Small values of $\nabla \cdot F_c$ could be accommodated by a small imbalance between $\nabla \cdot F_{rad}$ and $\nabla \cdot F_{mech}$, but large values of $\nabla \cdot F_c$ are not so easily accommodated and they seem to require the approximate validity of Equation (39).

There are sound reasons for supposing that $\nabla \cdot F_c$ is, in fact, of comparable magnitude to $\nabla \cdot F_{rad}$. Suppose, for example, that the radiating efficiency of the solar atmosphere, at a fixed gas pressure, has maxima below 1×10^5 K and beyond 1×10^6 K with a broad minimum at intermediate temperatures and that in this minimum region $\nabla \cdot F_{mech} > \nabla \cdot F_{rad}$. What must then happen, according to the arguments of Athay and Thomas (1956), is that the temperature will increase until some mechanism is capable of removing the energy deposited by $\nabla \cdot F_{mech}$. The rate of temperature rise will be limited only by thermal conduction, but the thermal conduction does not get rid of the energy. It only returns it to lower layers where it must be removed by non-radiative processes. Thus, we expect that the temperature will rise in the inner corona until $\nabla \cdot F_{rad}$ balances $\nabla \cdot F_{mech}$, and that it will rise as steeply as possible. Clearly, the upper limit to the rate of temperature rise is set by the condition $\nabla \cdot F_c = \nabla \cdot F_{mech}$ or $\nabla \cdot F_c = \nabla \cdot F_{rad}$. The condition $\nabla \cdot F_c > \nabla \cdot F_{mech}$ would violate the second law of thermodynamics, since it would be possible in such a case to extract more energy from the system than is supplied. The expected value of $\nabla \cdot F_c$ will be given by $\nabla \cdot F_c = e \nabla \cdot F_{mech}$ where e is the maximum efficiency with which the heat cycle $\nabla \cdot F_c = \nabla \cdot F_{K.E.}$ can be established. Hence, unless e is small compared to unity we expect that $\nabla \cdot F_c$ will be of the same order as $\nabla \cdot F_{mech}$ and $\nabla \cdot F_{rad}$.

Delache (1968, 1969) has pointed out that a system of shockwaves heating the transition region will, in general, supply momentum as well as heat to the transition region. It is clear, therefore, that Equations (38) and (39) are an oversimplification of the detailed problem and that ultimately Equation (37) must be solved in order to properly include all of the interactions between the energy terms.

All of the above arguments are qualitative of course, and must therefore be accepted only as possibilities. The details of the energy balance must await quantitative calculations using proper forms for each of the energy terms. Unfortunately, we do not yet know the physics well enough for any one of the terms in Equation (37). The details of shock wave dissipation in the sharp transition region, of radiative energy loss, of the magnetic field configurations (hence the thermal conductivity) and of the hydrodynamics of a system driven by thermal conduction are each difficult problems in themselves. Nevertheless, we can confidently expect much progress in these problems and in the solution of Equation (37) within the next decade.

C. ARE THE SPICULES DRIVEN BY F_c

We have argued in the preceding section (following Kuperus and Athay, 1967) that the thermal conduction through the transition region drives large scale motions in the upper chromosphere and low corona, including the spicules.

Spicules are believed to occur exclusively at the borders of super granules (Beckers, 1968) where the magnetic fields are systematically stronger than average and tend to be vertical. The increased vertical field strength is believed to result in an increase in F_{mech} (Osterbrock, 1961; Kuperus, 1965) and, most likely, in an increase of $\nabla \cdot F_{mech}$. Also, because of the vertical field orientation F_c is expected to be high. Thus, the association of spicules with regions of strong, vertical magnetic field is consistent with the suggestion that spicules are driven by the thermal conduction.

To suggest that spicules are driven by the thermal conduction is, of course, somewhat meaningless without a proposed mechanism for converting the heat energy of thermal conduction into kinetic energy. Such a conversion might be effected by using the conducted heat to raise the gas pressure (by raising the temperature in a restricted region) then by using the increased gas pressure to drive the spicule jet. Such a picture is not altogether implausible.

It is inherent to the nature of the conductive model of the transition region that $\nabla \cdot F_c$ will be large at the base of the transition region. As seen from Equation (24), the layer between $T=0$ and 1×10^5 K is only 2 km thick if F_c is constant. If we dissipate F_c in such a way that $\nabla \cdot F_c = $ const, we find

$$\frac{d}{dh}\left(\kappa T^{5/2} \frac{dT}{dh}\right) = c,$$

or

$$\kappa T^{5/2} \frac{dT}{dh} = c_1 h + c_2 \tag{40}$$

where c_1 and c_2 are constants. Integration of Equation (40) gives

$$\kappa \frac{2}{7}(T^{7/2} - T_1^{7/2}) = \frac{c_1}{2}(h^2 - h_1^2) + c_2(h - h_1). \tag{41}$$

At $h = h_1$, $\kappa T^{5/2} dT/dh = 5 \times 10^5$ erg cm^{-2} sec^{-1}, and at $h = h_2$ we set $\kappa T^{5/2} dT/dh = 0$. Thus, we have

$$c_1 h_1 + c_2 = 5 \times 10^5 \tag{42}$$

and

$$c_1 h_2 + c_2 = 0 \tag{43}$$

We assume that at $h_2 T \ll T_1$ so that equation (41) becomes

$$-\frac{2}{7}\kappa T_1^{7/2} = \frac{c_1}{2}(h_2^2 - h_1^2) + c_2(h_2 - h_1). \tag{44}$$

We then find from Equations (42), (43) and (44) $h_1 - h_2 = 4$ km. It is evident, therefore, that any reasonable model of the dissipation of F_c requires that that portion of F_c remaining at $T = 1 \times 10^5$ dissipate in a shallow layer immediately below the transition region, i.e., in a very thin shell at the top of the chromosphere.

We propose, therefore, that at the top of the chromosphere a large amount of thermal energy is deposited in a layer a few kilometers thick. Vertical mixing by convection and diffusion will spread the effect into the deeper layers to some extent. However, a heat flux of 5×10^5 erg cm^{-2} sec^{-1} is sufficient (ignoring radiative losses) to heat a cool gas of density 10^{11} cm^{-3} to 2×10^5 K at a rate of approximately 1 sec per kilometer of thickness. Thus, even a 1000 km layer would require only about 16 minutes to reach 2×10^5 K. If the layer remained at a density of 10^{11} cm^{-3} the gas pressure at $T = 2 \times 10^5$ K would exceed the coronal gas pressure by about a factor of 30. There appears to be no difficulty, therefore, in providing the driving force for vertical motions if, in fact, the conducted energy is not disposed of by radiation as we have asserted.

The upper chromosphere and low corona are normally stable against vertical motions. In this case, however, the high pressure buildup in the upper chromosphere could easily overcome the latent stability and drive vertical motions. The situation is very analogous to a Raleigh-Taylor instability in which a dense fluid overlying a less dense fluid is pulled downward by the force of gravity. An upward pressure gradient in the high chromosphere replaces gravity as the driving force behind the denser fluid. It is perhaps more than a coincidence, therefore, that the spicules and the jets of a Raleigh-Taylor instability bear a certain resemblance to one another.

Typical spicule velocities observed after the spicules have grown into extensions of over about 2 arc sec beyond the solar limb are about 25 km/sec. This is of the same order as, but perhaps somewhat smaller than, the sound velocity in the upper chromosphere. It is not at all clear whether spicules grow with a constant velocity or whether they have significant accelerations. Thomas and Athay (1961) have shown, for example, that spicule statistics are not sufficiently good to distinguish between a uniform velocity and a steady deceleration at the solar gravitational deceleration rate. Thus, we should use some caution in relating a spicule 'velocity' to the chromospheric sound velocity. Nevertheless, the average observed velocity for spicules suggests that such a relationship may indeed exist. If it does, there are perhaps a variety of theories that would predict such a relationship, including the preceding suggestion that spicules may be the jets of an instability analogous with the Raleigh-Taylor instability.

The growth rate of a small perturbation in a Raleigh-Taylor instability is given by Taylor (1950)

$$\eta^2 = ak \frac{\varrho_2 - \varrho_1}{\varrho_2 + \varrho_1} \tag{45}$$

where a is the acceleration directed from the higher density ϱ_2 toward the lower density ϱ_1 and k is the horizontal wavenumber. For $a = g_0 = 2.7 \times 10^4$ cm sec^{-2}, $k^{-1} = 6 \times 10^8$ cm, which is 2π times the scale height in the upper chromosphere, and

$\varrho_2 \gg \varrho_1$, we find $\eta = 0.04$ sec^{-1}. Thus, in its early stages a Raleight Taylor jet in the upper chromosphere would grow by a factor e in 25 sec. Specifically, the jet would grow from a length of 0.5 H to 1.35 H, where H is the upper chromospheric scale height, in 25 seconds. For $H = 10^8$ cm, we obtain an average growth velocity of 34 km sec^{-1}. This is sufficiently near the average spicule velocity to suggest that the growth rate of a Raleigh-Taylor jet is not inconsistent with the spicule velocity.

Equation (45) is valid in a vertical magnetic field provided k is not too large (Chandrasekhar, 1961). In a horizontal field of strength B_x

$$\eta^2 = ak \frac{\varrho_2 - \varrho_1}{\varrho_2 + \varrho_1} - \frac{B_x^2 k_x^2}{2\pi(\varrho_2 + \varrho_1)}, \qquad (46)$$

where k_x is the horizontal component of k. For $k_x = k$ and $\varrho_2 = 2 \times 10^{-14}$ η becomes imaginary when $B_x > 8.5$ G. Thus, even a modest magnetic field would suppress the growth of an instability and this could explain the absence of spicules in the interior regions of supergranule cells.

In broad outline, therefore, the suggestion that spicules may be the jets of a Raleigh-Taylor like instability driven by the thermal conduction through the transition region is not altogether implausible and has enough attractive features to warrant that it be seriously considered. Several alternative spicule theories have been proposed, some of which also have attractive features. However, to the author's knowledge this is the only theory that relates the spicules fundamentally to the transition region itself. It is for this reason that only this theory is considered here.

An initial study of the characteristics of thermally driven flow in an idealized, isothermal chromosphere has recently been completed by Bessey and Kuperus (1970). The results strongly suggest that spicule-like flow could be induced by localized heating in the upper chromosphere.

References

Alfvén, H., 1941: *Arkiva Mat. Astron. Fys.* **27A**, No. 25.
Athay, R. G., 1959a: *Paris Symposium Radio Astron.* (ed. by R. N. Bracewell), Stanford Univ. Press.
Athay, R. G., 1959b: *Astrophys. J.* **129**, 164.
Athay, R. G., 1965: *Astrophys. J.* **142**, 755.
Athay, R. G., 1966a: *Astrophys. J.* **145**, 784.
Athay, R. G., 1966b: *Astrophys. J.* **146**, 223.
Athay, R. G., 1967: *Astrophys. J.* **150**, 365.
Athay, R. G., 1969: *Solar Phys.* **9**, 51.
Athay, R. G., 1970: *Solar Phys.* **11**, 347.
Athay, R. G. and Canfield R. C., 1970: *Extended Atmosphere Stars*, NBS Special Publ. **332**, 65.
Athay, R. G. and Menzel, D. H., 1956: *Astrophys. J.* **123**, 285.
Athay, R. G., Menzel, D. H., Pecker, J.-C., and Thomas, R. N., 1955: *Astrophys. J. Suppl.* **1**, 505.
Athay, R. G. and Roberts, W. O., 1955: *Astrophys. J.* **121**, 231.
Athay, R. G. and Skumanich, A., 1968: *Solar Phys.* **3**, 181.
Athay, R. G. and Thomas, R. N., 1956: *Astrophys. J.* **123**, 299.
Beckers, J., 1968: *Solar Phys.* **3**, 367.
Bessey, R. J. and Kuperus, M., 1970: *Solar Phys.* **12**, 216.
Burton, W. M., Jordan, C., Ridgeley, A., and Wilson, R., 1969:

Chromosphere-Corona Transition Region, High Altitude Observatory, Boulder.
Chandrasekhar, S., 1961: *Hydrodynamic and Hydromagnetic Stability*, Oxford, pp. 462, 466.
Chiuderi, C., Chiuderi Drago, F., and Noci, G., 1970: *Solar Phys.*
Delache, Ph., 1967: *Ann. Astrophys.* **30**, 827.
Delache, Ph., 1968: *J. Quant. Spect. Rad. Transfer* **8**, 317.
Delache, Ph., 1969: *Chromosphere-Corona Transition Region*, High Altitude Observatory, Boulder.
Dubov, E. E., 1960: *Izv. Krymsk. Astrofiz. Observ.* **22**, 101.
Dumont, S., 1967: *Ann. Astrophys.* **30**, 861.
Dunn, R. B., 1965: Air Force Cambridge Research Lab., 65-398, Environmental Res. Paper 109.
Dupree, A. K. and Goldberg, L., 1967: *Solar Phys.* **1**, 229.
Giovanelli, R. G., 1949: *Monthly Notices Roy. Astron. Soc.* **109**, 372.
Goldberg, L., Muller, E. A., and Aller, L. H., 1960: *Astrophys. J. Suppl.* **5**, 1.
Hinteregger, H. E., Hall, L. A., and Schmidtke, G.: 1965, *Space Research V*, North Holland Publ. Co., Amsterdam.
House, L. L., 1964: *Astrophys. J. Suppl.* **8**, 307.
Ivanov-Kholodnyi, G. S. and Nikol'skii, G. M., 1961: *Soviet Astron.* **5**, 31. (Russian **38**, 45.)
Jager, C. de, 1959: *Paris Symposium, Radio Astron.* (ed. by R. N. Bracewell), Stanford Univ. Press.
Jager, C. de and Kuperus, M., 1961: *Bull. Astron. Inst. Neth.* **16**, 71.
Jefferies, J. T., 1960: *Astrophys. J.* **132**, 775.
Jordan, C., 1965: Thesis, Univ. of London.
Kopp, R., 1968: Air Force Cambridge Research Lab., 68-0312.
Kopp, R. and Kuperus, M., 1968: *Solar Phys.* **4**, 212.
Kundu, M. R., 1964: *Solar Radio Astronomy*. University of Michigan, Ann Arbor.
Kuperus, M., 1965: *Rech. Astron. Observ. Utrecht* **17**, 1.
Kuperus, M. and Athay, R. G., 1967: *Solar Phys.* **1**, 361.
Martyn, D. F., 1948: *Proc. Roy. Soc. (London)* **A193**, 44.
Osterbrock, D. L., 1961: *Astrophys. J.* **134**, 347.
Orrall, F. Q. and Zirker, J. B., 1961: *Astrophys. J.* **134**, 72.
Piddington, J. H., 1954: *Astrophys. J.* **119**, 531.
Pottasch, S. R., 1960: *Astrophys. J.* **131**, 68.
Pottasch, S. R., 1964: *Space Sci. Rev.* **3**, 816.
Schatzman, E., 1949: *Ann. Astrophys.* **12**, 203.
Schirmer, H., 1950: *Z. Astrophys.* **27**, 132.
Shklovskii, I. S., 1944: Thesis, Moscow State University.
Shklovskii, I. S. and Kononovich, E. V., 1958: *Soviet Astron.* **2**, 32. (Russian **35**, 37.)
Spitzer, L. Jr., 1962: *Physics of Fully Ionized Gases*, Interscience, New York.
Taylor, G. I., 1950: *Proc. Roy. Soc. (London)* **A201**, 192.
Thomas, R. N. and Athay, R. G., 1961: *Physics of the Solar Chromosphere*, Interscience, New York.
Uchida, Y., 1963: *Publ. Astron. Soc. Japan* **15**, 376.
Ulmschneider, P., 1967: *Z. Astrophys.* **67**, 193.
Ulmschneider, P., 1969: *Chromosphere-Corona Transition Region*, High Altitude Observatory, Boulder.
Weymann, R., 1960: *Astrophys. J.* **132**, 452.
Withbroe, G., 1970: *Solar Phys.* **11**, 208.
Woolley, R. v. d. R. and Allen, C. W., 1950: *Monthly Notices Roy. Astron. Soc.* **110**, 358.
Zirin, H., 1968: *Astrophys. J.* **154**, 799.
Zirin, H. and Dietz, R. D., 1963: *Astrophys. J.* **138**, 664.

5. CORONAL MAGNETIC FIELDS

GORDON NEWKIRK, JR.

High Altitude Observatory, National Center for Atmospheric Research*, Boulder, Colo., U.S.A.

1. Introduction – The Importance of Coronal Magnetic Fields

In his introductory lecture Dr. Evans examined several important, unsolved problems of the solar corona. Perhaps, the central one is the role played by magnetic fields. These fields appear to determine the flux of energy and matter from the lower levels of the solar atmosphere into the corona. The field configuration influences the density distribution of the coronal gas and undoubtedly determines the forms we observe there. Magnetohydrodynamic and plasma instabilities, which give rise to solar flares according to several theoretical models, originate in the magnetic fields permeating the corona above active regions. Transient events in the corona such as radio bursts, cosmic rays, and rapid changes in the form of the corona may all have a magnetic origin. Prominences appear to condense out of the corona and also take on the forms observed under the influence of magnetic fields. Finally, the same magnetic fields which control the rotation of the corona are swept out into interplanetary space by the solar wind to become an important component of the ambient medium of the earth. There the fields and their attendant particles influence the state of the earth's van Allen belts, the magnetosphere, the ionosphere, and a wide variety of phenomena in the earth's atmosphere.

Clearly, it would be impossible to discuss all these varied aspects of solar magnetic fields. We shall restrict ourselves to a review of the fields present in the corona below $\sim 3\,R$ and their consequences. As you will see, much remains to be learned about this important component of the solar corona.

2. The State of Our Knowledge

The study of coronal magnetic fields has had to rely almost exclusively upon calculation. Perhaps, a brief review of such calculations is in order. They all begin with measures of the photospheric field and then either in rectangular coordinates (Schmidt, 1964) over a volume small with respect to the sun or in spherical coordinates over a large volume (Newkirk *et al.*, 1968; Schatten *et al.*, 1969; Altschuler and Newkirk, 1969) calculate the potential field or modified potential field present. We shall examine these two methods in some detail.

H. Schmidt was the first to publish a program for the calculation of magnetic fields in the corona assuming that no currents flow above the photosphere. The comparison of these calculations with observations, of course, would allow us to examine if

---

* The National Center for Atmospheric Research is sponsored by the National Science Foundation.

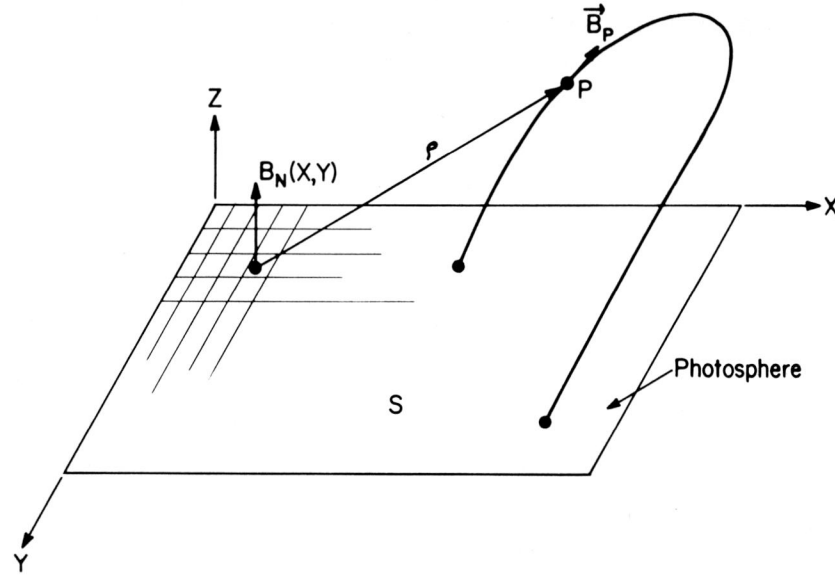

Fig. 1. Geometry for the calculation of the current free magnetic field according to the Schmidt program.

currents are appreciable in the low corona. He starts with the observed distribution of the normal magnetic field $B_N(X, Y)$ over an area S of the photosphere (Figure 1) and equates the normal field to a distribution of monopoles $\sigma(X, Y) = k B_N(X, Y)$. Thus, the potential at any given point P is

$$\psi_p = \int_S \frac{\sigma(X, Y)}{\varrho} \, dS \tag{1}$$

and the field is uniquely determined as the gradient of this scalar potential. Field lines can thus be calculated from an array of foot points covering S. For areas on the sun small with respect to the total, this technique gives a unique solution for the field at P and has proved quite successful in describing the shapes and magnitudes of fields above active regions.

To calculate the field above $\sim 1.3 \, R$ from the center of the sun, spherical geometry must be used as well as the observed distribution of the photospheric field over the *entire* sun. Moreover, we cannot use a distribution of monopoles over the photosphere since the normal field at any one point on the surface is influenced by all the other monopoles, i.e.,

$$B_N(R, \theta, \phi) \neq k \sigma(R, \theta, \phi) \tag{2}$$

and we have no unique way of prescribing σ to obtain a given $B_N(R, \theta, \phi)$. Since the field is assumed to be current-free (i.e., the gradient of a scalar potential),

$$\mathbf{B} = -\nabla \psi ; \tag{3}$$

and since $\nabla \cdot \mathbf{B} = 0$, the expansion of the field in terms of functions, which individually and in linear combination are solutions of Laplace's equation

$$\nabla^2 \psi = 0, \tag{4}$$

should permit the calculation of the coronal fields consistent with the photospheric fields. Such an orthonormal set of functions is the associated Legendre polynomials $P_n^m(\theta)$. Expanded in these spherical harmonics (Chapman and Bartels, 1940) the potential is

$$\psi(r, \theta, \phi) = R \sum_{n=1}^{\infty} \sum_{m=0}^{n} \left(\frac{R}{r}\right)^{n+1} (g_n^m \cos m\phi + h_n^m \sin m\phi) P_n^m(\theta) \tag{5}$$

where r, θ, and ϕ are the radius, colatitude, and longitude of any point in space, and R is the solar radius. At any point on or exterior to R the components of the field are uniquely determined and are

$$B_r = -\frac{\partial \psi}{\partial r} = \sum_{n=1}^{\infty} \sum_{m=0}^{n} (n+1) \left(\frac{R}{r}\right)^{n+2} (g_n^m \cos m\phi + h_n^m \sin m\phi) P_n^m(\theta),$$

$$B_\theta = -\frac{1}{r}\frac{\partial \psi}{\partial \theta} = -\sum_{n=1}^{\infty} \sum_{m=0}^{n} \left(\frac{R}{r}\right)^{n+2} (g_n^m \cos m\phi + h_n^m \sin m\phi) \frac{dP_n^m(\theta)}{d\theta},$$

$$B_\phi = -\frac{1}{r \sin \theta}\frac{\partial \psi}{\partial \phi} = \frac{1}{\sin \theta} \sum_{n=1}^{\infty} \sum_{m=0}^{n} m \left(\frac{R}{r}\right)^{n+2} \tag{6}$$

$$\times (g_n^m \sin m\phi - h_n^m \cos m\phi) P_n^m(\theta).$$

This expansion is the equivalent of fitting the surface field to a set of multipoles located at the center of the sun. In practice, the observations of the line-of-sight field $B_L(\theta, \phi)$ over the photosphere are used to determine the coefficients g_n^m and h_n^m by least-mean-squares. The total number of coefficients used is, of course, not infinite as in (6) but is limited by the size of the computer memory to a principal index N.

These two techniques have several important limitations. Inherently, the Schmidt program can be used only for heights small with respect to the dimension of S and at points well removed from the borders of S. The harmonic expansion, while applicable to larger coronal distances, can incorporate only a rather coarse resolution grid of the surface fields and cannot be used to calculate the fine details of the field above an active region, for example. The Schmidt program requires observations near the center of the disk for $B_N(X, Y)$ while the harmonic expansion demands the field over the entire sun – a distribution which is frequently a rather uncertain average of the temporally varying fields at the surface. Neither technique includes the influence of electric currents which flow in the corona as a result of the solar wind and transient effects. Short of a complete solution of the dynamical, thermal, and magnetic equations such as has been carried out by Pneuman and Kopp (1970), which we shall discuss later, we have to introduce some contrivance to simulate the effect of the wind. A technique of electrostatics first applied to this problem by Schatten (Schatten *et al.*, 1969) is to introduce a zero potential or source surface above the photosphere at some $R_w > 1.6 - 2.5\,R$. This potential surface forces the field lines to become radial at that radius and they are presumed to remain radial, except for the 'garden hose' spiraling, beyond. Actually, the field lines close above R_w and, strictly speaking, any comparison of field lines and the corona should be restricted to $r \leqslant R_w$. The techniques for solving for the Legendre coefficients with this zero potential surface are essentially the same as described above. We note that physically this technique replaces the volume currents which flow in the corona as a result of the interaction of the solar wind and the field by a set of surface currents on $r = R_w$. The exact value of R_w is chosen empirically either to match the shape of the calculated field lines to the shape of the corona or to bring correspondence between the frequency spectrum

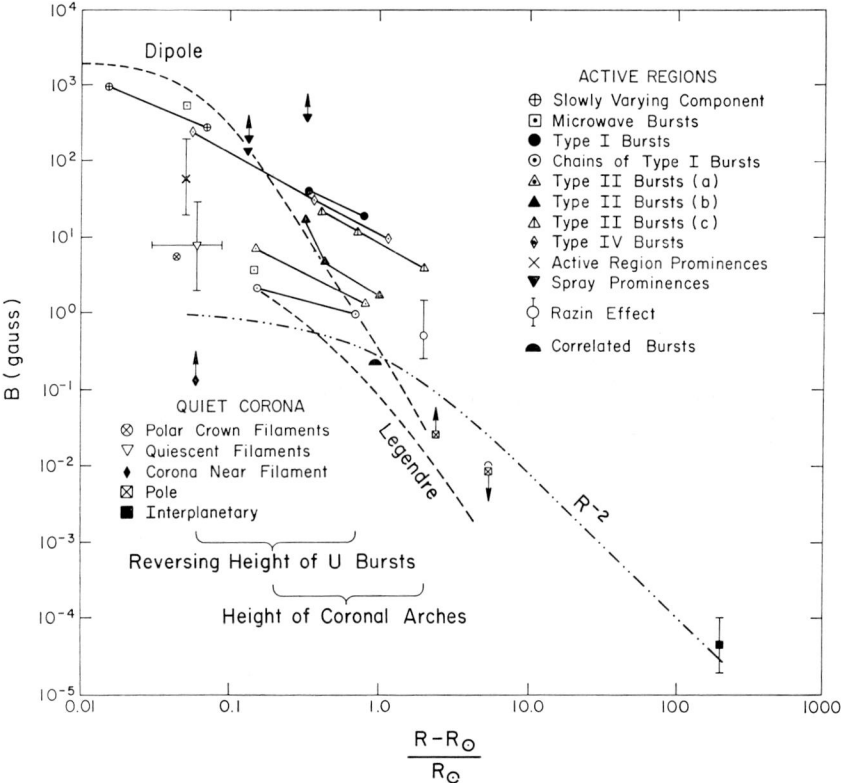

Fig. 2. Summary of measures of magnetic fields in the solar corona compared to (1) r^{-2} extrapolation from interplanetary space, (2) the field above a typical active region using the harmonic expansion method with $N = 9$, and (3) a simple dipole potential model for an active region. Except for the Razin effect and the correlated burst measurements, all other references are to be found in Newkirk, 1967.

of the variation of the magnetic field at 1 AU and that at R_w. Although the values of r_w so determined differ slightly, the discrepancy is probably not significant given the approximate nature of the model itself.

Measurements to check these calculations are unfortunately scarce (Newkirk, 1967; Takakura, 1966) except where direct detection of the Zeeman splitting in prominences can be made (Severny and Zirin, 1961; Rust, 1966; Harvey, 1969; Tandberg-Hanssen, 1970). Since reviewed several years ago, the analyses of the Razin effect in a single radio burst (Boischot and Clavelier, 1967; Ramaty and Lingenfelter, 1968; Bohlin and Simon, 1969) and of the weak polarization in some correlated bursts (Kai, 1969a) have added a few more measures of the absolute magnitude of the field at coronal heights (Figure 2). In the absence of any event by comparison between observed and calculated coronal fields, we compare the observations with three simple models: (1) an r^{-2} extrapolation from interplanetary space, (2) the Legendre polynomial field above a plage for the surface fields of November 1966; and (3) a potential dipole model of a plage region. This comparison suggests several cautions: (1) the Legendre approximation will not yield accurate results near active regions (a fact well known) and (2) radio bursts at $\sim 2\,R$ may well represent events in which a transient field disturbance is injected into the corona and may be unsuitable as a measure of the quiet magnetic field.

The only detailed comparison between observed and calculated magnetic fields in coronal space now at our disposal is in prominences (Harvey, 1969; Rust, 1966) which shows, in general, an agreement between the shapes of the fields and currently accepted ideas for the occurrence of prominences within the fields (Rust and Roy, 1970).

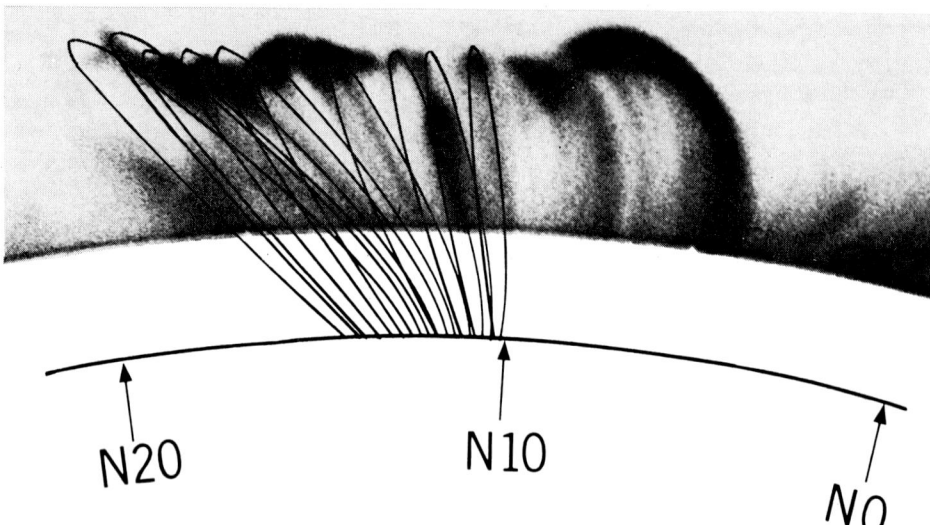

Fig. 3. Comparison of the shapes of bright coronal loops visible in 5303 Å (shown here as a negative) with the calculated potential field based on observations of the photospheric field near central meridian (Rust and Roy, 1970).

However, in active prominences, particularly, the measured fields are in excess of those calculated by a factor of five. The origin of this discrepancy is unknown. More recent measurements suggest that the discrepancy can be almost completely attributed to inaccurate measures of the surface fields (Rust and Roy, 1970). Similarly, comparison of the shapes of bright coronal emission regions and those of the calculated magnetic fields gives some confidence that the potential field is at least a good first approximation and that the coronal loops, arches, etc. constitute magnetic tubes of abnormally high density (Figure 3).

3. Influence on the Shape of the Corona

This brings us to the important question of the influence of magnetic fields on the shape of the solar corona. Although there may be some effect of magnetic structures of dimensions of the order of 30,000 km or less on such features as polar plumes (Saito, 1965; Newkirk and Harvey, 1968; Ivanchuk, 1968), in general only large scale magnetic fields will have a major influence on the corona. The investigation of the relation between the magnetic field and the density structure of the corona has followed two lines. One is to compare the calculated fields with the known shape of the corona (Newkirk *et al.*, 1968; Schatten, 1968; Altschuler and Newkirk, 1969; Newkirk *et al.*, 1970; Newkirk and Altschuler, 1970). Whether the comparison is intended as a prediction (Schatten, 1970a) or as a *post facto* analysis, the method is basically the same. The second line of attack is to use a simplified distribution of the field and to solve the hydromagnetic and solar wind equations simultaneously (Pneuman, 1968; Pneuman, 1969; Pneuman and Kopp, 1970) to determine the resultant distribution of material, the field, the velocity structure, and the energy flow in the modified corona.

To begin with the first approach let us first examine the pattern of calculated coronal fields present during a typical period as seen against an Hα spectroheliogram (Figure 4). The magnetic fields may be conveniently divided into Diverging Fields which are found in close association with plages, Low Magnetic Arcades and High Magnetic Arcades. Perhaps, most striking is the existence of magnetic arches connecting widely separated active regions. Such arches may well be the lines of communication which give rise to nearly simultaneous radio bursts in separated active regions (Wild, 1969a). In view of the close correlation between the positions of plages and coronal density enhancements, it is not surprising to find a similar correlation between such enhancements and the Diverging Field patterns.

A major conclusion of this type of analysis, in which the three-dimensional structure of the corona and the field are compared, is that coronal streamers appear to form over the High Magnetic Arcades. This is illustrated by the superposition of the K-coronameter isophotes of a streamer, identified at the 12 November 1966 eclipse, and the coronal magnetic map (Figure 5). This substantiates the idea long used in theoretical models (Kuperus and Tandberg-Hanssen, 1967; Pneuman, 1968, 1969) *that streamers develop above the neutral line separating large scale adjacent regions of opposite polarity.* (These features were also illustrated by a computer-drawn movie

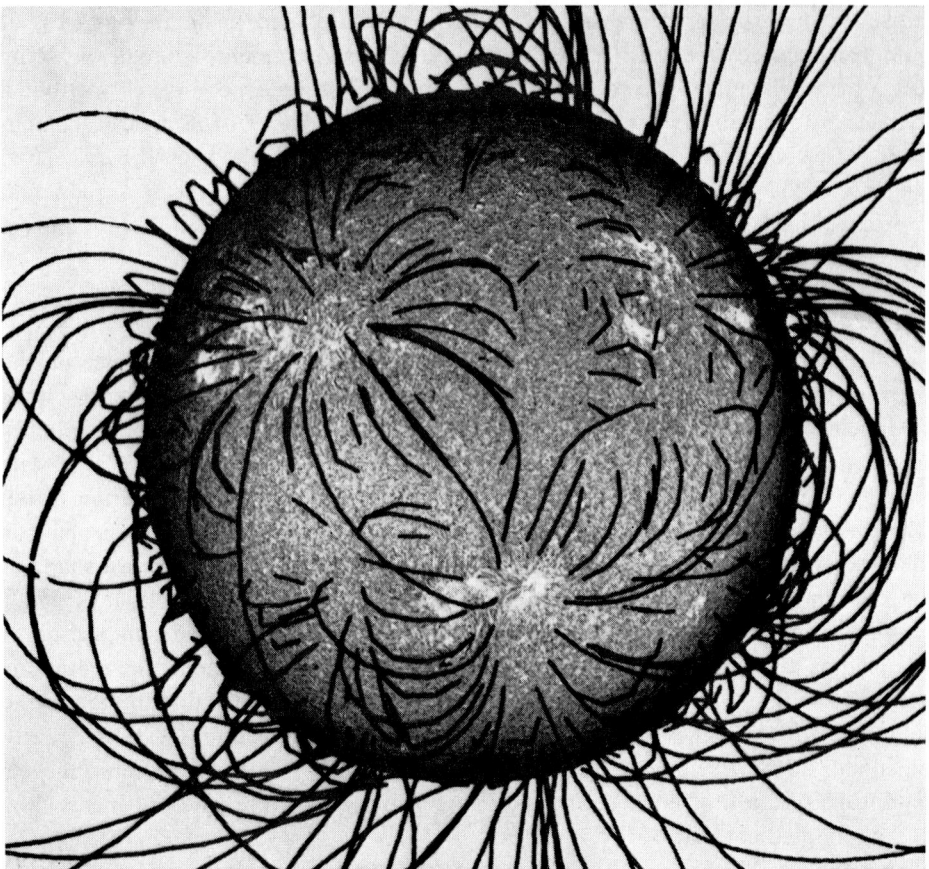

Fig. 4. Superposition of coronal field map (least-mean-square fit to B_L, $R_w = \infty$, corrected for magnetograph saturation) and the corresponding $H\alpha$ filtergram (Sacramento Peak Observatory).

which shows the true complexity of the coronal fields with an illusion of three dimensions created by the rotation of the sun.)

A comparison of the congruence of the shapes of small scale features in the corona with the magnetic field lines is almost inevitably restricted to an evaluation of their *projected* positions and appearances. Again we return to the 1966 eclipse and find that the agreement is quite good – we find open rays, arches, loops, etc. in the corona where they are indicated in the field. A similar conclusion is reached by examination of the most recent eclipse (Figure 6) as well as older data (Figure 7). *Thus, we conclude that much of the fine structure visible in the corona is simply a mapping of magnetic tubes in the approximately potential field.*

As an example of the more theoretical approach we cite the work of Pneuman (1968, 1969) and Pneuman and Kopp (1970). Here a simple distribution of field as well as pressure equilibrium are assumed at the base of the corona. The hydromag-

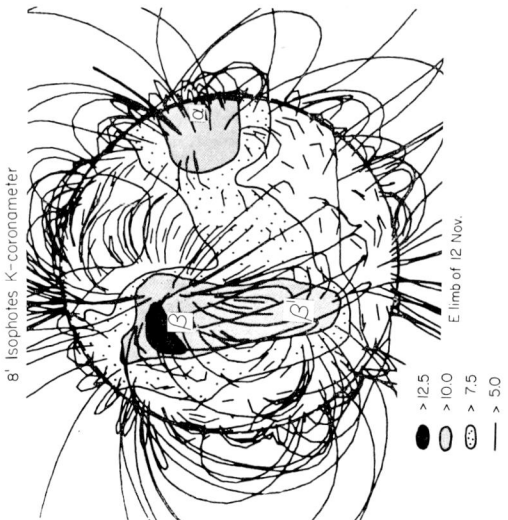

Fig. 5. Comparison of K-coronameter isophotes at 1.5 R, coronal magnetic field map, and the eclipse corona of 12 November 1966. The central meridian of the magnetic map corresponds to the east limb at the time of the eclipse and the line-of-sight proceeds from right to left across the map. Corresponding arches and rays can be easily located in the field and in the corona.

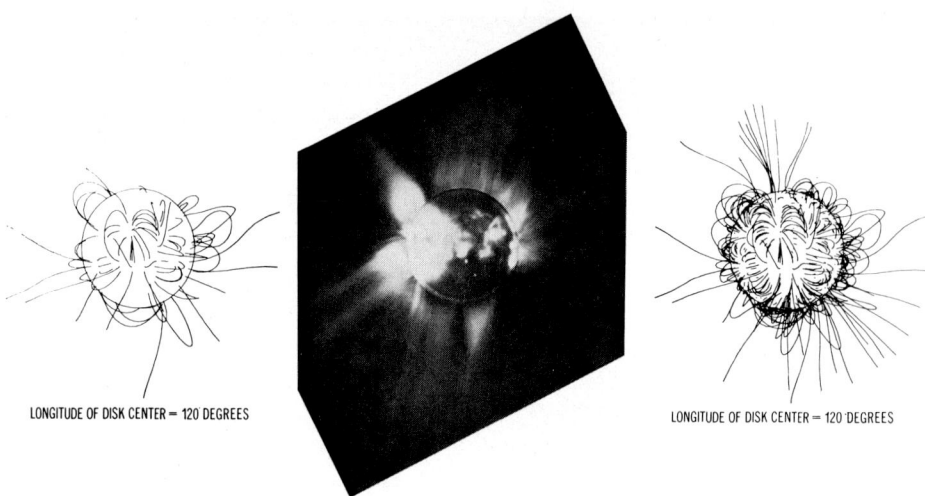

Fig. 6. Comparison of the solar corona of 7 March 1970 (outer corona HAO; X-ray corona seen on the disk courtesy Vaiana *et al.*, 1970, American Science and Engineering) with the corresponding coronal magnetic maps.*

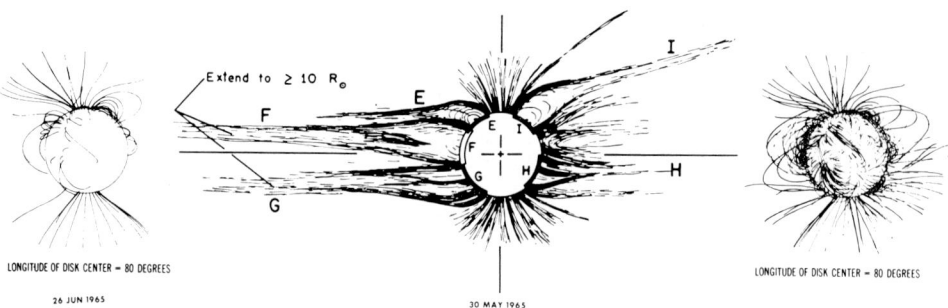

Fig. 7. Comparison of the solar corona of 30 May 1965 (drawing from Bohlin, 1968) with the corresponding magnetic maps. Note particularly the similarity between (1) the magnetic and coronal arches in streamer I and (2) the polar magnetic field and polar plumes. (See footnote Figure 6.)

netic and energy transport equations are then solved iteratively to arrive at a model for the streamer which includes such parameters as:

(1) the profile of the boundary between streamer and interstreamer,
(2) the axial enhancement,
(3) the temperature profile,
(4) the velocity structure.

At least for those parameters which can be measured, the agreement with our knowledge of the structure of the corona is impressive (Figure 8). Such models are also

---

* In this and subsequent coronal magnetic displays, the map on the right shows only field lines originating at foot points where $B_L \geqslant 0.16$ G while at the left all field lines originating where $B_L \geqslant 10\%$ of the maximum line-of-sight field present at the surface are displayed.

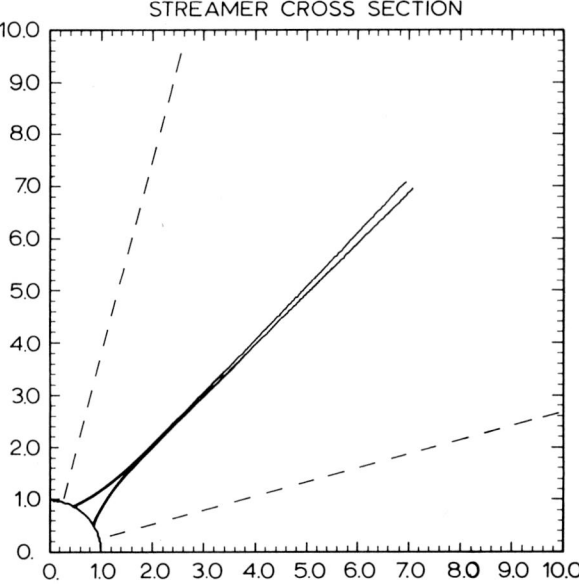

Fig. 8. The boundary between streamer and interstreamer regions for a representative streamer as calculated by Pneuman and Kopp.

important because they allow the construction of realistic ideas of how the visible structures we see in the corona are molded by the magnetic field and influence the structure and dynamics of the interplanetary medium.

4. Influence on Coronal Rotation

In addition to influencing the distribution of material and the expansion velocity of the solar corona, large scale magnetic fields clearly determine the rotation and transfer of angular momentum of the interplanetary medium. Here we must distinguish the corotation of a feature, such as a coronal streamer or a sector boundary in the field, from the angular velocity of the ions comprising the feature. Observational evidence for the tangential velocity of the corona at 1 AU derives from the orientation of comet tails (Brandt, 1967) and direct detection from space probes (Hundhausen, 1968). Both techniques yield a tangential velocity of 4–10 km/sec, which would require rigid rotation of the corona out to $\sim 15\,R$ is conservation of angular momentum occurred in the remainder of interplanetary space. Theoretical analyses (Pneuman, 1966; Weber and Davis, 1967; Modisette, 1967; Brandt et al., 1969) show this concept to be vastly oversimplified – coronal ions should actually lag behind the solar surface at all heights; however, they receive significant angular momentum from the solar magnetic field far out into the interplanetary medium. Except for measurements at 1 AU, we have no data on the rotation of the inner corona for comparison with these calculations.

The rotation of *structures* in the corona can be largely independent of the motions of the individual ions. Present information (Hansen *et al.*, 1970) shows that the low coronal enhancements rotate with the large scale magnetic structures on the surface (Wilcox and Howard, 1970) rather than with active regions. Moreover, these data suggest that the rate of rotation at a given latitude may *increase* with height in a similar manner to that found in the photosphere (Livingston, 1970). This apparently anomalous phenomenon can be explained (Pneuman, 1971) by the confinement of coronal gas to loops in the magnetic field having their foot points anchored at different latitudes and having different rates of rotation.

5. Coronal Magnetic Fields and Transient Events

Thus far we have discussed the large scale magnetic field and its influence as if the field were constant in time. Clearly, this is not the case and we shall examine several types of transient events which appear intimately connected with magnetic fields.

One such phenomenon is associated with solar cosmic rays. In addition to accounting for the acceleration of the particles, the observations require (Fan *et al.*, 1968): (1) a more or less direct channel for the escape of the particles from the flare region into interplanetary space, (2) storage and/or continuous generation of particles at the sun for a period of many days. An examination of the coronal magnetic field associated with a proton flare (Valdez and Altschuler, 1970) (Figure 9) suggests that the channel of direct escape may be found in the Diverging Fields associated with every active region and that storage may occur in some of the closed loops connected to the active region. That a proton flare may be associated with a permanent disruption in the large scale fields is shown by comparing Figures 9 and 10 in which the latter shows the same region one solar rotation later. The proton flare occurred in the interval. Note that the previously closed magnetic loops are open after the event. The fact that the open field lines appear in the current-free approximation indicates that a readjustment of the surface fields has occurred.

Radio occultation observations, either of natural sources (Dennison, 1970) or of satellite-borne transmitters (Levy *et al.*, 1969) give evidence for impulsive changes in the large scale magnetic field near the sun. In the first example, transient changes in the scatter width of the occulted Crab nebula with a time scale down to minutes may be interpreted as due to the interposition of streamers or other features of magnetic origin in the corona within the line-of-sight. A comparison of such observations with the calculated coronal fields has yet to be made. In the second example (Schatten, 1970b) (Figure 11), transient changes in the Faraday rotation are interpreted as a 'magnetic bottle' intruded into the line-of-sight at $10 R$ by an observed flare. The observed direction of Faraday rotation is consistent with the calculated fields in the low magnetic arch assumed to be the cause of the disturbance. Since the density and field both exceed the ambient value by an order of magnitude, we appear to have an example of a true intrusion of a magnetic bottle rather than a minor perturbation in a previously existing field.

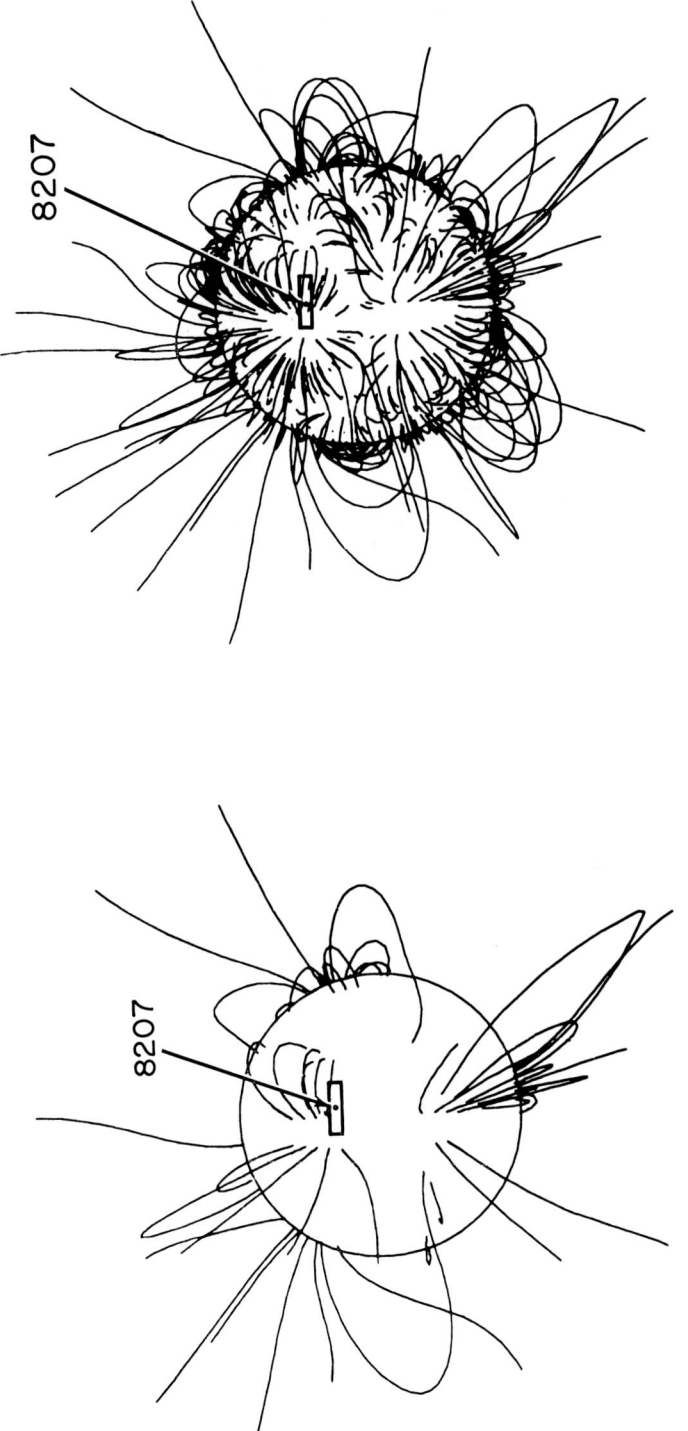

Fig. 9. Coronal magnetic maps for the Proton Flare of 16 April 1966 (Valdez and Altschuler, 1970) based on surface data taken *before* the occurrence of the flare. The location of the flare is marked with a rectangle. (See footnote Figure 6.)

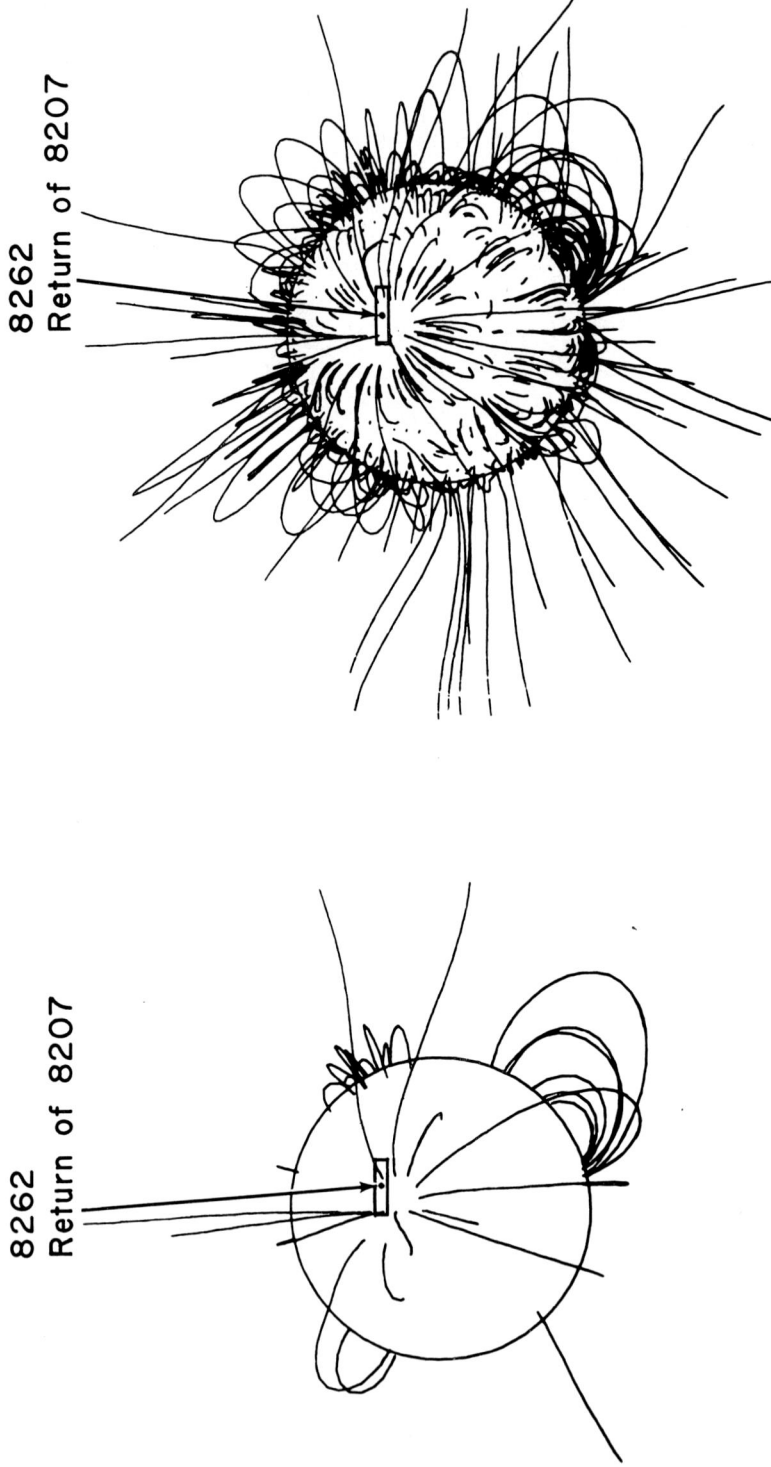

Fig. 10. Coronal magnetic maps for the Proton Flare of 16 April 1966 (Valdez and Altschuler, 1970) based on surface data taken *after* the occurrence of the flare. The location of the flare is marked with a rectangle. (See footnote Figure 6.)

Comparisons of radioheliograph observations of various radio bursts with the corresponding magnetic maps have been made only recently and thus no detailed analyses can be reported. However, a brief examination of some of the data suggests that we have some exciting discoveries in store for us.

One of the most energetic of radio events is the Type IV burst believed to be due to synchrotron radiation from mildy relativistic electrons. A comparison between radioheliograph observations (Wild, 1970) of several of these bursts and the calculated magnetic fields in the corona (Smerd and Dulk, 1970; Newkirk, 1970) shows that

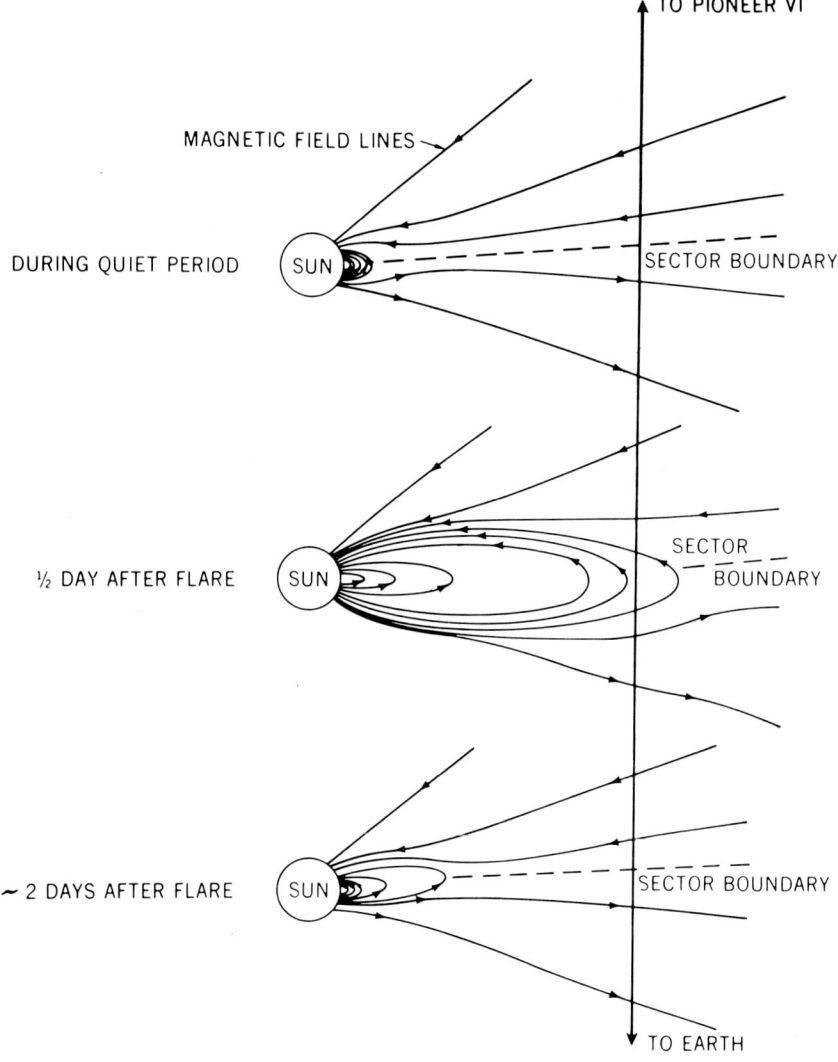

Fig. 11. Inferred geometry of the 'magnetic bottle' envisioned by Schatten (1970b) to account for the post flare transient change in Faraday rotation observed by Pioneer VI.

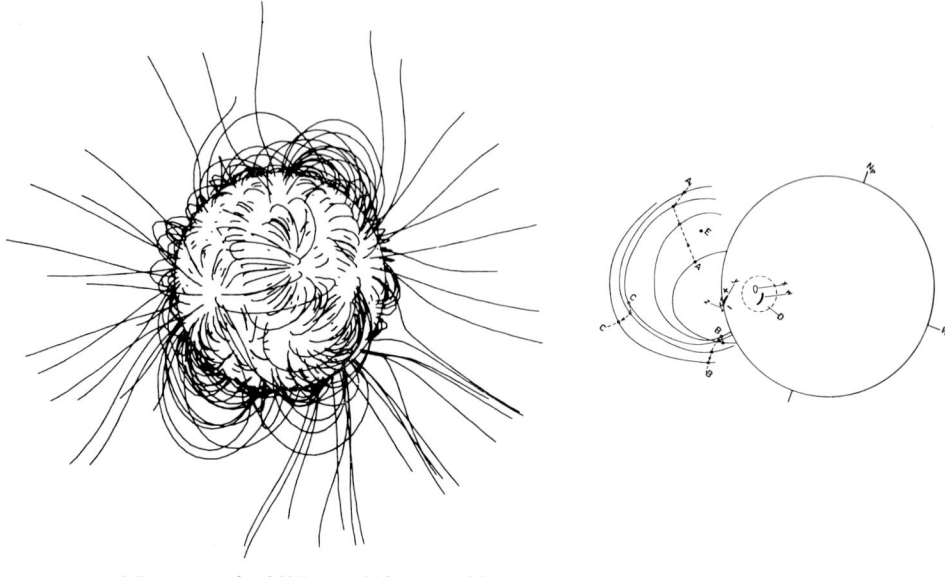

LONGITUDE OF DISK CENTER = 200 DEGREES

Fig. 12. Comparison of the shape of an expanding loop Type IV radio burst (Wild, 1969b) with the corresponding coronal magnetic arches. Perhaps, this is an example of a magnetic arch which has been filled with high energy particles although other interpretations are possible. (Field lines originating at foot points where $B_L \geq 0.16$ G are displayed).

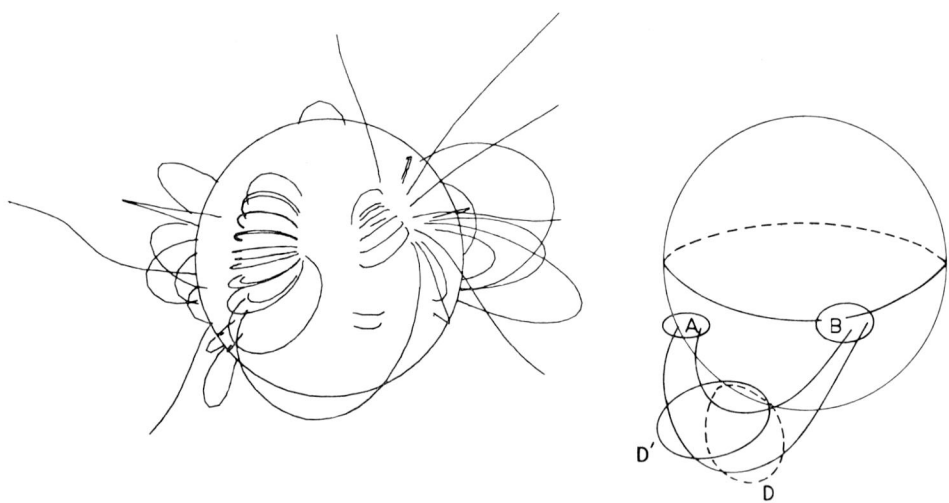

LONGITUDE OF DISK CENTER = 100 DEGREES

Fig. 13. Comparison of the magnetic arch inferred by Dulk (1970) from the radio bursts of 29–30 August 1969 (right) with the calculated magnetic map (left). Field lines originating at foot points where $B_L \geq 10\%$ of the maximum line-of-sight field present at the surface are displayed.

various subclasses of the Type IV bursts appear to be influenced by the field in different ways. In one (Figure 12, right) we see a loop of Type IV emission (Wild, 1969b), which appears to originate from the flare (X) under the low magnetic loops. Smerd and Dulk (1970) interpret this event as the impulsive expansion of the low magnetic loops which have trapped the high energy particles. Alternatively, we might imagine a shock disturbance propagating *along* the Diverging Fields which proceed to the east from the general region of the flare.

From the detailed analysis of another burst Dulk (1970) has inferred the presence (Figure 13, right) of a magnetic loop in which the particles are trapped to produce the Type IV burst (D) at the top. That such a loop is a fairly permanent feature of the field is suggested by the comparison with the magnetic map (Figure 13, left).

In another expanding Type IV burst (Kai, 1969b) (Figure 14) the expansion appears unambiguously to have occurred outward *along* the field lines. Here we may have evidence that the initiating shock wave, which is responsible for the acceleration of the particles, has been guided by the field and that the moving burst is really the moving shock front.

A final moving Type IV (Riddle, 1970) was accompanied by the expulsion of a spray prominence (McCabe and Fisher, 1970). Both the prominence and the burst appear to have been conducted out along the magnetic field (Figure 15). Riddle has suggested that this particular radio event may represent a vortex ejected from below.

Magnetic fields also appear directly responsible for guiding other types of radio disturbances. Kai (1969a) has reported a Type II burst which proceeded from a flare in only one direction and was followed by the disruption of a filament and a slowly rising Type IV. Inspection of Figure 16 strongly suggests that the channeling of the disturbance was, indeed, magnetic.

It has been suggested (Wild, 1970) that correlated radio bursts occur when high magnetic arches connecting widely separated regions conduct a triggering disturbance back and forth between the burst locations. Figure 17 suggests that this is actually the case although the channels for the southern group of correlated bursts are not nearly so obvious as those for the northern group.

6. Future Problems

Many future problems have, hopefully, been revealed during the course of this discussion. I mention only a few. Clearly, the interaction of the magnetic field with the density, temperature, and velocity structure of the corona requires more sophisticated treatment than has been possible by the potential field or the two stream magnetohydrodynamic models. The quantitative comparison of observations of radio burst events and detailed models including the magnetic field can be expected to yield new knowledge regarding the mechanisms of the bursts as well as of magnetic fields in the corona. Perhaps, the most pressing need is for actual observations of magnetic fields in the corona. Although radio observations can be expected to contribute much, the high angular resolution obtainable only at optical wavelengths suggests that

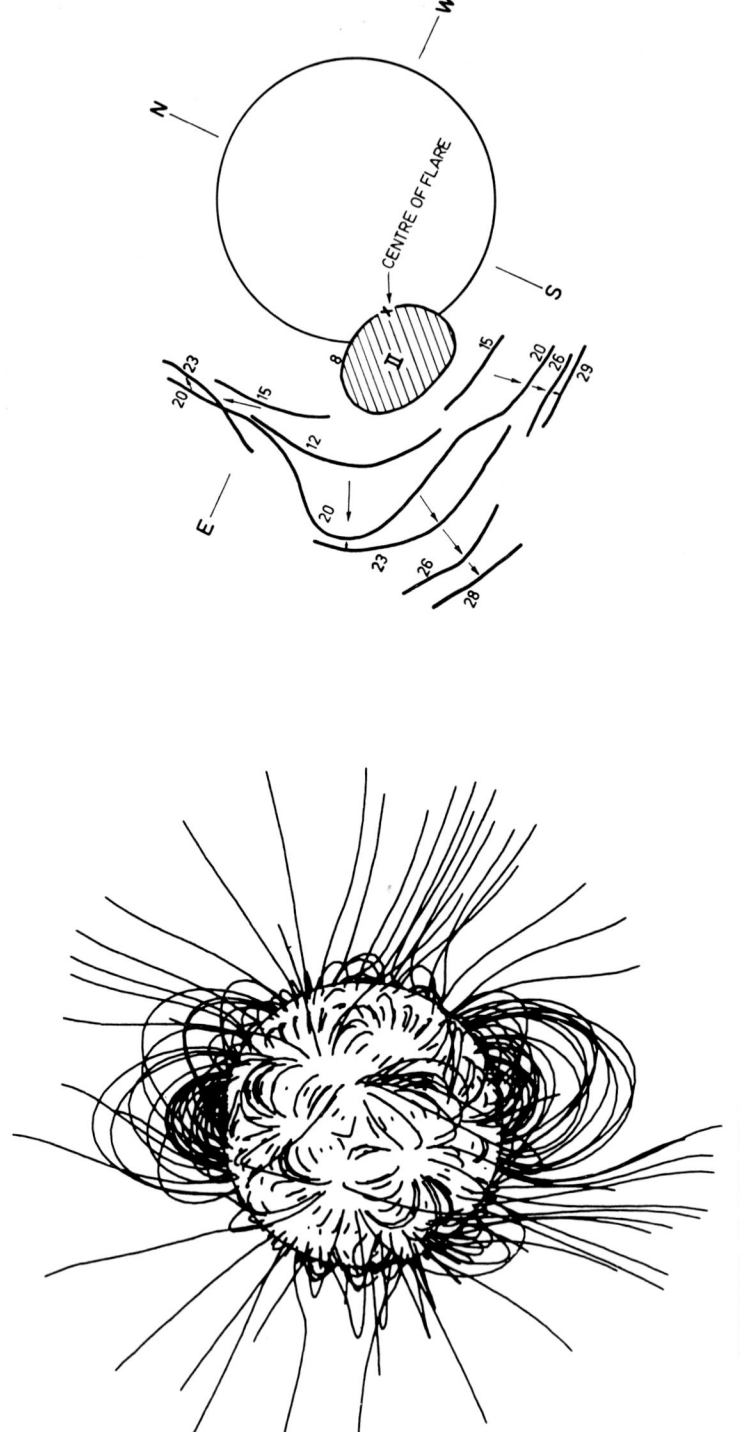

LONGITUDE OF DISK CENTER = 230 DEGREES

Fig. 14. Comparison of the successive positions of a rising Type IV burst (Kai, 1969b) with the corresponding coronal magnetic map. (See comment Figure 12.)

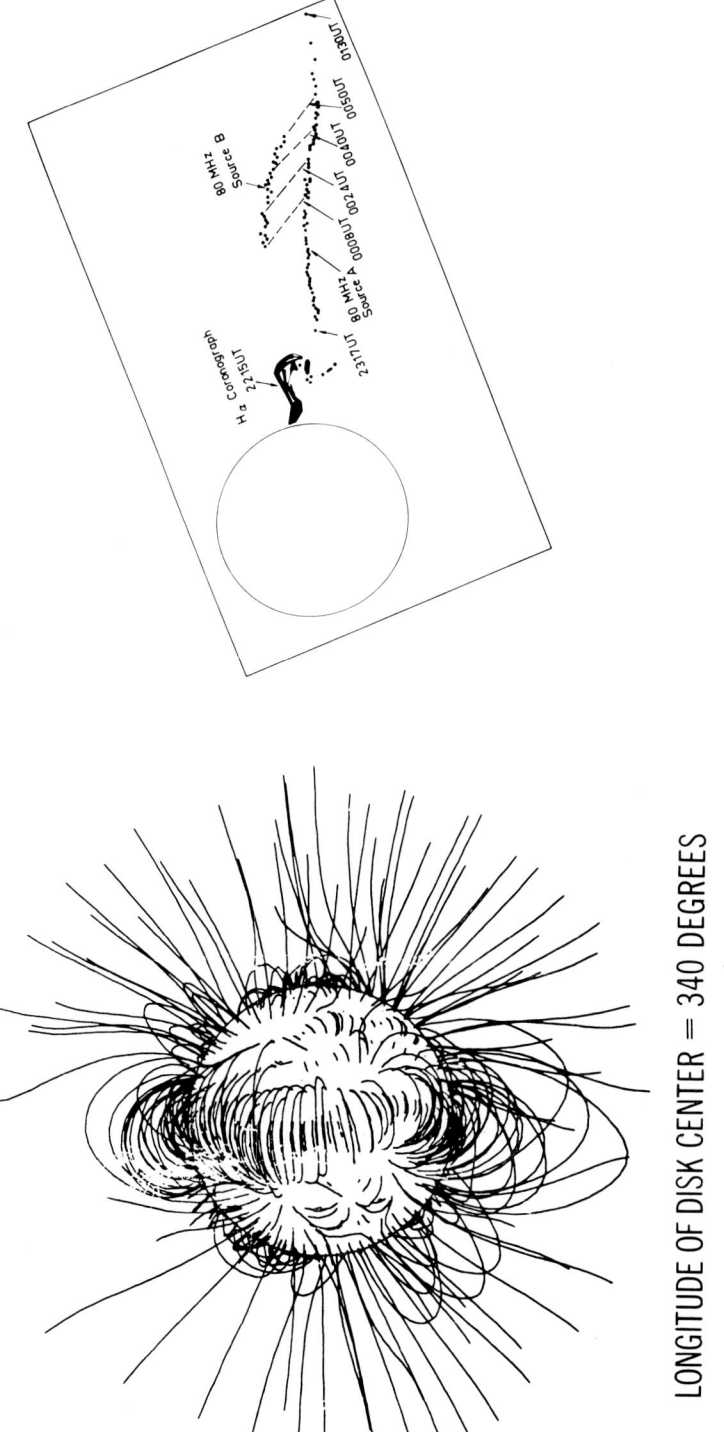

LONGITUDE OF DISK CENTER = 340 DEGREES

Fig. 15. An ejected spray prominence (McCabe and Fisher, 1970) and an associated moving Type IV burst (Riddle, 1970) appear to move out along magnetic lines of force. (See comment Figure 12.)

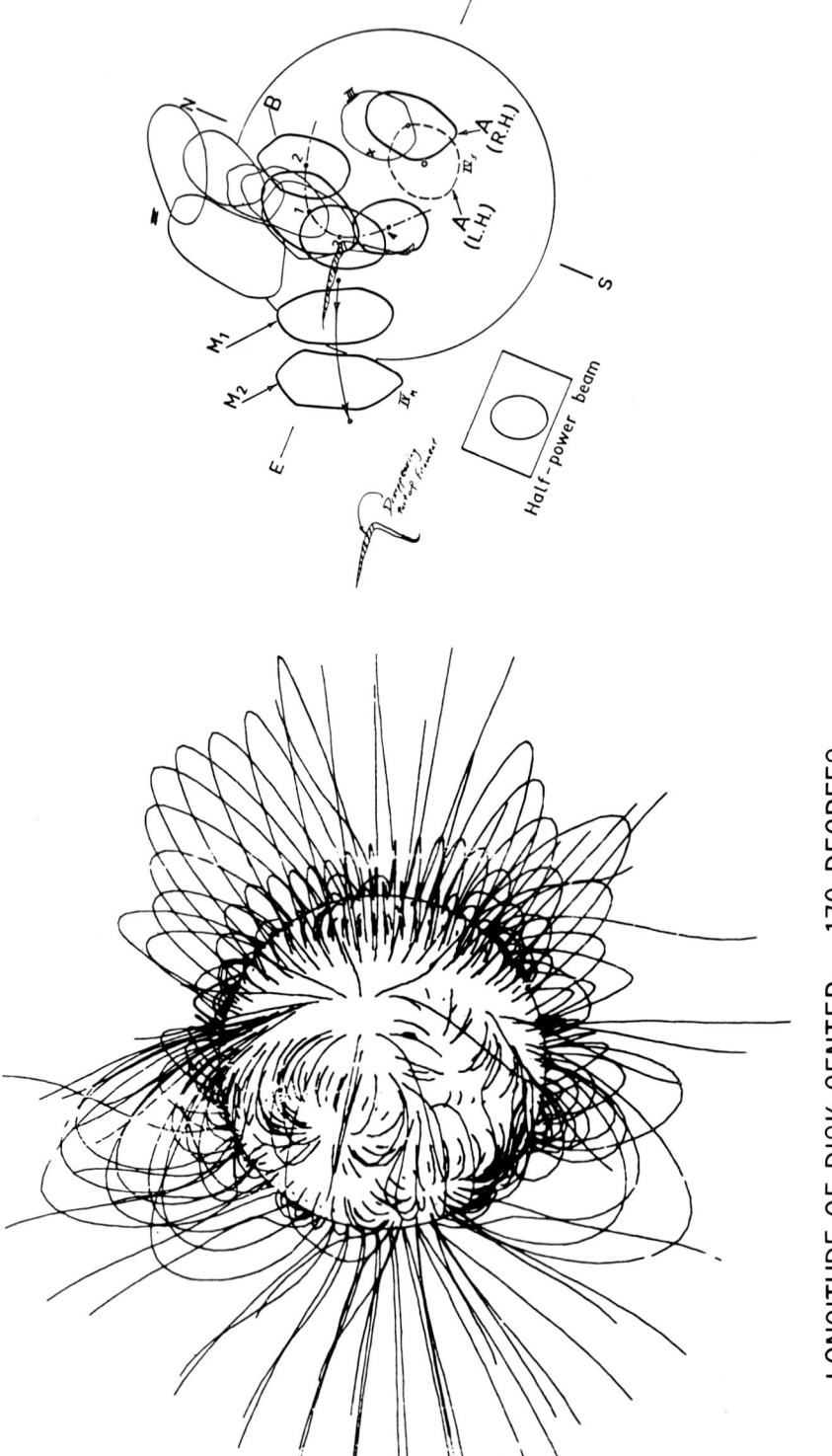

Fig. 16. Channeling of a directed shock, which gave rise to a complex of Type II, Type IV, and Type III radio events and a disappearing prominence (Kai, 1969a) by the coronal magnetic field. (See comment Figure 12.)

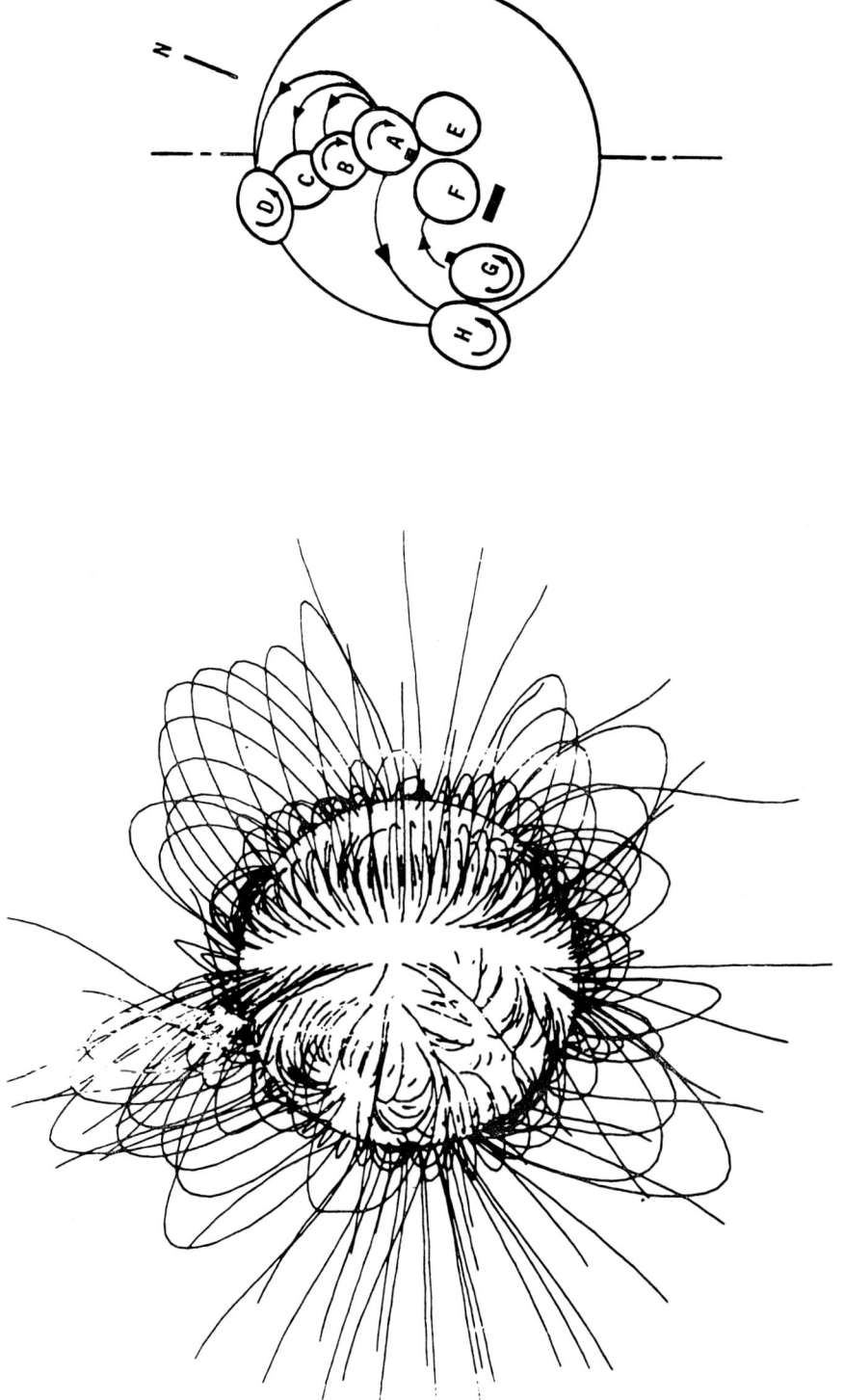

Fig. 17. Correlated radio bursts (Wild, 1970) between various centers (*A*, *B*, etc.) connected by curved lines in the radio heliogram (right) appear to be connected by magnetic arches in the magnetic map (left). (See comment Figure 12.)

emission line polarization observations (Hyder, 1966; Eddy *et al.*, 1967; Eddy and Malville, 1967; Hyder *et al.*, 1968; Charvin, 1970) have the greatest potential.

References

Altschuler, Martin D. and Newkirk, Gordon, Jr., 1969: *Solar Phys.* **9**, 131.
Bohlin, J. D., 1968: Ph.D. Thesis, Department of Astro-Geophysics, University of Colorado.
Bohlin, J. D. and Simon, M., 1969: *Solar Phys.* **9**, 183.
Boischot, A. and Clavelier, B., 1967: *Astrophys. Letters* **1**, 7.
Brandt, J. C., 1967: *Astrophys. J.* **147**, 201.
Brandt, J. C., Wolff, C., and Cassinelli, Joseph, 1969: *Astrophys. J.* **156**, 1117.
Chapman, S. and Bartels, J., 1940: *Geomagnetism*, Oxford University Press, London.
Charvin, P., 1970: *IAU Symp.* **43**.
Dennison, P. A., 1970: private communication.
Dulk, G. A., 1970: CSIRO, RPP 1395.
Eddy, John A., Firor, John W., and Lee, Robert H., 1967: presented at 124th meeting of AAS, Williams Bay, Wisconsin.
Eddy, John A. and Malville, J. McKim, 1967: *Astrophys. J.* **150**, 289.
Fan, C. Y., Pick, M., Pyle, R., Simpson, J. A., and Smith, D. R., 1968: *J. Geophys. Res.* **73**, 1555.
Hansen, Richard T., Hansen, Shirley, F., and Loomis, Harold G., 1970: *Solar Phys.* **10**, 135.
Harvey, J. W., 1969: Ph.D. Thesis, Department of Astro-Geophysics, University of Colorado.
Hundhausen, A. J., 1968: *Space Sci. Rev.* **8**, 690.
Hyder, C., 1966: Atti del Convegno sui Campi Magnetici Solari, Rome Observatory, Rome 1964.
Hyder, C. L., Mauter, H. A. and Shutt, R. L., 1968: *Astrophysics J.* **154**, 1039.
Ivanchuk, V. I., 1968: *Problems in Cosmic Physics*, 129.
Kai, K., 1969a: Proc. *Astron. Soc. Australia* **1**, 186.
Kai, K., 1969b: *Solar Phys.* **10**, 460.
Kuperus, M. and Tandberg-Hanssen, E., 1967: *Solar Phys.* **2**, 39.
Levy, G. S., Sata, T., Seidel, B. L., Stelzried, C. T., Ohlson, J. E., and Rusch, W. V. T., 1969: *Science* **166**, 596.
Livingston, W. C., 1969: *Solar Phys.* **9**, 448.
McCabe, M. K. and Fisher, R. R., 1970: *Solar Phys.* **14**, 212.
Modisette, J. L., 1967: *J. Geophys. Res.* **72**, 1521.
Newkirk, Gordon Jr., 1967: *Ann. Rev. Astron. Astrophys.* **5**, 213.
Newkirk, Gordon Jr., 1970: *IAU Symp.* **35**.
Newkirk, Gordon Jr. and Altschuler, Martin D., 1970: *Solar Phys.* **13**, 131.
Newkirk, Gordon Jr., Altschuler, Martin D., and Harvey, J. W., 1968: *IAU Symp.* **35**, 379.
Newkirk, Gordon Jr. and Harvey, J. W., 1968: *Solar Phys.* **3**, 321.
Newkirk, Gordon Jr., Dupree, Robert G., and Schmahl, Edward J., 1970: *Solar Phys.* **15**, 15.
Pneuman, G. W., 1966: *Astrophys. J.* **145**, 242.
Pneuman, G. W., 1968: *Solar Phys.* **3**, 578.
Pneuman, G. W., 1969: *Solar Phys.* **6**, 255.
Pneuman, G. W., 1971: *Solar Phys.*, in press.
Pneuman, G. W. and Kopp, Roger, 1970: *Solar Phys.* **13**, 176.
Ramaty, R. and Lingenfelter, R. E., 1968: *Astrophys. J.* **5**, 531.
Riddle, A. C., 1970: *Solar Phys.* **13**, 448.
Rust, D., 1966: Ph.D. Thesis, Department of Astro-Geophysics, University of Colorado.
Rust, D. and Roy, J. R., 1970: *IAU Symp.* **43**.
Saito, K., 1965: *Publ. Astron. Soc. Japan* **17**, 1.
Schatten, K. H., 1968: *Nature* **220**, 1211.
Schatten, K. H., 1970a: *Nature* **226**, 251.
Schatten, K. H., 1970b: *Solar Phys.* **12**, 484.
Schatten, K. H., Wilcox, J. M., and Ness, N. F., 1969: *Solar Phys.* **6**, 442.
Schmidt, H. U., 1964: NASA Symposium on Physics of Solar Flares, 107.
Severny, A. and Zirin, H., 1961: *Observatory* **81**, 155.

Smerd, S. and Dulk, G. A., 1970: *IAU Symp.* **35**.
Takakura, T., 1966: *Space Sci. Rev.* **5**, 80.
Tandberg-Hanssen, E., 1970: *IAU Symp.* **43**.
Vaiana, G. S., Krieger, A. S., and Van Speybroeck, L. P., 1970: *IAU Symp.* **43**.
Valdez, J. and Altschuler, M. D., 1970: *Solar Phys.* **15**, 446.
Weber, E. J. and Davis, L. Jr., 1967: *Astrophys. J.* **148**, 217.
Wilcox, J. M. and Howard, R., 1970: *Solar Phys.* **13**, 251.
Wild, J. P., 1969a: *Proc. Astron. Soc. Australia* **1**, 181.
Wild, J. P., 1969b: *Solar Phys.* **9**, 260.
Wild, J. P., 1970: *Proc. Astron. Soc. Australia.*

6. THE EXTENDED CORONAL MAGNETIC FIELD

JOHN M. WILCOX

Space Sciences Laboratory, University of California, Berkeley, Calif. 94720, U.S.A.

Abstract. The coronal magnetic field should contain many field lines connecting the photosphere to interplanetary space. A sharp boundary separates two adjacent sectors of opposite polarity. The large-scale structure of the corona is related to the photospheric sector pattern. The corona may frequently contain transient magnetic loops reaching out to five to ten solar radii.

As has already been mentioned in the introductory talk there is a good correspondence between the large-scale photospheric magnetic field and the large-scale interplanetary magnetic field observed with spacecraft near the earth. This situation means that the coronal magnetic field problem is bounded, although the outer bound is perhaps at a rather large distance from the region of the corona most often discussed. In the first portion of this paper we shall indicate the nature of the large-scale patterns in the photospheric field and then indicate some implications and related observations for the extended coronal field.

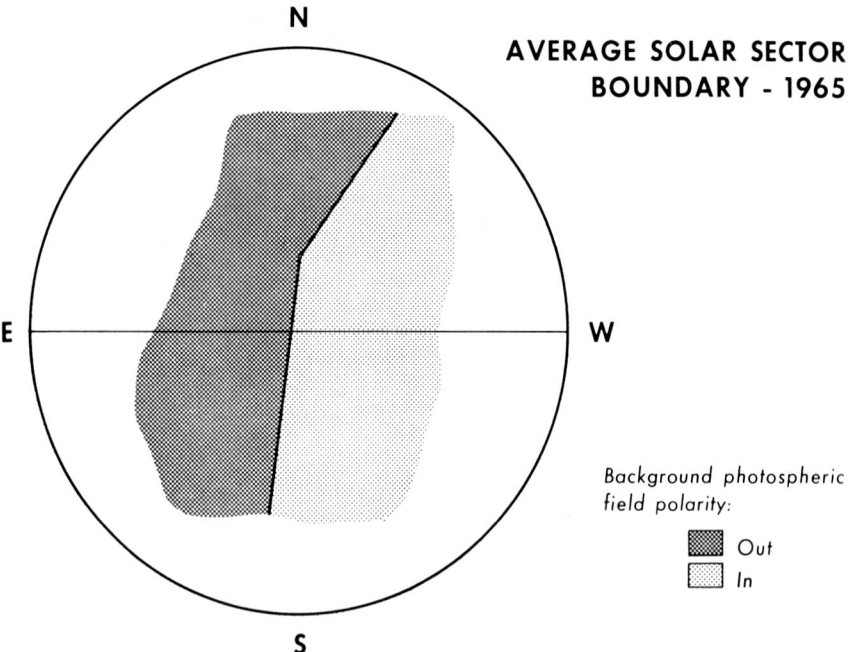

Fig. 1. A schematic of the average position of a solar sector boundary during 1965. On each side of the boundary the weak background photospheric magnetic field is predominantly of a single polarity in equatorial latitudes on both sides of the equator (after Wilcox *et al.*, 1969).

The classical Babcock (1961) model of solar magnetism utilizes the stretching and field amplication effects of differential rotation to explain a number of the observed solar magnetic phenomena, including the observation that the polarity of a bipolar magnetic region in the northern solar hemisphere is opposite from the polarity of a bipolar region in the southern hemisphere. An additional large-scale pattern in the photospheric magnetic field having rather different properties has recently been discovered (Wilcox and Howard, 1968). Figure 1 is a schematic showing some of the main properties of this solar sector pattern. A boundary exists approximately in the north-south direction. On one side of the boundary the large-scale weak photospheric field is predominantly directed out of the sun, and on the other side of the boundary this field is predominantly directed into the sun. This pattern exists over a wide range of latitudes on both sides of the equator. The boundary rotates in an approximately rigidly rotating coordinate system, since it is very little influenced by the shearing effects to be expected from differential rotation. The solar sector pattern thus differs from the Babcock model in two fundamental respects: (1) The solar sector pattern rotates in an almost rigidly rotating coordinate system while the Babcock model

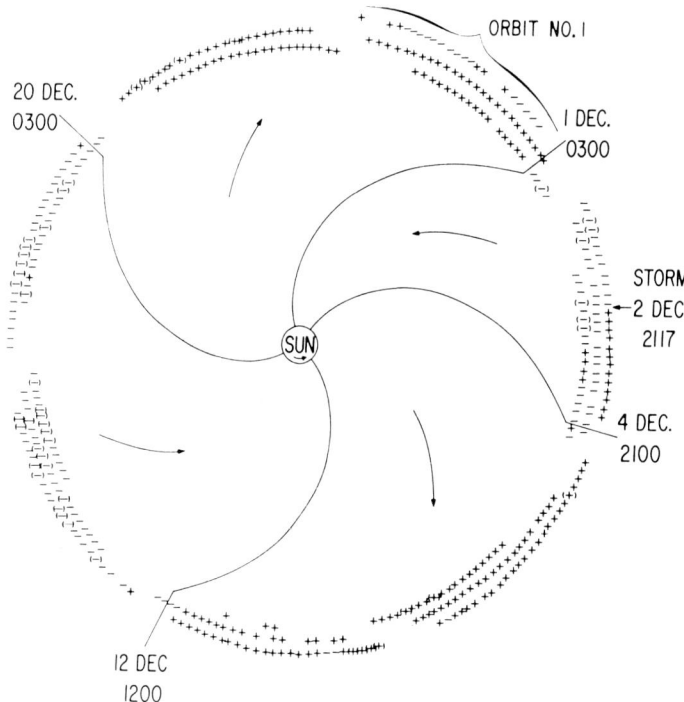

Fig. 2. The plus signs (away from the sun) and minus signs (toward the sun) at the circumference of the figure indicate the direction of the measured interplanetary magnetic field during successive 3-h intervals. The inner portion of the figure is a schematic representation of a sector structure of the interplanetary magnetic field that is suggested by these observations. The deviations about the average streaming angle that are actually present are not shown (after Wilcox and Ness, 1965).

depends on differential rotation to produce the observed effects, and (2) the solar sector pattern has the same polarity on both sides of the equator while the Babcock model (and observations) show that bipolar magnetic regions have opposite polarities on either side of the equator. Yet the two patterns coexist on the sun.

The solar sector pattern is the source (Wilcox and Ness, 1965) of a corresponding interplanetary sector pattern, an example of which is shown in Figure 2. Some of the solar magnetic field lines are carried outward by the radially flowing solar wind plasma. The combination of the radial plasma flow and the solar rotation leads to an Archimedes spiral shape for an average interplanetary field line. Thus a solar sector boundary is transported into interplanetary space in the form of an Archimedes spiral. The interplanetary sector pattern rotates with the sun so that a complete pattern sweeps past the earth every 27 days (the solar rotation period). The pattern shown in Figure 2 was approximately stationary in time for one year (1964) near the minimum of the 11-yr sunspot cycle.

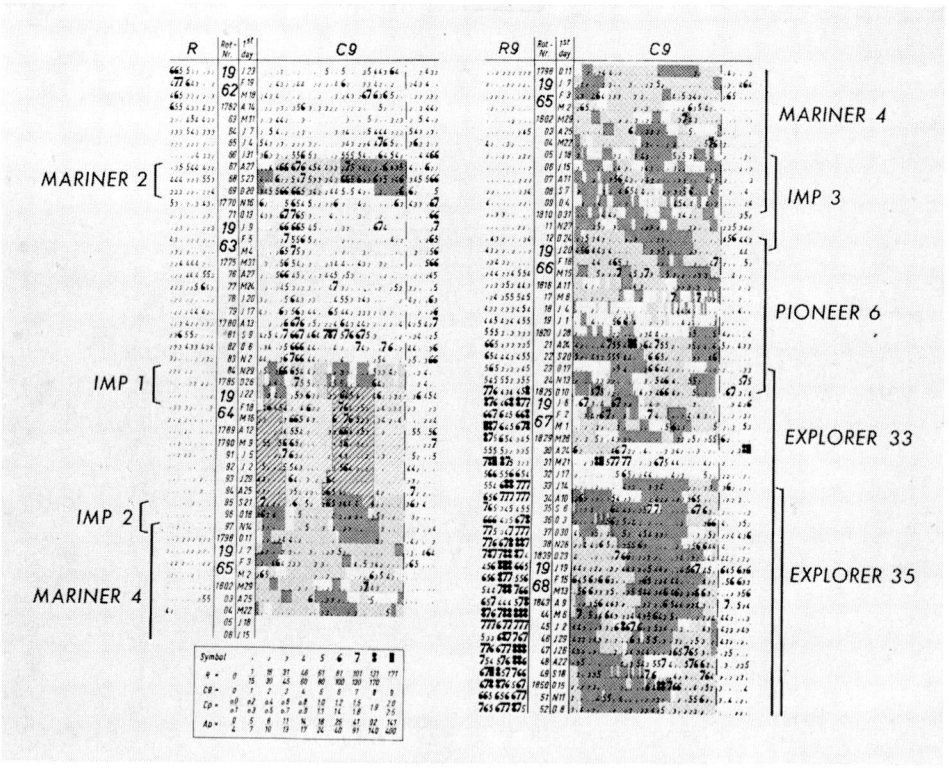

Fig. 3. Observed sector structure of the interplanetary magnetic field, overlayed on the daily geomagnetic character index C9, as prepared by the Geophysikalisches Institut in Göttingen. Light shading indicates sectors with field predominantly away from the sun, and dark shading indicates sectors with field predominantly toward the sun. Diagonal bars indicate an interpolated quasi-stationary structure during 1964 (after Wilcox and Colburn, 1970).

The evolution with time of the sector pattern (Wilcox and Colburn, 1970) is shown in Figure 3, which is basically a 27-day calendar. The top row represents the first 27-day rotation, the second row is the next 27-day rotation and so on. The shaded regions indicate the polarity of the interplanetary magnetic sector pattern. The

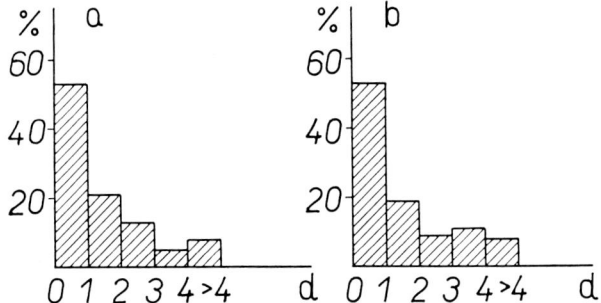

Fig. 4. Histograms of frequency distributions of the time difference between the central meridian passage of spot groups and the position of solar sector boundaries for the groups: (a) with flares of importance 1 + or greater; (b) with a number of flares equal or greater than 10 (after Bumba and Obridko, 1969).

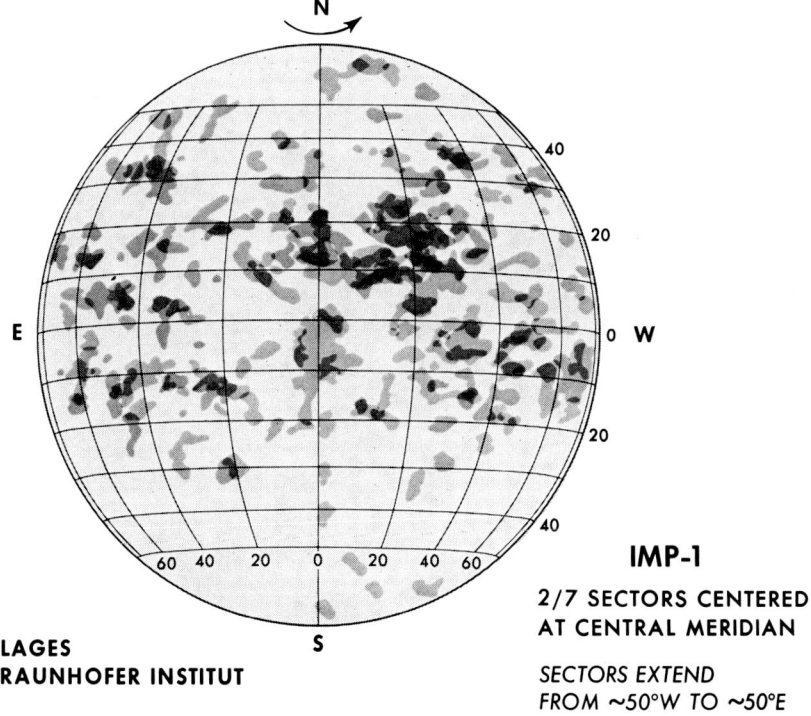

Fig. 5. Superposed-epoch analysis of calcium plage structure obtained from the daily Fraunhofer Institute maps of the sun. The sectors are approximately centered at central meridian, so that the leading edge of the sector is at about 50°W and the trailing edge of the sector about 50°E longitude (after Wilcox and Ness, 1967).

stationary pattern with four sectors per solar rotation can be seen during the year 1964. With the rise of solar activity in 1965 the sector pattern begins to change. Usually one solar rotation is quite similar to the preceding rotation, but in the course of several rotations an appreciable change in the sector pattern may occur. *From the discussion so far we see that the coronal magnetic field should contain many field lines connecting the photosphere to interplanetary space. A sharp boundary separates two adjacent sectors of opposite polarity.*

We will next establish that the region to the west of a sector boundary (before the boundary in the sense of solar rotation) is a quiet region while the region just east of the boundary (after the boundary) is an active region. First we may examine the location of flares. Do they occur at random with respect to sector boundaries? We see in Figure 4 that the region close to a sector boundary is the most likely site for a flare. Figure 4 is a histogram of flare occurrence as a function of distance from a sector boundary, where distance is measured in terms of days of rotation (one day equals 13° longitude). The results by Bumba and Obridko (1969) have been extended by Vladimirsky (private communication) using a larger body of observations.

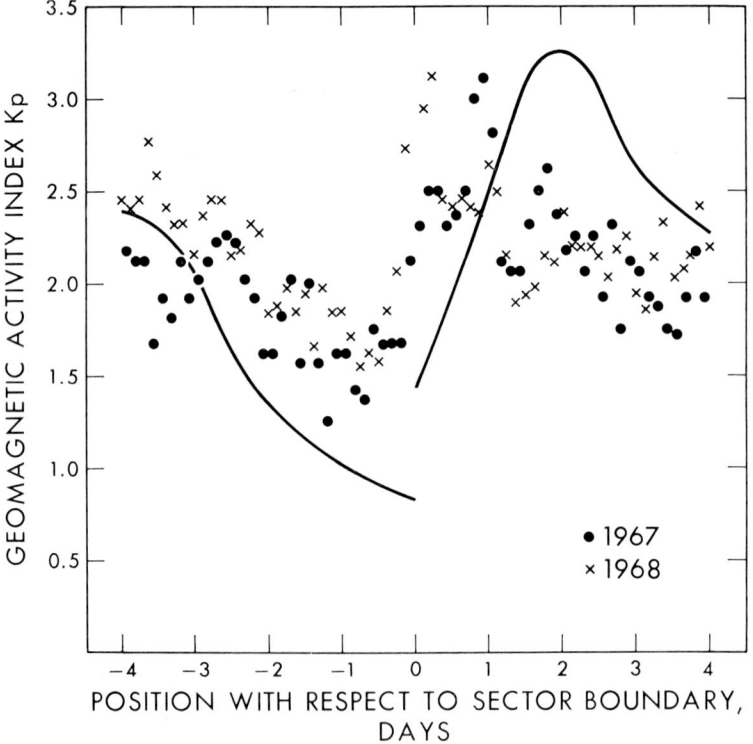

Fig. 6. Superposed epoch analysis of the magnitude of the planetary magnetic 3-hour-range indices K_p as a function of position with respect to a sector boundary. The abscissa represent position with respect to the sector boundary, measured in days, as the sector pattern sweeps past the earth. The solid line represents similar results obtained near solar minimum, the dots represent results in 1967 and the × represent results during 1968 (after Wilcox and Colburn, 1970).

Vladimirsky confirms these results and shows that the most likely position for a flare is just eastward of (after) a sector boundary. Is this a particular property of flares or does it extend to other solar activity? Figure 5 shows the average position of plages in the sectors observed near solar minimum (Wilcox and Ness, 1967). In Figure 5 the preceding boundary of a sector is at about 50°E and the following sector boundary is at about 50°E. We can see that the plages are more numerous in the areas just after the sector boundary.

Does this same property exist in the extended coronal field near the earth? To answer this question we may use the earth's magnetic field as a probe, since this field is influenced by interplanetary conditions. Figure 6 shows the average response of geomagnetic activity as a sector boundary sweeps past the earth (Wilcox and Colburn, 1970). The abscissa labelled zero represents the time at which a sector is observed near the earth and the graph shows the situation four days before and four days after this time. We see that in the days before the sector boundary geomagnetic activity is monotonically decreasing, with an almost discontinuous increase near the boundary

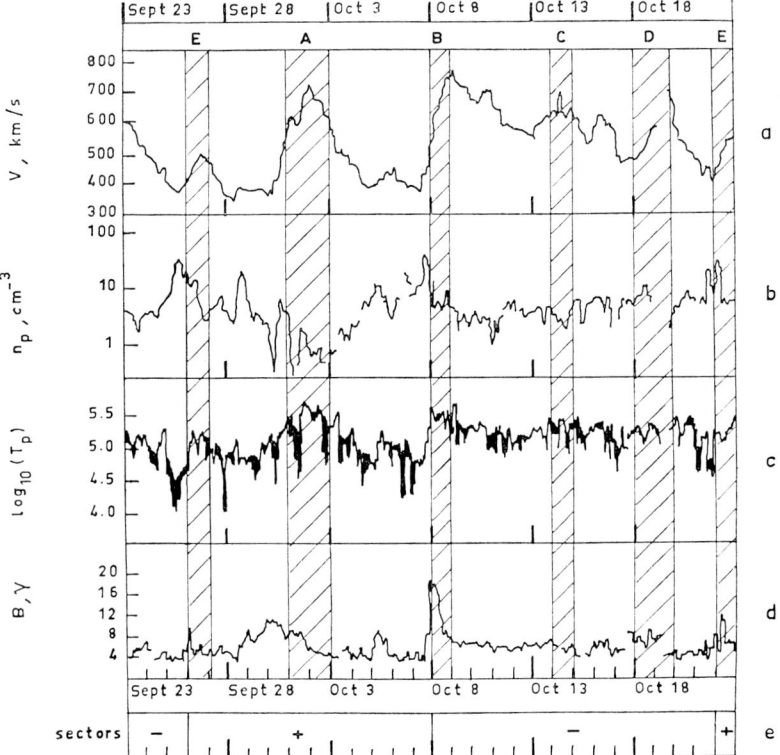

Fig. 7. Solar wind activity during solar rotation 1768. The vertical hatched regions represent CMP of coronal enhancements. a, solar wind velocity, b, proton density, c, temperature (upper and lower limits), d, interplanetary magnetic field magnitude, and e, sector polarity pattern (after Couturier and Leblanc, 1970).

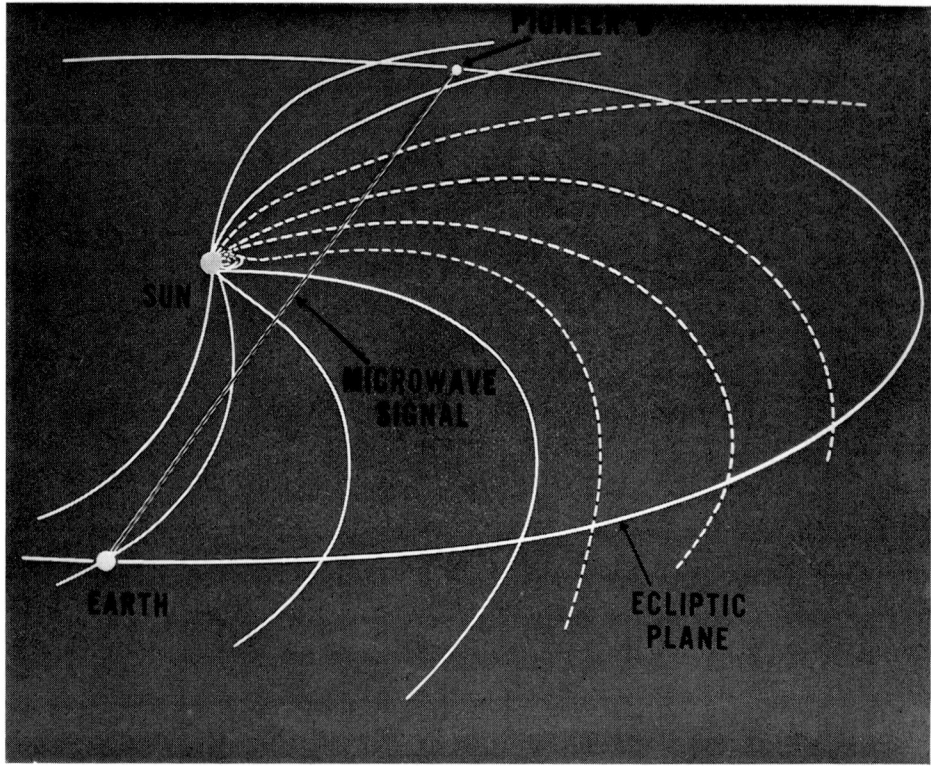

Fig. 8. Positions of the Pioneer 6 spacecraft, the sun and the earth at the time of observations of the Faraday rotation of the spacecraft telemetry signal.

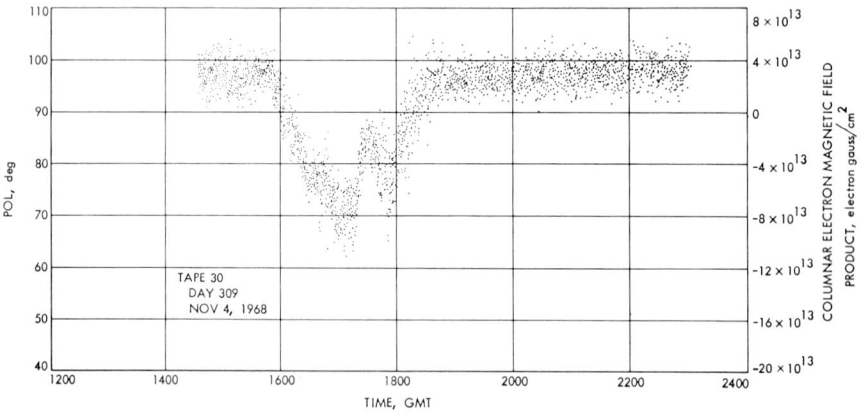

Fig. 9. Observed polarization of the telemetry radio signal from Pioneer 6 as a function of time (after Stelzried et al., 1970).

and a peak shortly thereafter. Thus we see again that the region just before the boundary is quiet and the region after the boundary is active.

Having established this boundary situation in the photosphere and at the distance of the earth we may inquire if the same effect exists in the corona. The vertical hatched regions in Figure 7 (Couturier and Leblanc, 1970) represent coronal enhancements observed with the Nançay radio interferometer at 169 MHz (1.77 m). These are

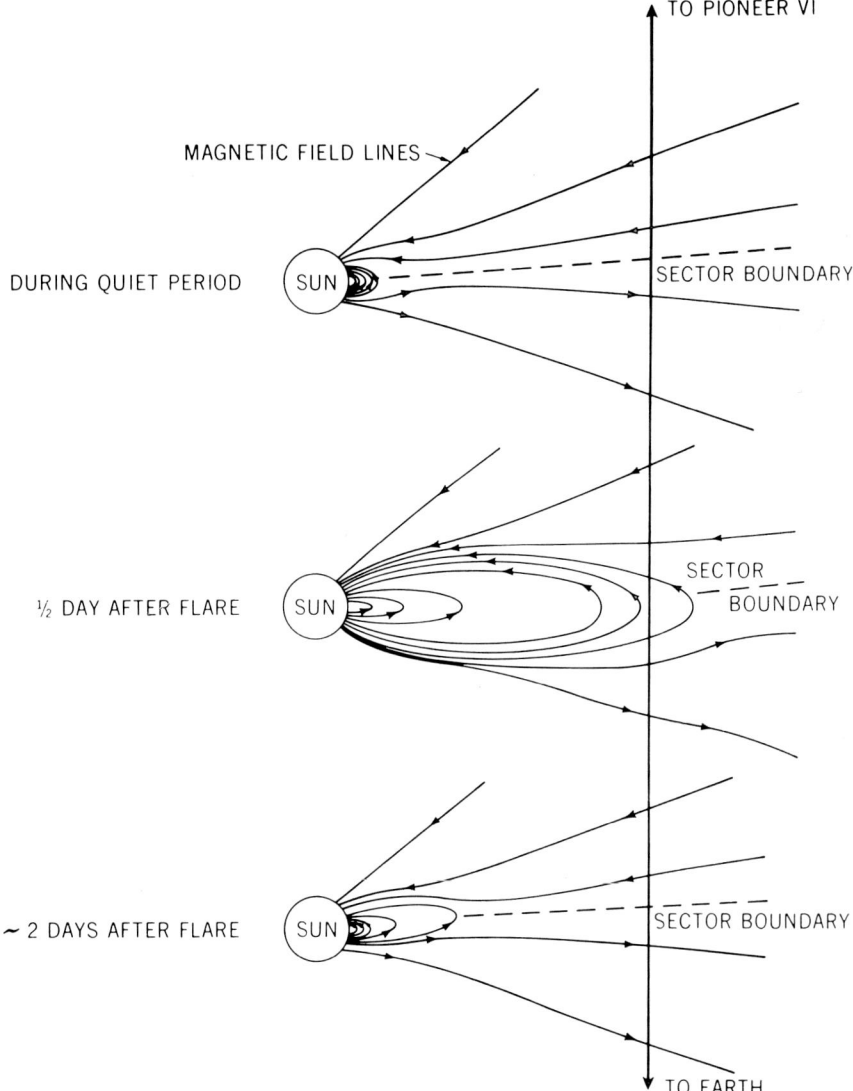

Fig. 10. View from the north of the magnetic field model proposed to explain the transient events observed with Pioneer 6. The line-of-sight between Pioneer 6 and the earth is shown to the right of each panel (after Schatten, 1970).

regions of enhanced electron density and temperature at altitudes between 0.2 and 0.5 solar radii above the photosphere. At the bottom of Figure 7 the interplanetary sector polarity is indicated. It can be seen that just after each sector boundary a coronal enhancement occurs. *Thus we see that the large-scale structure of the corona is related to the photospheric and interplanetary sector patterns.*

We examine now some unique observations of the coronal magnetic field in the regions 5 to 10 solar radii that were obtained when the Pioneer 6 spacecraft was occulted by the sun. The Faraday rotation of the microwave telemetry signal from the spacecraft was observed (Stelzried *et al.*, 1970) during the time in which the line-of-sight from earth to Pioneer had its closest approach to the sun in this region. The positions of Pioneer, sun and earth are shown in Figure 8. The Faraday rotation observation gives the product of the line-of-sight magnetic field multiplied by the density along the column from earth to Pioneer. Most of the rotation occurs near the shortest distance to the sun where the densities and field magnitudes are largest. Occasionally the observed Faraday rotation showed a change of 30° within a time interval of 2 hours, as shown in Figure 9. This observation has been interpreted (Schatten, 1970) in terms of loops of magnetic field transported outward from the sun by plasma ejected from flares. The three sketches in Figure 10 show the situation before, during and after the event shown in Figure 9. The Faraday observations in Figure 9 agree in direction and in magnitude with observations of an active region in the photosphere assumed to be the source of the flare. The flares producing events of this kind are not particularly large, and we may therefore assume that often the corona contains such magnetic loop structures. It is rare to find evidence of such loops in the observations of the interplanetary field near the earth. It therefore seems that most of the loops return to the sun after having reached perhaps a distance of 10 solar radii. *Thus the corona may frequently contain transient magnetic loops reaching out five to ten solar radii.*

Acknowledgements

This work was supported in part by the Office of Naval Research under contract N00014-69-A-0200-1016, by the National Aeronautics and Space Administration under grant NGL 05-003-230, and by the National Science Foundation under grant GA-1319.

References

Babcock, H. W.: 1961, *Astrophys. J.* **133**, 572.
Bumba, V. and Obridko, V. N.: 1969, *Solar Phys.* **6**, 104.
Couturier, P. and Leblanc, Y.: 1970, *Astron. Astrophys.* **7**, 254.
Schatten, K. H.: 1970, *Solar Phys.* **12**, 484.
Stelzried, C. T., Levy, G. S., Sato, T., Rusch, W. V. T., Ohlson, J. E., Schatten, K., and Wilcox, J. M.: 1970, *Solar Phys.* **14**, 440.
Wilcox, J. M. and Colburn, D. S.: 1970, *J. Geophys. Res.* **75**, 6366.
Wilcox, J. M. and Howard, R.: 1968, *Solar Phys.* **5**, 564.
Wilcox, J. M. and Ness, N. F.: 1965, *J. Geophys. Res.* **70**, 5793.
Wilcox, J. M. and Ness, N. F.: 1967, *Solar Phys.* **1**, 437.
Wilcox, J. M., Severny, A., and Colburn, D. S.: 1969, *Nature* **224**, 353.

7. INVESTIGATIONS ON CORONAL MONOCHROMATIC EMISSIONS IN THE OPTICAL RANGE

A. DOLLFUS

Observatoire de Paris, 92 Meudon

Abstract. Improved techniques were designed for the observations of monochromatic emissions in the Solar corona, and used between 1955 and 1963 with the Pic-du-Midi Coronograph during the last maximum of solar activity.

The analysis of these observations proved that, in the lower part of the corona, the electron density is heterogeneous and featured, as known by the eclipse plates, but with slow variations in time. The field of ionisation (or temperature) is far more variable or transient; the monochromatic emissions of different energies, illuminated in the coronal regions having the suitable ionisation temperature, show that this field of temperature has very strong gradients, complex structures and rapid variations with time.

1. Classification of the Ionisation Levels in the Corona

The solar corona is emitting some monochromatic radiations in its regions of highest electron density. These emissions are featuring streamers, arches and patches of light varying slowly with time.

Spectroscopic observations carried on during eclipses or with the coronographs displayed at least 30 of these radiations; they correspond to different ionisation potentials. Those of these radiations having potentials of about the same levels, display configurations of the same type. But the features are very different for radiations departing in ionisation potentials. The study of these configurations enables the classifying of coronal radiations by categories, and predictions can be derived for the ionisation range corresponding to a given emission.

With the first spectrograph of his coronograph, at Pic-du-Midi, B. Lyot, used in 1936 and 1937 a circular slit surrounding the image of the solar disc at a constant angular distance, and recorded successively a large number of spectra for different coronal emissions [1]. The same technique was extended to the ultraviolet during the total eclipse of February 25, 1952 [2], and was later used by other authors than Lyot himself. In 1957, we tried to gather all these materials for a classification of the emission lines by increasing ionisation. The result was that these lines definitely range themselves in 4 categories. Table I gives this classification in 4 groups [3].

In this table, we report the interpretation of the lines given by B. Edlèn and his computed potential ionisation level [4]. It is striking to realize that the empirical classification corresponds accurately to 4 ionisation groups in the Edlèn's identifications.

Each group is designated by the wavelength of its most prominent emission line in the visible range.

Finally, it appears that the different lines in each group have close potential ionisations and behave in similar manner. As a result, we can describe the ionisation

TABLE I

Group 6374 Å	Group 5303 Å	Group 6702 Å	Group 5694 Å
3454.3	3388.0 Fe XIII 325	3329.6 Ca XII 589	4412.4 A XIV 682
3534.0	3642.8 Ni XIII 350	3600.9 Ni XVI 455	5444.5 Ca XV 814
3800.7	4232.2 Ni XII 318	4086.5 Ca XIII 655	5694.5 Ca XV 814
3987.3 Fe XI 261	5116.1 Ni XIII 350	4351.4 (Co XV)	
4566.6	5302.9 Fe XIV 355	4358.9	
6374.5 Fe X 233	10746.8 Fe XIII 325	6701.9 Ni XV 422	
7891.9 Fe XI 261	10797.9 Fe XIII 325	7059.6 Fe XV 390	
		8024.2 Ni XV 422	

potential field in the corona, by limiting the study to 4 characteristic prominent lines only, each belonging to one of these categories. The corresponding observational possibilities are very powerful. We concentrated our analysis on the 4 following radiations:
– the red line $\lambda 6374$ Å (Fe X), corresponding to the ionisation potential 233 eV.
– the green line $\lambda 5303$ Å (Fe XIV) trigged for the potential 355 eV.
– The deep red line $\lambda 6702$ Å (Ni XV) responding to 422 eV.
– the yellow line $\lambda 5694$ Å (Ca XV) illuminated at 814 eV.

2. Direct Photography of the Coronal Emissions for 4 Ionisation Levels

In order to collect the proper observations during the last maximum of solar activity, we designed in 1954, a new bi-refringent filter, selecting the coronal radiations 5303 of Fe XIV and 6374 Å of Fe X. This monochromator was constructed by the French Company OPL, and mounted on the Pic-du-Midi coronograph in 1955. It was described in 1956 [5]. The basic principle is the Lyot's type bi-refringent filter, improved by the use of the Evans' split-elements [6]. Adapted to the 20 cm coronograph, this monochromator covers a field larger than the apparent diameter of the Sun, enabling a complete study of the coronal emissions all around the sun in a single exposure [7]. The selectivity is 1.5 Å; monochromatic emissions as faint as 10×10^{-6} times the brightness of 1 Å of the solar disc are recorded; the exposure time is of the order of 1 minute. In addition to the selection of 5303 and 6314 emissions, the filter is able to transmit the coronal line Ca XV at 5694 Å, providing special adjustment. Furthermore, it can be adapted to the radiations Hα (6563 Å) and Hβ (4861 Å) of hydrogen and D_3 (5875 Å) of helium emitted in the chromosphere and prominences.

Then, a second bi-refringent filter was designed in 1956 and adapted to the Pic-du-Midi coronograph in 1957. This new monochromator is selecting the coronal radiations 6374 Å of Fe X, 6702 of Ni XV, and 5694 of Ca XV. The radiations 6374 and 5694 Å, respectively emitted in coronal regions of maximum and minimum ionisation (233 eV and 814 eV), can be observed simultaneously in the field with the same adjustment. The coronal emission 6702 (422 eV) can be observed simultaneously with prominences in Hα. The filter was described in 1957 [8].

Finally, a simultaneous use of both filters in successive sequences enabled the observer to record photographic pictures of the corona with each of the lines corresponding to the 4 groups of ionisation potentials displayed within the visible spectral range. Figure 1 shows the kind of pictures obtained; this is an image in $\lambda 5303$ Å.

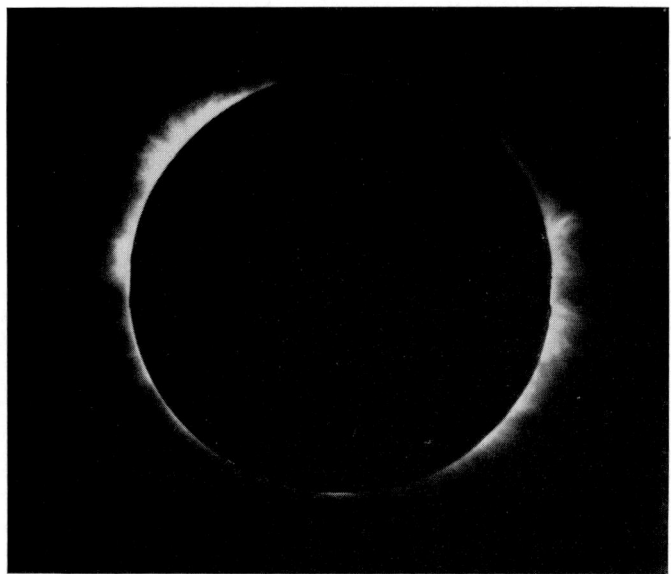

Fig. 1. Example of monochromatic picture of the corona. Image taken in emission $\lambda 5303$ of Fe XIV. 14 March 1959 at $14^h 35^m$ UT.

Figure 2 displays sequences of images successively obtained in Hα, at 233 eV ($\lambda 6374$), at 355 eV ($\lambda 5303$), at 422 eV ($\lambda 6702$) and at 814 eV ($\lambda 5694$). Each frame is a portion of full images of the disk similar to Figure 1.

The pictures are provided with sensitometric calibrations in absolute scale enabling photometric determinations of the luminance for the different emissions as a ratio to the luminance of the solar disc's center [9].

Furthermore, we collected direct photographs of the corona in white light, by the technique achieved by Lyot in 1932 and later refined. Figure 3 gives an example of a white light corona picture. The streamers displayed with this technique are due to the light scattered in the corona by the electrons and are characteristic of the electron density distribution in the inner corona.

An extensive survey of the coronal structures was carried on by these techniques at Pic-du-Midi, between 1955 and 1963. J. L. Leroy and E. Maurice cooperated in these observations and in several instrument improvements. A certain number of plates were also collected by the Pic-du-Midi Observatory staff under the direction of J. Rösch.

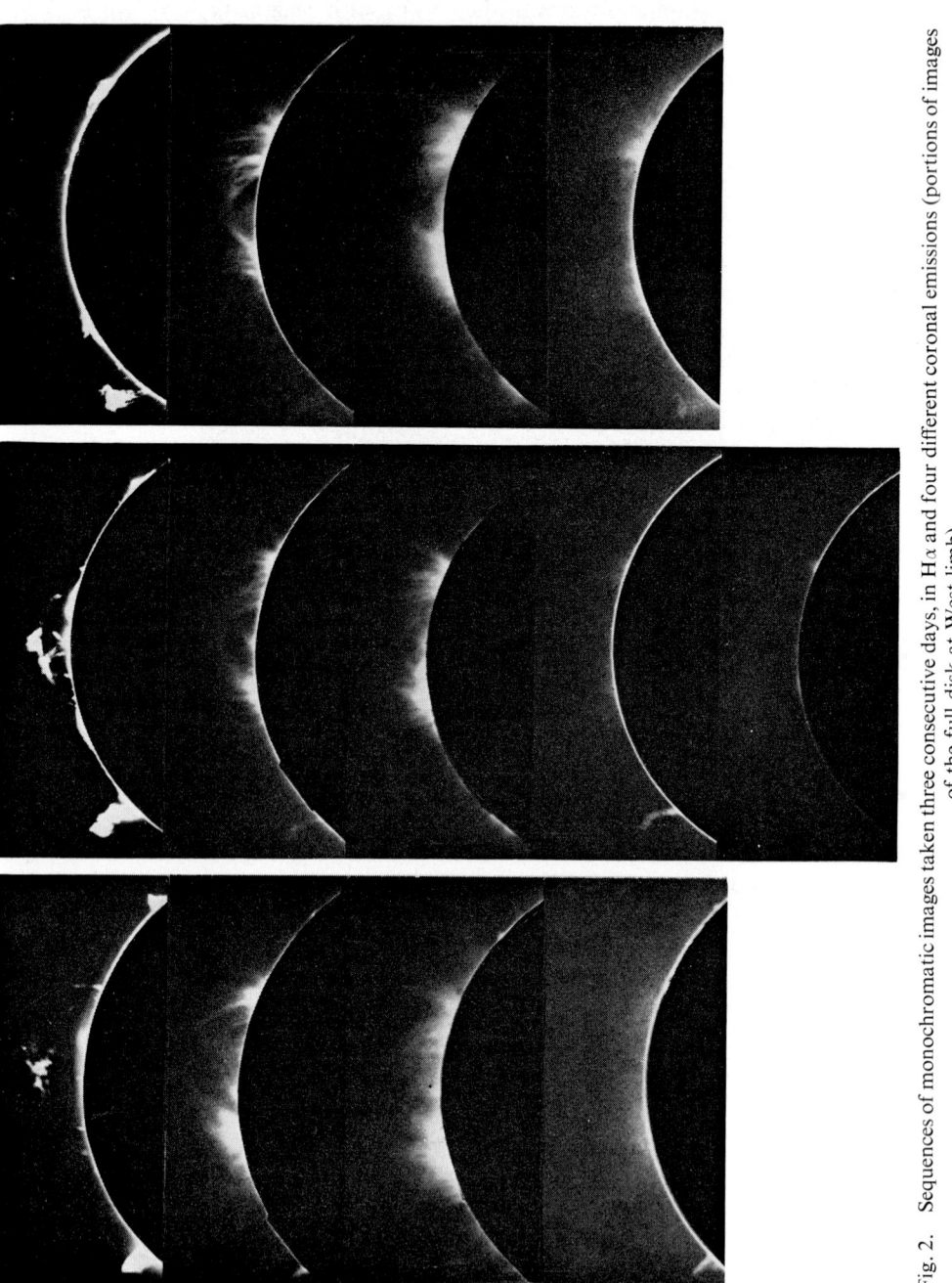

Fig. 2. Sequences of monochromatic images taken three consecutive days, in Hα and four different coronal emissions (portions of images of the full disk at West limb).

Left column: 20 June 1960   from top to bottom: Hα: 16ʰ30ᵐ UT   6374: 08ʰ12ᵐ   5303: 08ʰ03ᵐ   6702: 17ʰ30ᵐ
Central column: 21 June 1960   from top to bottom: Hα: 09ʰ30ᵐ UT   6374: 10ʰ57ᵐ   5303: 08ʰ15ᵐ   6702: 07ʰ26ᵐ   5694: 10ʰ00ᵐ
                                                      Hα: 09ʰ45ᵐ UT   6374: 07ʰ43ᵐ   5303: 07ʰ30ᵐ   6702: 09ʰ20ᵐ

Fig. 3. Example of white light picture of the Corona. Image taken with a red colour filter.
1 March 1963.

3. Basis of the Interpretations for the Coronal Pictures

For the analysis of the plates, we should consider the basic formula relating the intensity I_i of an emission line to the local temperature T, the transition energy W, and the local electron density N_e; we have also to consider the population N_z of the level, as a ratio to the total number N_0 of the atoms for the element.

The relation is expressed by the somewhat simplified relation:

$$I_i = K_1 T^{-1/2} e^{-W/KT} \frac{N_z}{N_0} N_e^\alpha \quad \text{with} \quad 1 < \alpha < 2 \qquad (1)$$

The exponent α is equal to 2 if the process is purely collisional and equal to 1 in a pure radiative process. Practically, for the coronal area concerned, α lies probably between 1.5 and 2.

The intensity I_e of the white light corona is simply proportional to the local electron density:

$$I_e = K_2 N_e \qquad (2)$$

In principle, the quantities I_i and I_e have to be integrated all along the line of sight in the corona; however, it is the purpose of the present observational survey to try to disentangle the perspective effect, or overlaps in corona structures, by collecting a large amount of pictures, and selecting carefully the small number of those for which major prominent features appear to be strikingly well isolated at the limb, thus ruling out the blurring by too long path integration, or limiting the possibilities of overlap misinterpretations.

4. Heterogeneities of the Ionisation Temperatures

With the observational material collected, it is easy to prove that the ionisation field is not uniform in the corona. In case of a homogeneous ionisation temperature, Equation (1) reduces to a term proportional to N_e^α, meaning that the emission is highly correlated with the electron density field. Such is not the case. Figure 4 displays 2 images obtained successively, the first one in white light showing the features of the electron density, the second with the radiation 5303 Å of Fe XIV corresponding to the

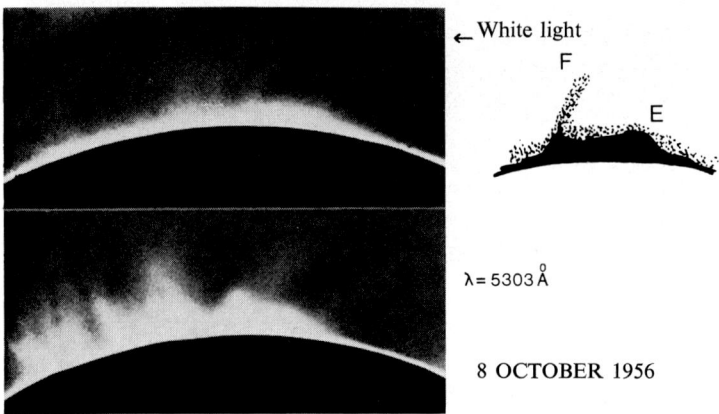

Fig. 4. Comparison of coronal features for the electron density (upper image in white light) and for the monochromatic emission 355 eV at 5303 Å (lower image through the monochromator). 8 October 1956.

ionisation potential 355 eV. The streamer F is bright in the white light image, indicating a high electron density; it is not visible in the 5303 Å image, proving that the local ionisation temperature departs from the value 355 eV necessary to illuminate this radiation. On the contrary, the streamers at left are bright in 5303 Å despite the low electron density proved by the weakness of the white light emission, involving a local ionisation temperature of exactly the value 355 eV necessary to strongly enhance the Fe XIV emission.

5. Gradients in the Ionisation Temperature Field

The 4 radiations observed produce images of coronal structures respectively enhancing 4 different levels of ionisation, corresponding to 233 eV, 355 eV, 422 eV and 814 eV. For these different levels, the features displayed are not similar. Furthermore, they have a tendency to avoid appearing in the same area. They often look like shells of filaments enveloping themselves. An obvious explanation is the existence of gradients in the ionisation temperatures. Each radiation is produced in the areas of the corona having exactly the temperature required for the trigging of its transition [10].

Examples of enveloping structures are given in Figures 5 and 6.

INVESTIGATIONS ON CORONAL MONOCHROMATIC EMISSIONS IN THE OPTICAL RANGE 103

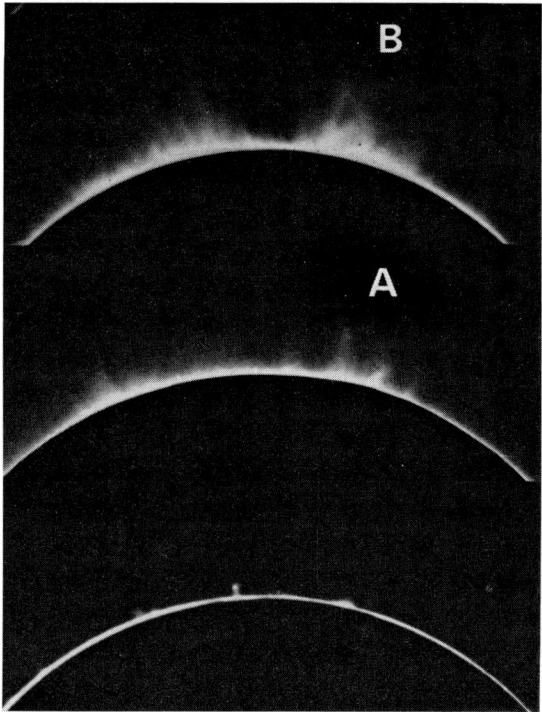

Fig. 5. Example of gradients in the coronal ionisation temperature field: Cool streamer: 19 May 1960 (West limb). From top to bottom: 5303 Å (355 eV); 6374 Å (233 eV); Hα (chromosphere).

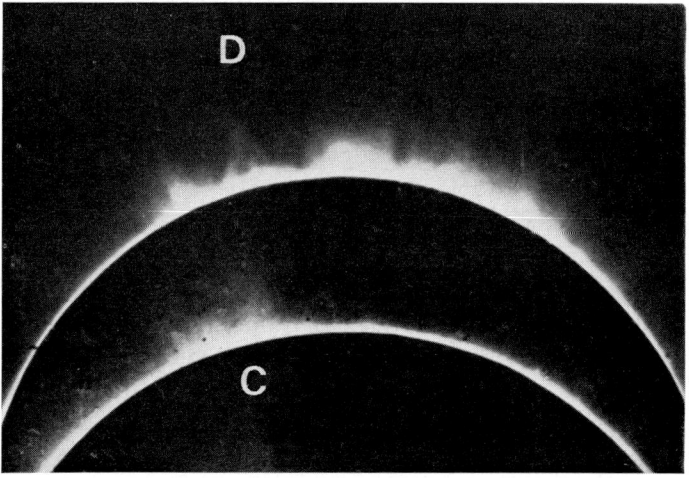

Fig. 6. Example of gradients in the coronal ionisation temperature field: Hot streamer: 27 October 1959 (East limb). Top: 5303 Å (355 eV); Bottom: 6702 Å (422 eV).

In Figure 5, obtained on May 19, 1960, the conic structure A seen in 6374 Å (at 233 eV) is outlined in the image at 5303 Å by a shell of ionisation 355 eV; this is the case of a low temperature nucleus with a radially increasing ionisation.

Figure 6 obtained on October 27, 1959, corresponds to the opposite configuration. The emission D at 5303 Å (355 eV) is enveloping a hotter center C enhanced at 6702 Å (422 eV) corresponding to an area of hotter temperature than average with outward decreasing temperature.

In the latter case, the gradient of ionisation is $422-365=57$ eV in 20" corresponding to 14000 km, meaning 5 eV for 1000 km. According to the ionisation temperatures computations of Burgess and Seaton, and more recently of C. Jordan (11), a reasonable figure for the temperature variations is about 5000 K for 1 eV. As a result, gradients observed in the corona are of the order of 25000° for 1000 km, or near 1% for 1000 km. Such local gradients of ionisation temperatures are very high indeed, and prove great inhomogeneities in the lower corona temperatures.

6. Local Variations in the Ionisation Temperature

Time variations in the emissions' intensities may be due to changes in the electron density, according to Equation (1). But we know, by the study of the white light pictures, that these variations are small and proceed slowly. Accordingly, the brightness variations observed are mainly due to changes in the local ionisation temperatures.

Figure 7 gives an example – among others – of a coronal area illuminated at 2 different ionisation levels successively (10). On May 29, 1960, in area G, the energy, initially high, seems to decrease continuously within a few hours:

– Initially, the temperature is too high to correspond to any of the two transitions under survey.

– At 10^h25^m, the decreasing ionisation reached the level generating the enhancement of the Fe XIV transition needing 355 eV, as materialized by the beginning of the illumination of the corresponding 5303 Å radiation.

– At 11^h25^m, the ionisation reached exactly 355 eV, because the brightness at 5303 Å reached its maximum. The transition of Fe X observed at 6374 Å with 233 eV, has not yet appeared at all at this time.

– At 12^h00^m, the emission $\lambda 5303$ Å of Fe XIV at 355 eV was substantially decreased, although simultaneously, the radiation $\lambda 6,374$ Å was increasing, with its maximum at 12^h20^m.

– At 13^h00^m, the illumination for 355 eV had completely disappeared, and now the illumination for 233 eV was decreasing, to vanish completely soon after on the last picture of the sequence.

The ionisation energy in this streamer decreased during the 3 hours of observation.

The reverse situation is observed in the same area the next day with a display of progressive increasing ionisation temperature with time. On Figure 8, the dome emitting $\lambda 6374$ Å (233 eV), at 8^h55^m on the right side of the upper right picture, is evolving in the successive 6374 images to feature itself into shells enveloping a display of 5303 Å emissions (355 eV), itself progressively featuring dome-like structures. The last image is taken in λ 5303 Å at 15^h00^m. The average increase of ionisation in this area is of the order of 30 eV per hour.

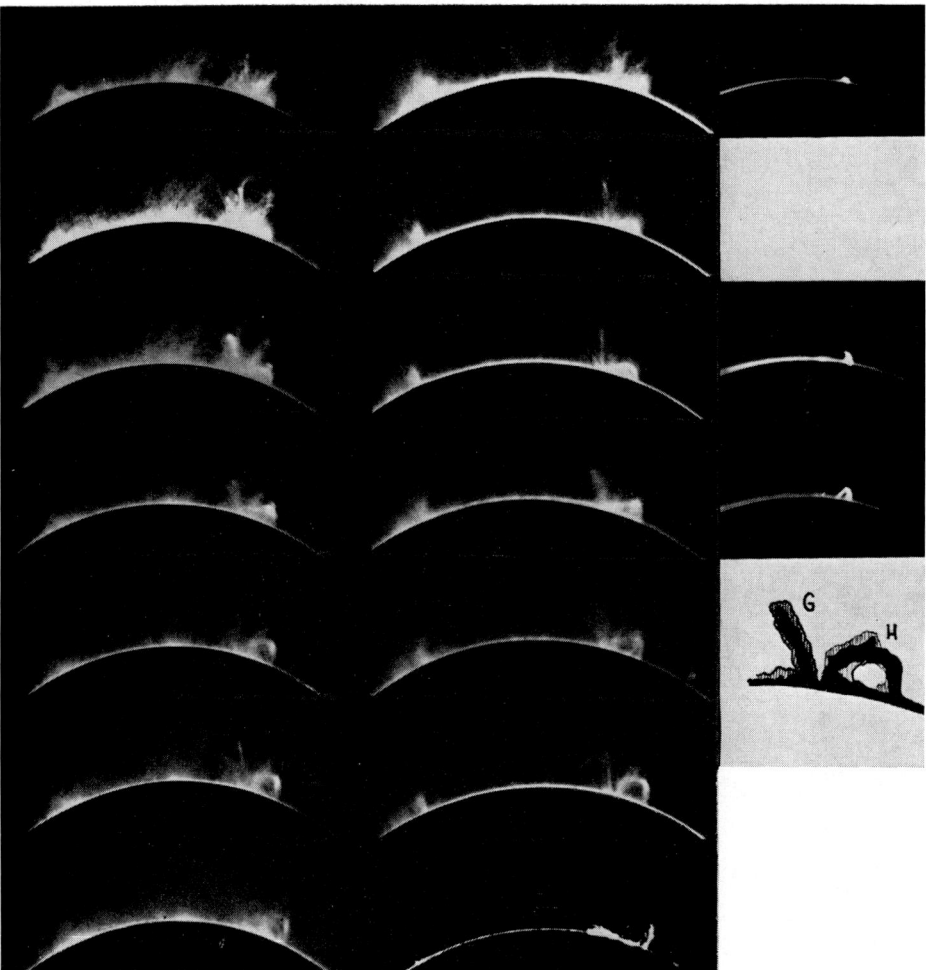

Fig. 7. Coronal emissions above an active region: Rapidly cooling streamer in G. Arch moving upwards on H. 29 May 1960 (East limb).

Emission 5303 Å	Emission 6374 Å	White light
09^h45^m UT	09^h30^m TU	09^h30^m UT
10^h25^m	10^h40^m	—
11^h25^m	11^h10^m	11^h10^m
12^h00^m	11^h20^m	12^h20^m
13^h15^m	13^h00^m	—
13^h40^m	13^h55^m	—
13^h45^m	Hα at 14^h99^m	—

For Figure 7, the ionisation level of area G is 355 eV at $11^h 25^m$, and 233 eV at $12^h 20^m$. The variation of ionisation is of the order of 122 eV in 55 min, corresponding to a variation of temperature of about 10000 K per minute. This rapid decrease gives a relative change of 1% in 2 min. Such very high variations correspond to a substantial fraction of the total energy stored locally in the corona.

However, the two observations discussed above correspond to regions above active centers.

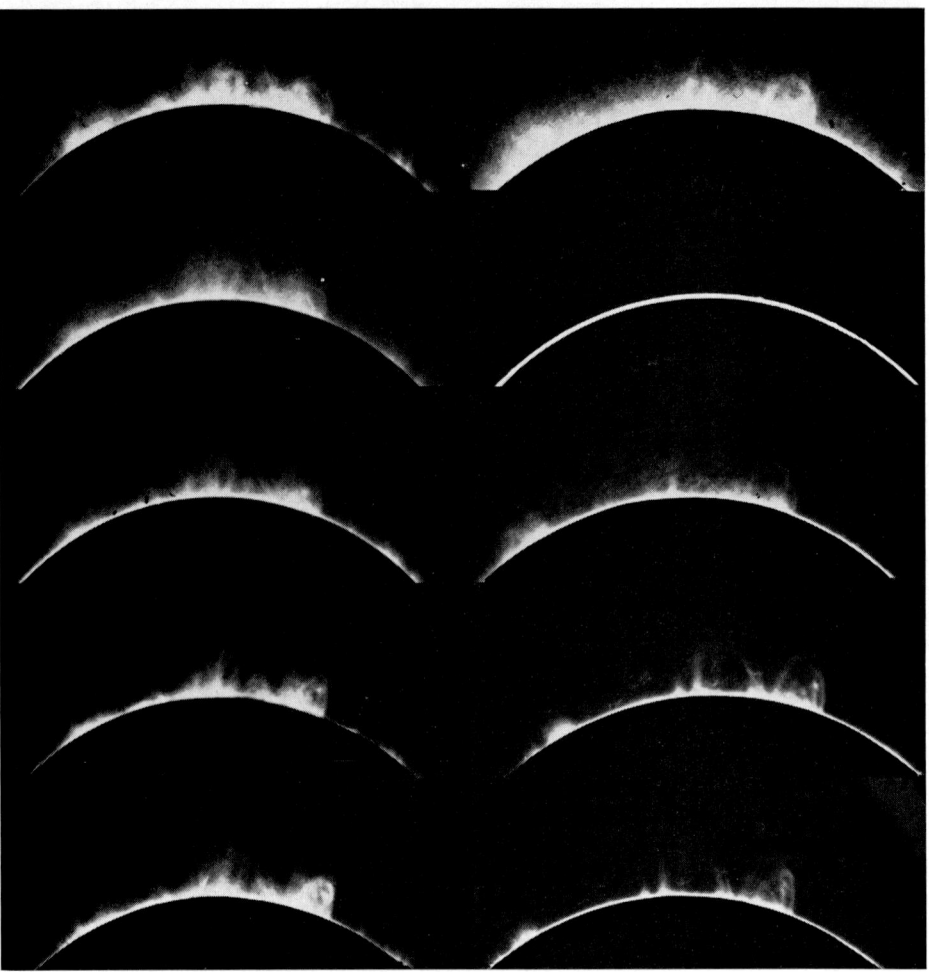

Fig. 8. Coronal emission above an active region: next day (see previous day Fig. 7) 30 May 1960.

Emission 5303 Å	Emission 6374 Å
$08^h 30^m$ UT	$08^h 55^m$ UT
$10^h 50^m$	Hα at $12^h 10^m$
$11^h 40^m$	$11^h 55^m$
$13^h 45^m$	$14^h 10^m$
$15^h 00^m$	$14^h 45^m$

7. Monochromatic Emissions and the Corona Magnetic Field

A striking result is that the coronal monochromatic features belong to the same types of configurations as prominences. Figure 2 shows, for consecutive days, an example of similarities between prominences and coronal features.

It is well known that the structures of prominences are correlated with the magnetic field's configurations; their streamers underline the lines of force of the field.

Accordingly, the monochromatic pictures of the corona can be used to suggest qualitative information about the solar magnetic fields' configuration; they are more widespread than the prominences and can be observed till 400 000 km towards the Sun's edge.

The behaviour of the monochromatic emissions with respect to the magnetic fields can be described as follows:

– According to the classical views, the corona electron density is featured by the magnetic field configurations in somewhat frozen permanent structures, varying slowly with time.

– The ionisation should vary rapidly in location and time through these configurations, giving localized enhancements of radiations appropriate to the local level of ionisation. Because the electron density is one of the factors governing the resulting emissions intensities (see Equation (1)), the overall patterns of the emission radiations are finally outlining some aspects of the magnetic configurations.

8. Motions of Coronal Material

The sequences of monochromatic pictures taken every few minutes during several hours, like those reproduced in Figures 7 and 8 show some variations in the coronal configurations. None of them are as fast as the motion of knots in solar prominences; most of the time, the illuminations seem to increase or vanish in the same location, without motions; other nearby streamers are animated independently by their own variations. The whole corona behaves somewhat like auroral displays.

During a full-day observing period, a monochromatic coronal structure is scarcely modified drastically; its individual streamers are rather modulated in its details. Propagations in the light enhancements might be indicated, though we were not able to disentangle clearly such processes. For day-to-day variations, we can recognize the overall structures despite the solar rotation, though somewhat modified (cf. Figures 7 and 8, also Figure 2). Once, a fan of streamers was modified during the survey of 6 hours in a given day, but remodelled nearly in the original configuration the following day [12].

As a whole, we are observing motions of the excitations enhancing the emissions, rather than motions of the coronal material itself. This result is a clear confirmation of the previews expressed by B. Lyot, after his pioneering survey in 1944 [13].

However, above active centers, it happens that the coronal material itself be displaced. Figure 7 shows in H a coronal arch, enveloping a loop-prominence and moving upwards simultaneously with the loop, the vertical velocity being of the order of 3.5 km/sec.

9. Spectroscopic Study of the Corona Motions

However, to disentangle the contributions of the true velocities in the coronal material with those resulting from the rapid variations of ionisation, the spectroscopic Döppler displacements can be used. The 20 cm Pic-du-Midi coronograph was equipped by B. Lyot with a spectrograph of the Littrow type giving a dispersion of 7 Å/mm. By increasing the focal length, it was possible to obtain large-size spectra of the corona lines on high contrast plates, with a dispersion of 3.2 Å per millimeter and a resolution of 20,000, on solar images having a diameter of 45 mm.

The radiation 6374 Å emitted by Fe x in the corona appears in a spectral range practically free from Fraunhofer lines and is particularly adapted to radial velocity measurements. The line 6383 Å of Neon was recorded throughout all the exposure time, as a wavelength reference. Figure 9 gives examples of such spectra. 35 spectra were measured, using the Evershed technique, by superposing a positive on a negative of contrast unity, reversed face to face, and shifting them until superposition of the two symmetric corona line images, and measuring the translation required. Such measurements gave radial velocities on about 100 areas in the corona, on segments of 0′4 length, covering a range of distances to the solar limb from 0′5 to 4′0.

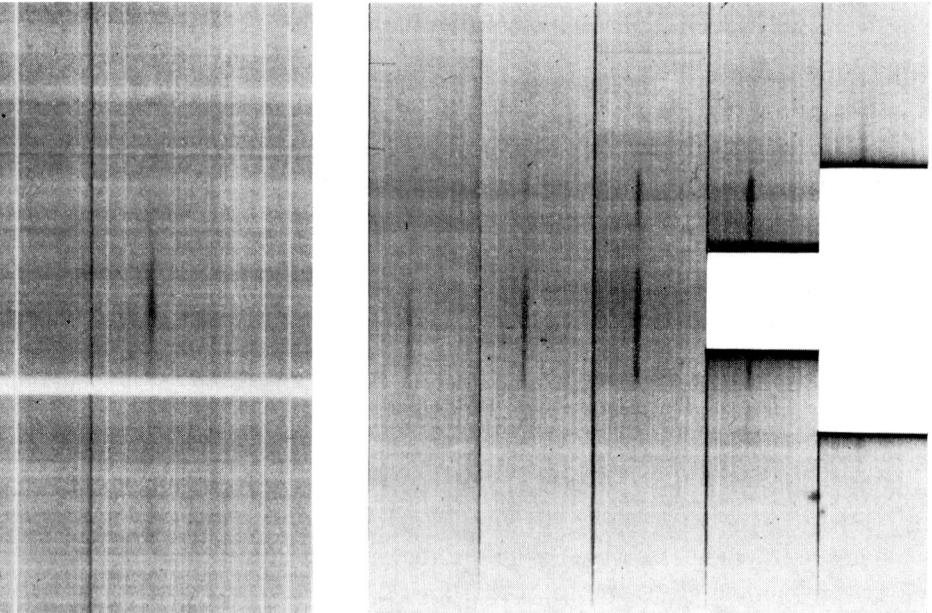

Fig. 9. High resolution spectra of $\lambda 6374$. Right: 11 July 1942. Successive sections by increasing steps of 1′ above East limb (taken by B. Lyot). Left: 26 July 1952. At 1.3′ above East limb. Neon line 6383 Å recorded simultaneously (taken by A. Dollfus).

After correction of the shifts due to solar rotation, it was possible to derive the velocity histogram shown in Figure 10. The motion of coronal material along the line of sight is one half of the time smaller than 0.5 km/sec. Exceptionally, bright knots crossed by the slit above active regions displayed localized radial velocities of nearly 4 km/sec.

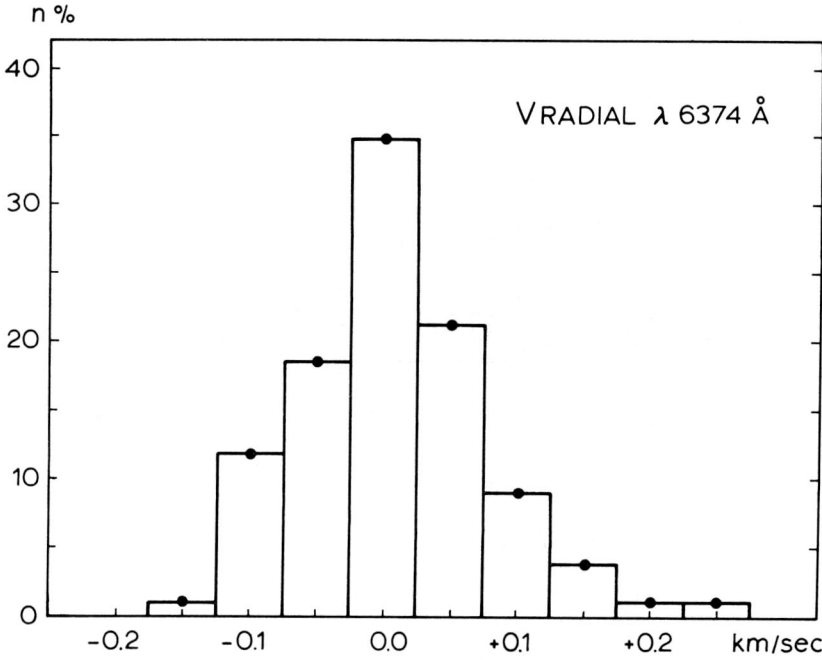

Fig. 10. Histogram of radial velocities in $\lambda 6374$ Å corona spectra.

As a conclusion, it is again confirmed that, except above some active regions, the coronal material is stable and varies very slowly with time. Most of the apparent variations observed in monochromatic lights are due to rapid local variations of the ionisation temperature.

10. Cinetic Temperature of the Coronal Emissions

Another interesting result is provided by careful analysis of the line profiles on high resolution spectrograms. As stated above, the 6374 Å line of Fe x is particularly suitably located in the spectrum for accurate measurements, because completely devoid of any blend with Fraunhofer lines. The spectra were recorded with a linear slit perpendicular to the direction of the solar radius, sections by sections, at distances increasing from 0'5 to 4'0 towards the edge of the Sun. (See Figure 9.)

Later, each section is measured by small segments of 0'4 of length, corresponding to small areas well located in the coronal image. The microphotometric tracings of the coronal line and the comparison Neon line, Figure 11, are converted in intensities after correction of the plates' photometric calibration. Then, the coronal lines are corrected from the resolution and aberrations of the spectrograph. If $B_1 = e^{-(\lambda/\gamma)^2}$ is the equation of the coronal profile, and $B_2 = (I/\beta\sqrt{\pi})e^{-(\lambda/\beta)^2}$ is the instrumental profile given by the Neon, the final profile results from the deconvolution giving: $B = \alpha \, e^{-(\lambda/\alpha)^2}$ with $\alpha^2 = \gamma^2 - \beta^2$ and $\alpha = \gamma/\alpha$. The average resolution measured on the plates was 0.3 Å and is 3 times smaller than the coronal profile, involving a small correction. The equivalent width is: $L = \alpha\sqrt{\pi}$.

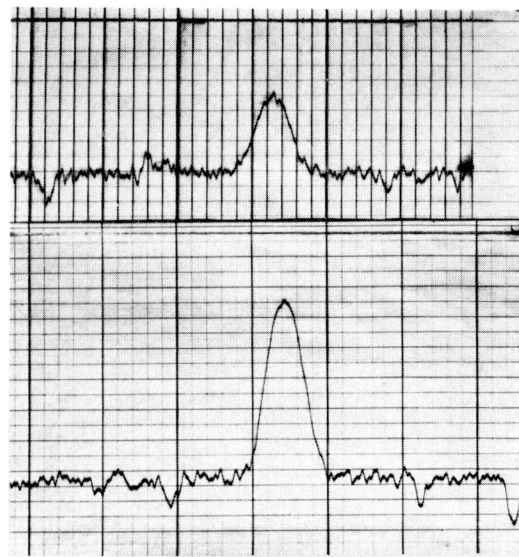

Fig. 11. Coronal line 6374 Å. Resolution: 20,000; Dispersion: 3,2 Å/mn; Solar diameter: 45 mn; High contrast plates.

The accuracy of the determination for L is limited by the fluctuations of the granulation for the emulsion. If the average lines intensities are $E = 30 \times 10^{-6}$ times the luminance of 1 Å at the center of solar disc, and the average scattered background brightness $B = 15 \times 10^{-6}$ times the luminance of the solar disk, the accuracy for L is of the order of 3% and varies as $\sqrt{1 + (LB/E)}$.

We recorded 7 of these large-size spectra in 1949 providing 42 profiles. 6 new spectra were added in 1952, giving 33 measures. 6 spectra previously taken by Lyot in 1942 were recorded and added 27 new profiles. The instrumental blurring in these 1942 spectra was given by Fraunhofer's lines profiles.

The brightness and the observed profiles are resulting from the integration along the line of sight for different coronal features; slight radial velocities may increase the line width; some measurements were discarded when complex structures seem to overlap in the field. Then the profile of the line is always symmetrical and follows reasonably the equation $B = E_0\, e^{-(\lambda/0.26)^2}$. The annual means of the measurements are summarized in Table II.

The equivalent width of the line remains always close to 0.94 Å, and seems to be completely independent from solar activity and altitude. Figure 12 displays the equivalent width as a function of the line's intensity. Dots stand for diffuse areas in the corona; crosses mean well identified streamers in which no apparent overlaps are occurring. The corresponding cinetic temperature is plotted at the bottom of Figure 12.

The striking result is that the measures are sharply picked around a well defined maximum, still sharper if we select only the well defined streamers (crosses) [14].

TABLE II

Distance to the solar limb	1942	1949	1952	Mean
0′5 to 1′4	0.956	0.945	0.895	0.933
1′5 to 2′4	0.913	0.972	0.925	0.936
2′5 to 3′4	0.931	0.990	0.910	0.940
3′5 to 4′0	0.920	1.020	—	(0.960)
Mean	0.934	0.970	0.907	0.940
Solar activity	30	135	50	

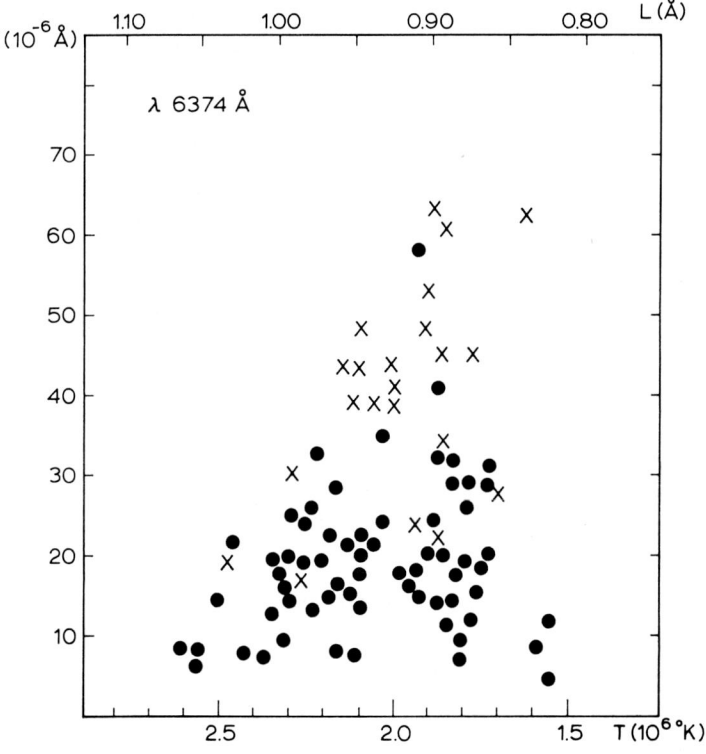

Fig. 12. Plot of the equivalent width L (or cinetic temperature T K) versus intensity (in 10^{-6} of 1 Å on solar disk) for the coronal line $\lambda 6374$. Dots: diffuse area. Cross: localized streamers.

A deduction is that, should the broadening of the line not be due to kinetic motions, but to large-scale turbulences for volumes smaller than the resulting power, such turbulences should be exceptionally constant at all distances and all areas in the corona, and this assumption is not straightforwardly realistic.

The more likely explanation, thermal broading, gives an average of about 2.0×10^6 K. When the high temperature in the corona is locally decreasing, the 6374 Å radiation begins to appear for about $T = 2.4 \times 10^6$ K; it reaches its maximum intensity for $T = 2.0 \times 10^6$ K; then its brightness decreases, and the emission vanishes when the temperature sinks lower than 1.7×10^6 K.

The maximum temperature of 2.0×10^6 K for the radiation 6374 Å is not in agreement with the latest developments of the di-electronic recombination theory.

C. Jordan [11] computed a temperature of the order of 1.2×10^6 K.

The discrepancy is puzzling; the ionisation processes or coefficients in the solar corona may still be more complicated than the present state of the art of the theory, including di-electronic processes.

11. Conclusions on the Behaviour of Coronal Emissions

As a whole, all the above-reported observations are consistent with the corona model in which the density is modulated by the magnetic field configurations and almost frozen in it with limited and slow variations in time. Motions are most of the time slower than 0.5 km/sec except above active regions.

The monochromatic emissions of different ionisation levels are trigged in the areas of this density field having the suitable ionisation temperature for the potential required. For Fe x of 233 eV, this temperature is 2.0×10^6 K, and the emission is reduced to the order of 100 times for departure of $\mp 20\%$ to this value. Therefore, the monochromatic emissions are materializing the areas of equal ionisation temperatures (isotherms) in the corona medium. As the monochromatic emissions are also influenced by electron density, the overall configurations have some connection with the electron density distribution, and consequently with the magnetic field configurations.

However, the ionisation temperature field is highly heterogeneous in the corona, giving rise to gradients as high as 1% in 1000 km. Furthermore, the local ionisations are rapidly varying with time and, above active centers, the amplitudes may reach 1% in only 2 min.

References

[1] B. Lyot: 1938, *L'Astronomie*, p. 193.
[2] B. Lyot and A. Dollfus: 1953, *Compt. Rend. Acad. Sci.* **237**, 855.
[3] A. Dollfus: 1957, *Compt. Rend. Acad. Sci.* **245**, 2011.
[4] B. Edlèn: 1954, *Monthly Notices Roy. Astron. Soc.* **114**, 700.
[5] A. Dollfus: 1956, *Rev. Opt.* **35**, 625.
[6] J. Evans: 1949, *J. Opt. Soc. Am.* **39**, 229.
[7] A. Dollfus: 1955, *Compt. Rend. Acad. Sci.* **241**, 1717.

[8] A. Dollfus: 1957, *Compt. Rend. Acad. Sci.* **245**, 32.
[9] E. Maurice: Thèse (3ème Cycle) Fac. Sciences, Paris, 13 Juin 1964.
[10] A. Dollfus: 1962, *Compt. Rend. Acad. Sci.* **255**, 3369.
[11] C. Jordan: 1969, *Monthly Notices Roy. Astron. Soc.* **142**, 499.
[12] A. Dollfus: 1957, *Compt. Rend. Acad. Sci.* **244**, 1880.
[13] B. Lyot: 1944, *Ann. Astrophys.* **7**, 31.
[14] A. Dollfus: 1953, *Compt. Rend. Acad. Sci.* **236**, 996.

8. CORONAL EVENTS OBSERVED IN 5303 Å

RICHARD B. DUNN

Sacramento Peak Observatory, Air Force Cambridge Research Laboratories, Sunspot, N.M., U.S.A.

Abstract. The observational and interpretational problems associated with photographing the solar corona in the wavelength 5303 Å are discussed. The events in the corona are discussed within three broad categories, (1) Slow, (2) Loops and Arches and (3) Fast Events. A classification scheme for coronal events is proposed.

1. Introduction

The green and red coronal lines, 5303 Å and 6374 Å, are the strongest lines in the visible spectrum. It is natural that Lyot (1944) should choose these lines to observe the changes in the corona outside of eclipse. During 1940-41 he developed the birefringent filter and ingeniously fitted it with three cameras that permitted simultaneous observations of wavelengths 5303 Å, 6374 Å, and Hα. Unfortunately, during 92 h of movies he did not see any large movements and did not corroborate the earlier eclipse observations of large motions itemized in his paper, or Waldmeier's visual observations of the motions of two coronal arches. We realize now that he was just unlucky and did not happen to be observing when there was a coronal event.

Movies of the coronal lines were not taken again until 1955, when this author joined the staff at Sacramento Peak and modified the small telescope used for taking prominences, which had been designed by Evans while at the High Altitude Observatory. The modification consisted of the addition of a coronal-quality lens in a focussing mount, a dust tube, and a 5303 Å and 6374 Å birefringent filter started by Dunn while at Harvard and finished by him at Sacramento Peak Observatory. Harry Ramsey, George Schnable and Howard DeMastus began surveying the corona in 5303 Å and obtained movies on those days when the sky was transparent. This program has never been extended to 6374 Å because we could not find a film that was satisfactory for both lines. Ramsey and Schnable left in the early 1960's and DeMastus has been operating this instrument unaided in recent years.

The early movies and the instrument were described by Evans (1957). Kleczek (1963) describes the regular structures in the green solar corona and recently Bruzek and DeMastus (1970) have described flare-associated expansions seen on some of the movies. Comments on the movies in the reviews and symposiums have been brief, presumably because of the sparsity of published papers.

2. Observational Parameters

2.1. Observational difficulties

For a number of reasons the movies are very difficult to make. One must understand these problems in order to interpret what he sees in the movies. The observational difficulties are as follows:

(1) *Faintness of Corona*. Even in the brightest areas of the corona the central intensity of the green line is only 200×10^{-6} of the intensity of the continuum of the solar disk. The faintest coronal features seen on the photographs are on the order of 10×10^{-6}. Thus the green line is faint. The transmission of the filters is low and this means the exposures are long, one or two minutes.

(2) *Scatter*. Because of the low relative intensity compared to the disk, one must use a coronagraph (Evans, 1953a; Lyot, 1939) to eliminate the scatter of the disk light into the image of the surrounding corona. In addition, some sort of filter must be used so that the ratio of intensity of the center of the green line to the local continuum is preserved. For instance, it would not do to permit 100 Å of continuum to pass through the filter, since the green line is only three-quarters of an angstrom wide. The optimum width of the filter is approximately the same as the width of the green line, or three-quarters of an angstrom. In the past only birefringent filters (Evans, 1953b) have been used for the corona. Dollfus (1956, 1957b) developed the most sophisticated of these filters which is still in use at Pic du Midi. Recently, Orrall at the University of Hawaii and Leroy at Pic du Midi have been using dielectric interference filters with considerable success. These filters have the advantage of simplicity and higher transmission compared with the birefringent filters that use conventional polaroids. In comparing filters one must always consider the overall intensity transmitted by the filter. This is proportional to the transmission, aperture, and angular field. A low value of transmission, for instance, can be compensated by a larger angular field.

(3) *Sky Brightness and Fluctuations*. Any ice particles, dust or insects between the coronagraph and sun cause the sky to brighten. This brightening and poor seeing account for the flashes over the entire scene that appear in many of the movies. With its two-angstrom filter the Sac Peak instrument is operated until the sky is 50×10^{-6} of the disk at a distance of six minutes of arc from the limb. A one-angstrom filter would permit observations in skies twice as bright.

(4) *Low Contrast*. With a sky of 50×10^{-6} and a two-angstrom filter, a 10×10^{-6} coronal feature would be less than 10% of the sky. High contrast films with gammas approaching four must be used for best results. The contrast is so low that it is difficult to see the region even with a low-power eyepiece.

(5) *Over-occulting*. The scene is always over-occulted by about 30 seconds of arc. Diffraction around the occulting disk makes one think the solar limb should be even with the occulting disk, but Figure 1 shows the true relationship between the edge of the sun and the occulting disk. Because of guiding errors and seeing, under-occulting accentuates the flicker of the background.

2.2. PARAMETERS OF THE SAC PEAK INSTRUMENT

The Sac Peak instrument has an objective with a 15-cm aperture and a 5-cm diameter image. The birefringent filter is not wide-field and is 2 Å wide with 6 polaroids (theoretical transmission about 9%). The only film that has ever shown the faint arches and streamers is Kodak IV-J exposed 1 to 2 min, depending on the brightness of the sky.

116   RICHARD B. DUNN

2.3. Interpretation of Coronal Scene

To avoid drawing erroneous conclusions we must keep in mind several points as we interpret and view the scenes.

(1) *Resolution*. During the long exposures for the corona seeing distortions and guiding errors are accentuated compared with the same scene photographed in Hα. Nevertheless, Figure 1 shows the coronal detail to be more diffuse than that seen in Hα. This relationship can be demonstrated by Lyot's movies, since he used identical and simultaneous exposures for Hα, the green and the red line. The prominences are 100 times over-exposed but still seem sharper than the coronal structure.

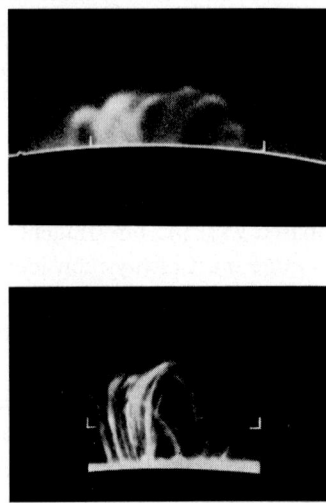

Fig. 1. Simultaneous pictures taken on November 22, 1956 show that the structure in the coronal loops in the top scene are more diffuse and more uniform than that in the Hα loops at the bottom. Although corresponding witness lines show the relative height of the occulting disk and the relative position of the scenes, detailed comparisons are difficult to make. Note small concentric arches on left. These faded later in the sequence. More scenes of this event are shown by Evans (1957).

(2) *Smearing*. Bruzek and DeMastus (1970) point out that since the exposures are 60 sec long a wave traveling at 1000 km/sec will be smeared by 60 000 km. Thus a thin shell traveling fast may not be visible in these pictures.

(3) *Time Resolution*. The time resolution of 60 to 90 seconds per frame defines an upper limit of velocities that can be seen on the films. Features traveling faster than 2000 km/sec would only show on one frame.

(4) *Radial Velocities*. Line-of-sight velocities greater than 100 km/sec will be shifted 2 Å and will not be seen through the Sac Peak filter.

(5) *Background*. The scenes are always superimposed on a strong scattered light background. The exposures are always set to record the sky so that the faintest coronal features show best. Occasionally the exposure on the movie changes abruptly

as the observer compensates for a sky change. All photometry of details must take this and the radial gradient of the sky into account.

(6) *Optical Thickness.* The corona is optically thin, so that the crossing of two loops looks like a bright dot and a spherical shell of gas may look like a loop.

(7) *Projection Effects.* Projection effects, complicated by the optically thin corona, can be very misleading.

(8) *Continuum.* The stronger prominences have enough continuum to show in these green line movies. Thus two angstroms of continuum can often equal the intensity of the green coronal line. This can be seen in Figure 2. The implication is that some of the bright corona in the tops of active loops may be continuum, and some of the "ejections" or "surge-like" features may be prominences showing in continuum. On the other hand, if a large cloud of high-density gas were ejected by a flare, later to condense into loop prominences, we would expect to see its continuum on these films. If a shell of continuum existed we would also expect to see it moving out at the time of a flare.

Fig. 2. Spike-like prominence structure shows in the green line picture at the top but is not visible in the 5303 Å spectrum. This structure is due to the filter transmitting the faint prominence continuum that is just visible in the spectrum taken near the Magnesium *B* lines. September 13, 1956.

3. Normal Undisturbed Corona

Coronal events are very rare. Nothing happens on the average movie. Lyot (1944) concluded, "In general, the corona changes in form and aspect, not like most of the prominences (that is, by relative movements of their parts), but by the appearance and disappearance and relative intensity variation of the elements which compose it. The coronal arches and coronal clouds, which often form complex ensembles, are born in one place along invisible-trajectories which existed beforehand and whose origin and mode of formation we cannot explain." This holds true for almost all the scenes of the corona. He goes on to describe "brightness variations", "weakening of arches" and other slow changes that occur in the scene. Dollfus (1957a) suggests the movements in the corona are explained by changes in excitation. In the Sac Peak movies one often sees slow motions and the development of a "hole" or patch (Evans 1957) high in the corona, or dark lanes radiating outwards in the corona. We would conclude that these areas do not emit 5303 but may show in other forbidden lines that are excited at lower or higher temperatures (Dollfus 1962).

4. Loops and Arches

4.1. "Open" and "Closed" Regions

Some coronal scenes look "open" as in Figure 3 and some look to be all loops and arches or "closed" as in Figure 4. The differences are presumed to be due to the magnetic field structure. The coronal structures are related to the magnetic field, and we say they "map" the magnetic field. This is misleading since the force-free calculations show that the field should uniformly permeate the entire area and should not lie only in loops. It would seem to be more accurate to say that the corona defines particular flux tubes in the corona. This is best illustrated by the coronal loop sys-

Fig. 3. "Open" coronal region December 10, 1966.

tems shown in Figure 5. Successively higher and higher loops glow and fade in place. With time the upper boundary of the system is seen to rise until the process is dissipated. We do not expect the field to change during this process since the loops do not expand. The loops are simply selecting different flux tubes to show the field at any one time. This implies a footpoint process. We would expect the same flux tubes to show from plotting the trajectories of prominences over a long time.

Fig. 4a. "Closed" coronal region. December 16, 1966. Note that the faint arches to the left of center are "banana" shaped at the end nearest the center of the picture.

The arches and loops may be very basic coronal structures since many scenes appear to contain nothing else. What is the difference between arches and loops? Is one a less energetic example of the other? There is no difficulty discovering the large concentric arches on Figure 4 and there is also no difficulty in recognizing the well-developed flare-associated loop system on Figure 1 or on November 2, 1969 and November 18, 1968 in Figure 5. The difficulty arises when the loops become older and hence grow to higher heights and are superimposed on other background features as on October 21, 1968 in Figure 5. The characteristics of loops and arches are listed as follows:

4.2. CHARACTERISTICS OF ARCHES

(1) Larger than loops, often extending half a solar radius (Kleczek, 1963).

(2) Connect active regions together. Kleczek expects the surplus north field of one region to connect to the surplus south of another (Kleczek 1963).

(3) Occur in concentric systems with as many as five arches in one system (Kleczek, 1963).

Fig. 4b. "Closed" coronal region. March 13, 1969. Note the "banana" shaped "dark" structure on the right. This may be due to projection effects of the nearby brighter regions.

(4) Generally uniform in intensity along their width and length, at least more so than prominences. The diameter of cross section is 8–12000 km, but can be thicker (Kleczek, 1963).

(5) Very stable (Kleczek, 1963). Certainly less active than the early stages of loops. However, loops tend to slow down as they grow higher.

(6) Shapes are more like a bridge than hairpin (Kleczek, 1963).

(7) Can be seen edge-on as an elliptical shape. This indicates that they are not spherical shells.

(8) No inherent radial velocity. Less than 1 km/sec (Lyot, 1944).

(9) Form 'sheaths' for Hα 'coronal rain' prominences.

(10) More diffuse than corresponding Hα features.

(11) Weak and stable X-ray emission. Presumably the ASE X-ray pictures of arches between active regions are the same as arches seen in the green coronal line (Van Speybroeck, 1970).

(12) Outline particular flux tubes of magnetic field.

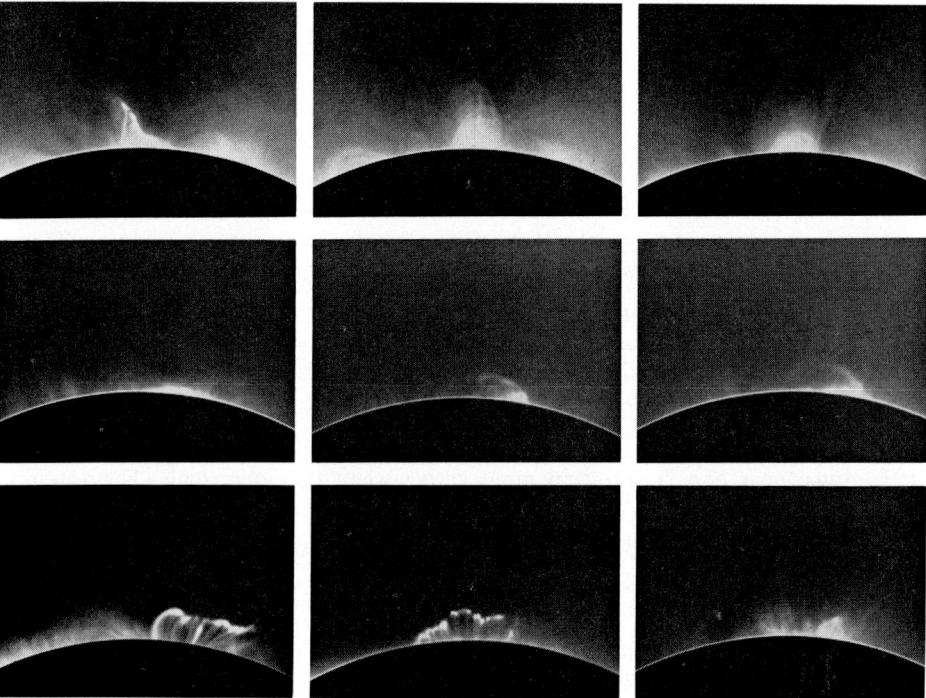

Fig. 5. Coronal Loops. *Top Row*: Growth of a loop system over three consecutive days. September 26–28, 1963. Note similarity to arches on the third day. *Middle Row*: Loop expansion on October 22, 1968, UT 16:30, 20:00, 21:42. *Bottom Row left to right*: Loops on November 2, 1969; Loops on November 18, 1968; Old diffused loop system on October 21, 1968.

(13) Evolution unknown. We believe they grow in place.

(14) Disrupted by flare-associated waves in the corona.

(15) Not flare-associated themselves.

4.3. CHARACTERISTICS OF LOOPS

(1) Smaller than arches, 50–100 thousand km (Kleczek, 1963).

(2) Always in a single active region (Kleczek, 1963).

(3) Occur in systems that at any one time show all loops at one height. Concentric loops are rare, but can occur. One shows to the left in Figure 1. It subsequently faded away.

(4) Less uniform along their length than arches, especially in the early stages of development. The diameter of cross section is given by Kleczek as 3–8 thousand km.

(5) Often show active brightening along their length.

(6) Often hairpin in shape (Kleczek, 1963).

(7) Can be seen edge-on as hairpins (Kleczek, 1963), so clearly are not spherical shells.

(8) Some radial velocity. Newkirk (1957) finds an average of 14 km/sec for the velocities of moving coronal material and 34 km/sec for the associated prominence material, always with the same sign as the coronal velocity. Karimov (1963) finds similar magnitudes but sometimes different signs.

(9) Form "sheaths" for associated Hα loop prominence (see Figure 1 and Evans, 1957). On large loops Hα breaks up into parallel filaments that show no sign of twisting (Newkirk, 1957).

(10) More diffuse than associated prominence. Figure 1 shows this to be true, however in this particular case the Hα picture was taken with a much shorter exposure than the coronal picture. The comparison should be made with simultaneous pictures with the same exposure length, as Lyot has done.

(11) Probably active X-ray event. (Vaiana *et al.*, 1968; Teske, 1970).

(12) Outline particular flux tubes of magnetic field.

(13) Evolution clearly observed. They grow in place, often growing uniformly along their length. They fade and then grow again at a higher elevation. This process slows down near end. They seem more active when they are small. The Hα loop prominence often associated with a coronal loop forms as a bright knot at the top and flows down both sides of the loop. Velocities of 100 km/sec for the Hα loops are common.

(14) Can show presence of flare-associated wave, although there are no observations of a wave disrupting a really well-defined loop system.

(15) Post-flare. The Hα loops identified with the coronal loops have been shown to occur after the flare (Bruzek, 1964 and others).

(16) Yellow coronal line (Ca xv 5694) is closely associated with the tops of coronal loops. We believe this to be a real effect that is not caused by the long line-of-sight distance one might expect from projection effects in the tops of loops.

4.4. Shape of arches and loops

The fact that arches are more "bridge-like" than the more "hairpin" loops would seem to be due entirely to the separation of the footpoints and to projection effects. As the footpoints separate, the lines of force will look less like hairpins. When the footpoints are well below the limb of the sun the part that projects above will look more like a "bridge". In both arches and loops I tend to favor the hypothesis that some coronal excitation mechanism selects a particular footpoint so a specific flux tube is activated. It is significant that these individual flux tubes are not "banana-shaped" in outline. In the recent ASE X-ray photographs (Van Speybroeck, 1970) there are apparently banana-shaped arches, going between active regions. Examination of Figure 3 suggests that the outline of the entire coronal arch *system* is banana-shaped. We immediately ask what is the cross section of the banana? It has to be either circular or a narrow, thin sheet best described by Goldberg as a "squashed banana." If it is circular then the arches within the concentric arch system will really be large tubes, each one shaped like the skin of the banana. This is unpalatable because the concentric arches are seen to project as ellipses on eclipse photographs and on these movies, and a banana skin would not project in this way. In addition the loops in Hα and in the corona do not look like a banana skin. Another less conclusive point is that the bits and pieces of the Hα prominence are seen to lie along the concentric arch. (See Figure 1). This picture is more compatible with a slender flux tube rather than a banana skin. So at this time I find it more acceptable, or should I say more palatable, to identify arch systems with a squashed or flattened banana and not with a concentric system of shells, or with the "arcades" seen in Newkirk's movies of field calculations. Projects that might help define the "squashed banana" are: (1) Studies of the outer dimensions of concentric arches as seen on eclipse pictures and on the green line movies. (2) Measurements of the shapes of loop systems during their evolution. (3) Tracings of the trajectories of prominence material in loop and "coronal rain" prominences. And (4) Study of the predicted shapes of arches by means of calculated current-free fields from photospheric magnetic observations.

This discussion suggests a host of questions: How uniform is the arch and loop along its length and width? Can we study its cross section to obtain the temperature and pressure gradient across the arch or loop? Is it really in hydrostatic equilibrium, as Meyer (1970) suggests, or is it much more uniform along its length? When their resolution is improved will the X-ray pictures reveal the individual loops within the banana? Is there a reflection point that cannot be seen in these movies because of over-occulting? In the X-ray pictures the reflection point does not appear to show as an especially bright region. Do arches seen in the green and red lines really coincide with each other and with the continuum arches? Recent eclipse observations suggest they do, but the correspondence is not clear in the studies of red and green line scenes taken simultaneously (Dollfus, 1962; Lyot, 1944). Will the magnetic field calculations show "squashed bananas" or only "arcades" (Altschuler, 1969)? Would the cross section of a "loop tunnel" (Bruzek, 1964) plotted throughout its evolution show a squashed banana?

4.5. Evolution of arches and loops

All arches and loops appear to form in place. We do not see them expand. The arches evolve so slowly that we do not know whether they form downward from the top like Hα loops, whether they show uniformly along their lengths, or whether they form upward from the bottom. From the movies we note that the loops can brighten and fade quite rapidly. This implies that the source of the coronal excitation mechanism can be switched on and off quite rapidly. Perhaps the decay of the loop or arch is a function of its size and the smaller loops fade rapidly while the larger arches and loops continue for days, even with their supply of energy turned off. What could be learned from a study of the lifetime and total energy in loops as a function of height and separation of footpoints? How does the energy in the small loop at the start of a growing loop system compare with that in the larger loops at the end of their growth? Can we study the footpoints by current-free field calculations using the Schmidt program and the X-ray pictures? Will the X-ray pictures show how the arch system evolves?

4.6. DIFFERENCES BETWEEN ARCHES AND LOOPS

I wonder about the significance of the differences between arches and loops listed in Sections 4.2 and 4.3. The differences in size and shape may simply be a function of the distance between the footpoints. Both may be squashed bananas. Both may be fairly uniform along their length. The radial velocities in coronal loops may be due to prominence material dragging the coronal material, as suggested by Karimov (1963), and are not inherent in the corona itself. Both arches and loops form sheaths for prominence material and both are more diffuse than the prominences. Both may be X-ray features.

Most solar astronomers would say that because of the high activity in the tops of the loops as seen in the Hα line and yellow line, and because the loops are post-flare phenomena, the two are decidedly different. On the other hand, as the loop region grows larger it becomes increasingly more difficult to find the yellow line, and the loop process seems to decay. Some of the very large clearly identified loop scenes are remarkably inactive.

I think the possibility exists that both arches and loops are formed by the same physical process but that the loops are more active and show yellow line and broad Hα in their tops simply because they are smaller and associated with stronger magnetic fields and with the sudden release of energy or perhaps with a sudden manipulation of footpoints that may accompany a flare. Perhaps the studies indicated in Section 4.5 would show a closer association.

5. Flare-Associated Waves in the Corona

There are about twenty cases of flare-associated waves observed in the corona. The original "whip" (September 12, 1956) shown by Evans (1957) is typical of these events. In 1956 I measured this whip, located the flare and made a time-distance plot that was never published. I offer these measurements here as an example of the difficulties of measuring fast-moving waves in the corona.

The event occurs in less than eight frames. The pictures are reproduced in Evans' paper. I treated each frame independently and outlined the arches as best I could. I then drew the center of each arch. These center lines are shown in Figure 6 as solid lines superimposed on dotted lines. The dotted lines represent the average position of the arches as sketched from several earlier scenes. Four points along the arch are shown as strong dots in the 19:04–19:08 portion of the figure. The time-distance plot of these dots is shown in Figure 7 along with the inferred velocity. (Number one is the lowest of the four).

There are a number of inferences that can be made from this event.

(1) The event is flare-associated since the time of the small flare shown in Figure 6 coincides with the start of the event. Because of similarities, we assume that this is the coronal manifestation of the flare-associated wave.

(2) The wave is confined to a narrow beam. In other words, it does not spread out over the entire corona but affects only the arches. There are some changes in one

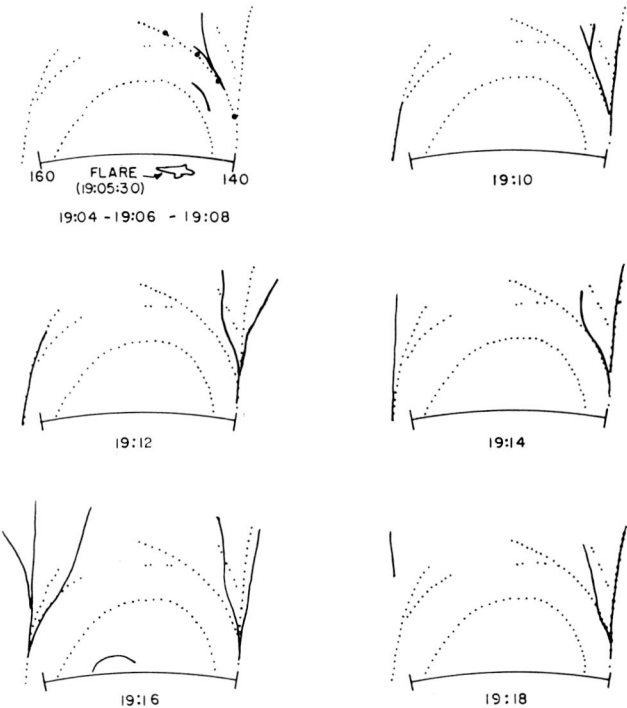

Fig. 6. Fast Event. Coronal "whip". September 12, 1956. Dotted arches in all scenes represent pre-event arches. Strong dots in right of 19:04–19:08 scene are the four points from bottom to top that were measured for the time-distance plot in Figure 7.

structure quite distant from the arches, but at least we do not see the entire corona blown apart. This beaming implies magnetic guiding or deep-seated photospheric phenomena. The latter is also implied in the activation of large surges.

(3) Either magnetic lines of force in the arch are broken or the top of the arch is stretched enormously compared with the visible arch. In Figure 6 one could reconnect the arches shown at 19:16 to form the old arch.

(4) The motion of the arch is underdamped, that is it oscillates once, or perhaps it bounces off the adjacent region. We know now from the Dec. 30, 1956 event that arches can oscillate, but will very rapidly spring back into their old position. We also know that arches can fall into new positions with no overshoot (see Section 6, Realign-

ment). I think this suggest the hypothesis that the magnetic lines of force were broken in the whip event and were locked into the plasma. It also suggests that there is material motion in the corona during the flare-associated wave events.

(5) The flare-associated wave has no difficulty in breaking or crossing the weaker magnetic lines of force.

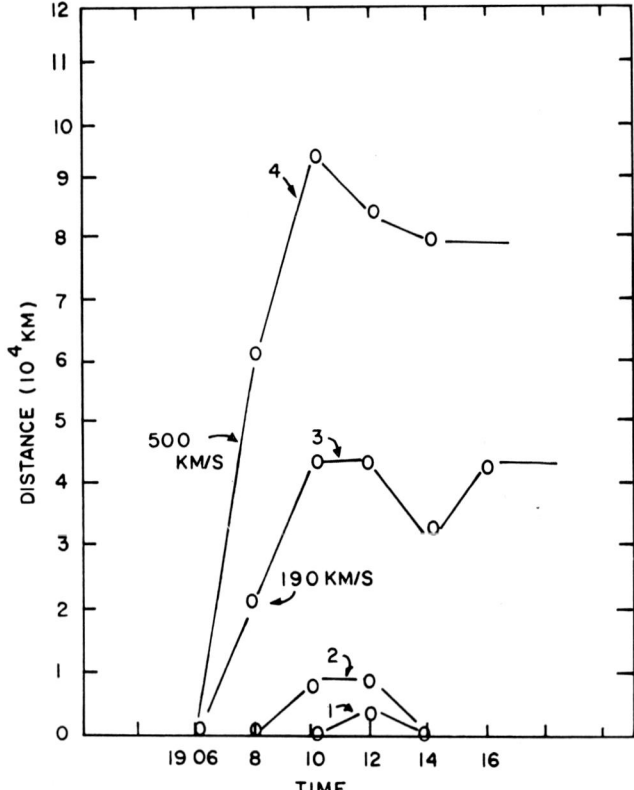

Fig. 7. Time-distance plot of coronal 'whip'.

(6) We must be seeing a disturbance that carries the material along with it. Hence the velocity of the arches as inferred from Figure 7 implies a lower limit to the velocity of the waves.

(7) There is no brightening of the green coronal line in the arch as the wave passes through it.

(8) There does not appear to be a large cloud radiating a continuum ejected at the time of the flare because if there were it would show in these pictures (see Section 2.3, paragraph 8). A part of an arch or possibly a shell can be seen at 19:06. This suggests the need for a continuum patrol that could utilize much shorter exposures.

(9) There were no loops after this particular event.

With the exception of (9) this description applies to the other fast events as well. Perhaps there are now enough data to study the theoretical relationship between the strength of the magnetic fields and the velocity of the wave in the hope of learning more about the flare process.

6. Classification of Coronal Events

Some of the events in the last 15 years of Sac Peak movies have been described by Evans (1957), Kleczek (1963) and Bruzek and DeMastus (B&D) (1970). Evans had only a few events available to him and these were described as "large and rapid variations in intensity", "high apparent velocities", "development of holes", "surge-like coronal knots" and "whip". At this time the concept of flare-associated waves was not developed, nor was it when Kleczek (1963) offered a diagram representing coronal "structures" and listed the various phenomena as "brightening (usually as associated with loops)", "emergence of a loop", "fast expansion", "splitting of loops", "merging of loops", "oscillation of arches" and "opening of an arch". Recently Bruzek and DeMastus (1970) have drawn attention to flare-associated coronal expansion phenomena. In their summary they offer the following categories:

(1) Slow Expansion Phenomena.
    (a) Long-lived coronal loop condensations.
    (b) Slowly expanding arches.
(2) Accelerated Expanding and Quasi-Exploding Arches.
    (a) Pre-flare generation and accelerated expansion of coronal arches.
    (b) Fast expanding and exploding arches emerging from the flare.
    (c) Expansion and disruption of already existing arch during flare occurrence.

At this point it seems that we do not have a really all-inclusive event classification scheme. Kleczek's loop "emergence" and the B&D "loop expansion" are the same. The "whip", "opening of an arch" and the B&D classification 2C are the same. No flare was visible with Kleczek's "oscillation of arches", but from our knowledge of flare-associated waves it probably was a flare-associated event where the magnetic field was strong enough to prevent a "whip" and "opening of an arch". In reviewing these papers and all the coronal movies, including the more recent ones, I suggest that there are really only three classes of events as seen on movies, and these are slow, loop and fast. I offer the scheme in Table I as including all the terms.

In the movies there is no difficulty in recognizing the large coronal loop systems. They are very distinctive, and I doubt that it is necessary to subdivide the loop events as I have in Table I to include all Kleczek's terms. The term "coronal loops" has been used at Sacramento Peak ever since the November 22, 1956 event described by Evans (1957). I suggest that, because of our more recent knowledge of Hα and coronal loops, this term should be adopted since it is more descriptive than "sporadic condensations".

In the fast events I offer only one new class of event and that is the Fast Realignment (FR). This occurred on three scenes, November 20, 1957, February 21, 1968 and March 12, 1969. The last two were obviously associated with a wave. Large arches

TABLE I
Coronal Events

*S*low Events (*S*)	
Brightenings (*SB*)	(Lyot)
Fadings (*SF*)	(Lyot)
Motions (*SM*)	(Lyot)
Hole (*SH*)	(Evans)
Expansions (*SX*) (Some flare associated)	(B & D–1b)
*L*oop Events (*L*) (Post Flare) (Active X-ray)	
Brightenings (*LB*)	(K)
Fadings (*LF*)	(K)
Splittings (*LS*)	(K)
Merging (*LM*)	(K)
Expansion (*LX*)	(K "emergence", B & D–1a)
*F*ast Events (*F*) (Flare-associated waves)	
Ejections (*FE*)	(Evan's "surge-like")
Accelerated Expansion (*FX*)	(K "fast-expansion", B & D–2a and 2b)
Oscillations (*FO*)	(K)
Realignment (*FR*)	(See text)
Disruptions (*FD*)	(B & D–2c, Evan's "whip", K "opening of arch")

are seen to snap into new positions almost instantaneously and with no overshoot. This could be an interesting event because it suggests that there was a sudden change in the magnetic field at one of the footpoints, which forced the arch to relocate or realign.

One more category of events is the coronal counterpart of an ascending or eruptive prominence. To my knowledge, none of the Sac Peak movies were taken as a prominence ascended. Waldmeier (1961) has visually observed such an event. One spray, March 7, 1959, was observed at Sac Peak in the corona. In this case the similarity between the Hα and coronal event suggests that continuum emission accounted for the activity in the corona.

This classification scheme is open to criticism because one cannot always show that there is an associated flare. Nevertheless, I think that with our recent knowledge of flare-associated waves and loops it is safe enough to adopt such a scheme. The classification emphasizes that Lyot's description of slow movements in the corona is correct unless a flare occurs, when one sees the effects of the flare-associated wave in the corona. After the flare one expects loop "tunnels", and the associated X-ray event.

7. Event Movie

Sac Peak intends to assemble a movie of the *Coronal Events*. This will include examples of all the classes and all the well-observed events. The archives contain 7 slow expansions and 18 loop scenes, including 6 loop expansions. There are 26 fast events categorized as follows: 11 expansions, 9 disruptions, 1 oscillation, 2 ejections and 3 realignments. Several of the fast events have slow expansions before or after the flare-associated event. A written commentary showing which scenes have been studied in detail together with the flare association will be included with this movie.

8. Conclusions

The physical process associated with arches and loops may be the same, if we take into account the stronger magnetic fields and closer footpoints of the loops compared to arches. We further conclude that the corona does not change except when a flare occurs when the flare-associated wave moves out through the corona and coronal loops may form. We suggest a classification of slow, loop, and fast coronal events. The associated wave or Bruzek's "explosion" is narrow in solid angle and has no difficulty crossing the magnetic lines of force.

Acknowledgements

I wish to thank Howard DeMastus for assembling the *Coronal Event* movie. He has singlehandedly kept this program going over the last several years. I also wish to thank A. Bruzek and J. Kleczek for their informal comments and discussion at Brighton, and Jack Zirker for his many stimulating conversations during the NATO Advanced Study Institute on the Physics of the Solar Corona at the Hotel Cavouri at Vouliagmeni, Greece. This paper would not have been written if it were not for the kind invitation of Dr. C. J. Macris to attend this splendid study institute.

References

Altschuler, M. D. and Newkirk, G.: 1969, *Solar Phys.* **9**, 131.
Bruzek, A.: 1964, *Astrophys. J.* **140**, 746.
Bruzek, A. and DeMastus, H. L.: 1970, *Solar Phys.* **12**, 447.
Dollfus, A.: 1956, *Rev. Opt.* **35**, 625.
Dollfus, A.: 1957a, *Compt. Rend.* **244**, 1880.
Dollfus, A.: 1957b, *Compt. Rend.* **245**, 32.
Dollfus, A.: 1962, *Compt. Rend.* **255**, 3369.
Evans, J.: 1953a, *The Sun* (ed. by G. Kuiper), University of Chicago Press, Chicago, p. 635.
Evans, J.: 1953b, *The Sun*, (ed. by G. Kuiper), University of Chicago Press, Chicago, p. 626.
Evans, J.: 1957, *Publ. Astron. Soc. Pacific* **69**, 421.
Karimov, M. G.: 1963, *The Solar Corona*, (ed. by J. Evans), Academic Press Inc., New York, p. 297.
Kleczek, J.: 1963, *Publ. Astron. Soc. Pacific* **75**, 9.
Lyot, B.: 1939, *Monthly Notices Roy. Astron. Soc.* **99**, 580.
Lyot, B.: 1944, *Ann. Astrophys.* **7**, 31.
Meyer, F.: 1970, *Physics of Solar Corona*, (ed. by C. J. Macris), Reidel, Dordrecht, Holland.
Newkirk, G.: 1957, *Ann. Astrophys.* **20**, 127.
Teske, R.: 1970, 'OSO Satellite', Univ. of Michigan, 05567-2-F.
Vaiana, G. S., Reidy, W. P., Zehnpfennig, T., Van Speybroeck, L., and Giacconi, R.: 1968, *Science* **161**, 564.
Van Speybroeck, L., Krieger, A., and Vaiana, G.: 1970, *Nature* **227**, 818.
Waldmeier, M.: 1961, *Z. Astrophys.* **53**, 198.

9. THE SOLAR CORONA IN THE ELEVEN-YEAR CYCLE

M. WALDMEIER

Swiss Federal Observatory, Zürich, Switzerland

Abstract. The corona changes its shape, structure and brightness according to the phase of the 11-yr cycle. The following discussion is based on coronal variations from 1958 (maximum of cycle No. 19) through 1969 (maximum of cycle No. 20). The material used has been gathered at the Astrophysical Observatory Arosa (monochromatic corona) and on nine total eclipses of the sun (white light corona).

1. The Total Intensities of the Lines 5303 and 6374 Å

The intensity of these two lines is measured along the sun's circumference at intervals of 5 degrees and at a distance of about 30" from the sun's limb. The sun of each set of 72 measurements is called the total intensity for the day under consideration. The yearly mean values of the total intensities of the lines 5303 (Fe XIV) and 6374 (Fe X) are given in Table I. The values are subdivided into the contributions from the northern (N) and the southern (S) hemisphere. Sunspots reached their maxima at

TABLE I

Total intensities of the coronal lines 5303 and 6374

Year	5303			6374		
	S	N	S+N	S	N	S+N
1958	613	677	*1290*	581	623	*1204*
1959	503	646	1149	565	629	1194
1960	358	486	844	*583*	460	1043
1961	297	424	721	*566*	469	1035
1962	145	303	448	*601*	508	1109
1963	56	205	261	618	463	1081
1964	20	*135*	155	699	566	*1265*
1965	21	*130*	151	536	437	973
1966	46	*312*	358	520	444	964
1967	265	*368*	633	390	*440*	830
1968	482	410	892	478	552	*1030*
1969	378	360	738	–	–	–

1957.9 and 1968.9 with a minimum between them at 1964.7. One of the most outstanding features of the sunspot activity in the years from 1958 to 1968 was the strong asymmetry between the two hemispheres, the northern being the more active one.

The total intensity of the line 5303 very closely follows the sunspot activity with maxima in 1958 and a minimum in the years 1964/65. The amplitude of this eleven-year variation is large, the total intensity at maximum being more than eight times that at minimum. As a result of the fact that cycle No. 19 was by far more active than cycle No. 20, the total intensity in 1958 exceeds that in 1968. Due to the mentioned asymmetry of sunspot activity in the years 1958 to 1967 the northern intensity of the line 5303 surpasses the southern one. From 1964 to 1966 N is more than six times larger than S.

The strong response of the line 5303 to sunspot activity has two reasons. The intensity of that line depends on density and temperature in such a way that both, an increase in density as well as an increase of temperature raise the line intensity. At sunspot maximum both density and temperature of the corona are higher than at sunspot minimum.

Quite different is the behaviour of the line 6374. Its total intensity shows but small variations, which at first sight seem not to be related to the sunspot cycle. A closer inspection however reveals that the total emission of this line has two maxima in each sunspot cycle, one coinciding with the sunspot maximum (1958, 1968), the other with the sunspot minimum (1964). The intensity of the red line increases with density, but decreases with temperature as the ionisation state Fe x, from which the red line originates, becomes more and more depleted when the temperature rises. Coronal density and coronal temperature are not independent of each other, but vary in the same sense: the higher the density the higher the temperature. As the influences of these two parameters upon the intensity of the red line are of opposite sign, they may more or less cancel out each other. This explains the almost invariable intensity of that line. The slight increase at sunspot maximum results from the increased density, that at sunspot minimum from the low temperature.

The N–S-asymmetry is observed in the red line too, but with opposite sign as in the green line, the southern hemisphere showing greater emission than the northern one. From white light pictures obtained at eclipses it follows that the density was larger in the northern hemisphere. Therefore the increased red line emission of the southern hemisphere has to be explained by a temperature, which is much lower than that in the northern hemisphere.

2. Shape and Structure of the Corona

From eclipse observations it is well known that the overall structure of the white light corona undergoes a marked change in the eleven-year cycle. At sunspot maximum the isophotes are almost circular, whereas at sunspot minimum they are elliptical, showing a larger extension of the corona at the equator than at the poles (Figure 1). The ellipticity of the isophotes expressed by the Ludendorff-coefficient, $a+b$, is given in Table II for the eclipses between 1958 and 1970, that of 1966 being omitted because of its irregularities. The ellipticity increases on the descending branch of the solar activity (1958–1964) and decreases on the ascending branch. The general appearance of the corona is dominated by its long streamers and these are in connection with the stationary prominences. At sunspot maximum the prominences are found all around the sun's limb and so are found the coronal streamers. This gives the corona a spherical appearance. At sunspot minimum prominences are concentrated in middle and low heliographic latitude and completely absent from the polar caps. The concentration of the streamers toward lower latitudes gives the corona an elongated appearance. The different aspect of the corona at maximum and minimum is strengthened by the fact that at sunspot minimum streamers are inclined to the equator, whereas at sunspot maximum they are more or less radial or even slightly inclined toward the poles.

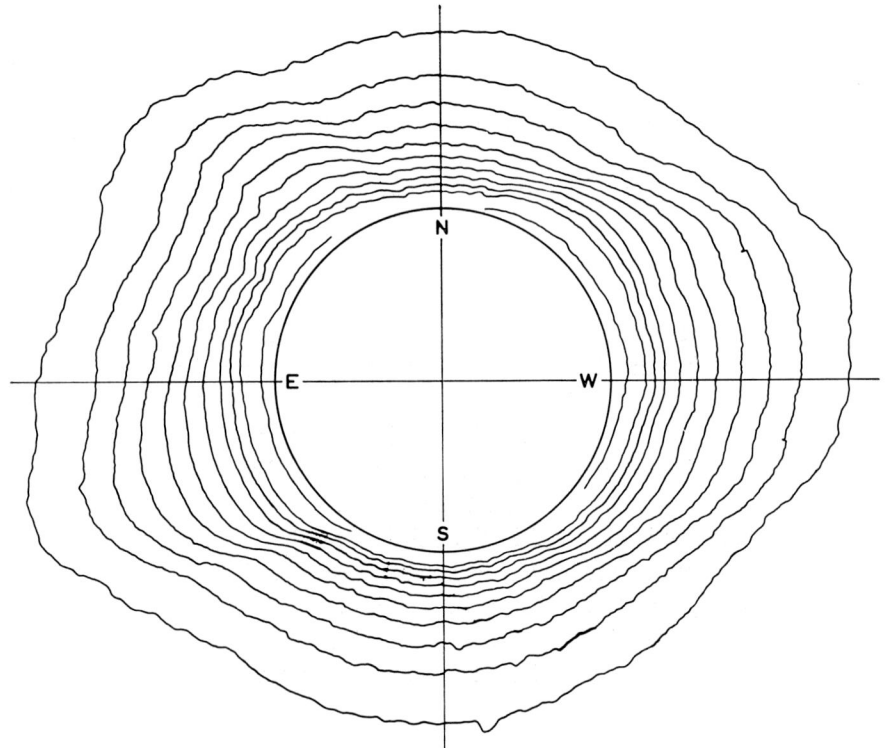

Fig. 1a. Isophotes of the strongly flattened corona of July 20, 1963.

TABLE II

The ellipticity of the corona

Eclipse	$a+b$
1958 Oct. 12	0.07
1959 Oct. 2	0.14
1961 Febr. 15	0.17
1962 Febr. 5	0.28
1963 July 20	0.29
1965 May 30	0.24
1968 Sept. 22	0.06
1970 March 7	0.00

When in the years around sunspot minimum polar regions are bare of long streamers, the short polar plumes – believed to represent magnetic lines of force – become visible. For the first time after the solar maximum of 1958 they were observed at the eclipse of February 15, 1961, in the northern polar cap. At the eclipses of 1962, 1963 and 1965 polar plumes were well developed in both hemispheres. At the eclipse

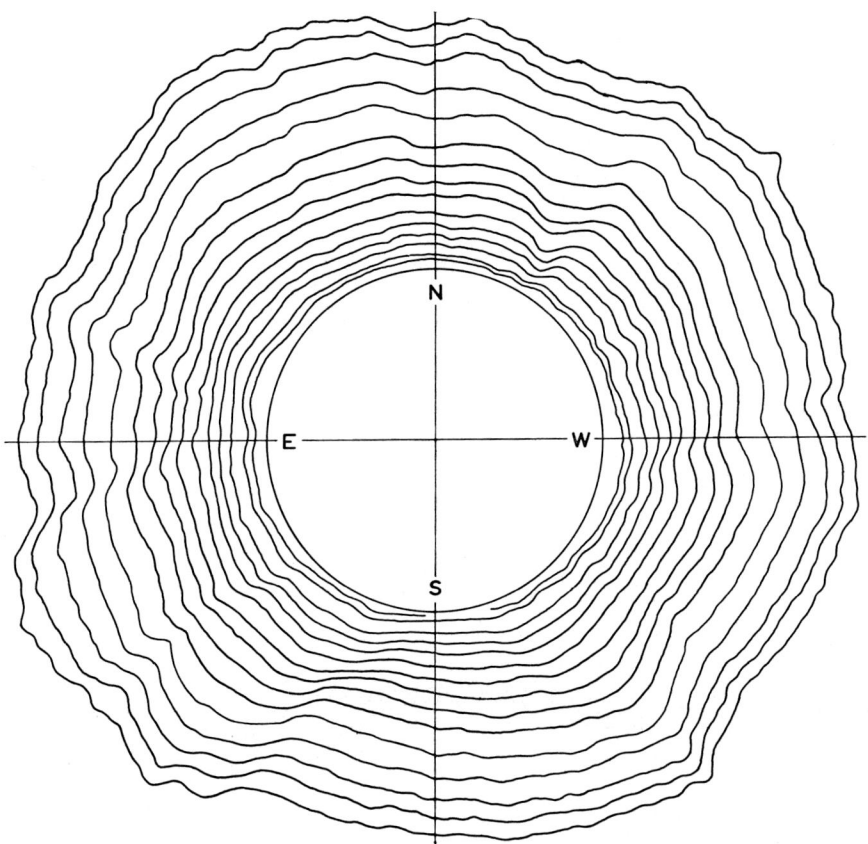

Fig. 1b. Isophotes of the circular corona of September 22, 1968.

of November 12, 1966, they had already disappeared from the northern polar cap, but were still visible in the southern hemisphere. Finally at the 1968 eclipse polar plumes had disappeared altogether. The different behaviour of the two hemispheres results from the fact that in the years under consideration the northern activity was ahead of the southern by about half a year.

The shape of the corona has also been observed in the light of the line 5303. Figure 2a gives isophotes of that line for nine different days around sunspot minimum

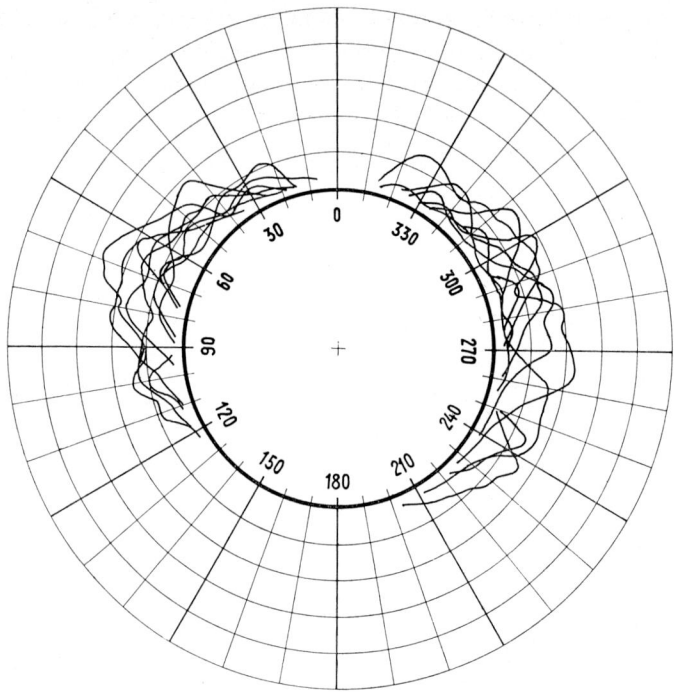

Fig. 2a. Isophotes of the coronal line 5303 Å around sunspot minimum (1963–1965).

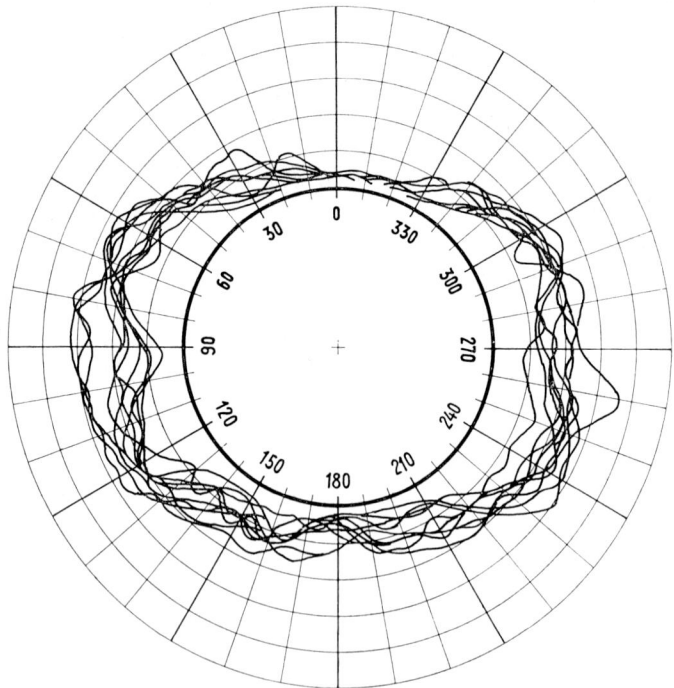

Fig. 2b. Isophotes of the coronal line 5303 Å around sunspot maximum (1968/69). N is at 0°, E at 90°. The distance from the sun's limb is for the first circle 0.12, for the second 0.24, and for the third 0.36 solar radii.

(1963–1965). The set of ten isophotes represented by Fig. 2b refers to days around sunspot maximum (1968/69). The isophotes give the largest distance from the sun's limb up to which the line was visible. In all heliographic latitudes the extension of the green corona is larger at sunspot maximum than at minimum. At maximum the isophotes of the monochromatic corona show – in contrast to the white light corona – a strong flattening. Around sunspot minimum the line 5303 actually was absent at the poles. The strong asymmetry which occurred in that year stands out clearly in the isophotes: in the southern hemisphere the corona's extension is smaller, and the region free of line emission around the pole is much larger than in the northern hemisphere.

3. The Zonal Structure of the Corona

The intensity of the line 5303 varies not only with time, but also with heliographic latitude. Figure 3 shows the latitude distribution of the 5303-intensity from 1958 through 1969. The line emission is concentrated in two zones. The main zone is found at low latitudes over the zone of plages and spots. This zone drifts from about 30° latitude at the beginning of a new cycle (1965) to about 10° at the end (1964). It is strong at sunspot maximum (1958) and weak at minimum (1964). Corresponding to the asymmetry of the solar activity the northern peak is the stronger one from 1958 to 1967. At higher latitudes the intensity shows a second, less pronounced maximum. It makes its first appearance shortly after sunspot maximum at about 60° and holds this latitude for the years of the descending branch of the sun's activity. With the beginning of the new cycle the polar zones migrate toward the poles, which are reached at sunspot maximum. Then the polar zones fade away and are replaced by new ones at 60° latitude. The polar zones stand out more clearly in the individual daily observations than in the yearly means.

Between the main and the polar zone there is a minimum of 5303-intensity. At that place the stationary polar filaments are found. These filaments lie along circles of constant latitude, often forming the "polar crown" of prominences. On opposite sides of the filaments or of the gap in the 5303 emission the magnetic field shows opposite polarities. As the polar zone of prominences moves toward the pole, the polar magnetic field disappears and is replaced by the opposite field of the low latitude side of the polar zone. This migration gives rise to the change of the polarity of the polar field around sunspot maximum.

The latitude distribution of the intensity of the line 6374 is given in Figure 4. We already mentioned that the time variations of the red line are small compared to the corresponding variations of the green line. The same is true for the local variations. On a general decrease of the intensity from lower to higher latitudes maxima over the main zone of activity are superimposed. These are produced by the increased density in these regions. Other regions of high red line emission, like the polar regions around sunspot minimum or the southern hemisphere in the years 1962 through 1966, where the green line is very weak or missing, are produced by low temperatures.

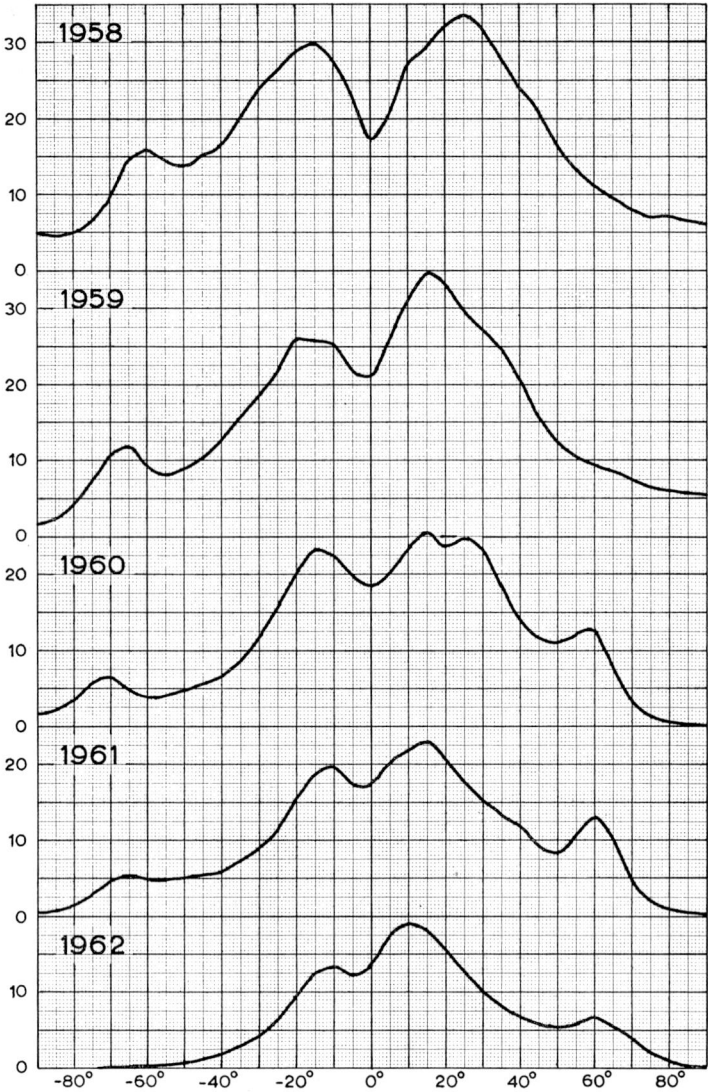

Fig. 3a. Latitude distribution of the line 5303 Å for 1958–1962.

4. Concluding Remarks

The intensity of a coronal emission line depends on electron-density N_e and temperature T. From the simultaneous measurements of two lines, e.g. of 5303 and 6374, N_e and T can be calculated provided that all numerical values involved in the processes of ionization and excitation are known. The intensity of the white light corona delivers N_e irrespectively of T, and the ratio of the green to the red line, $r = i_{5303}/i_{6374}$, furnishes T independently of N_e.

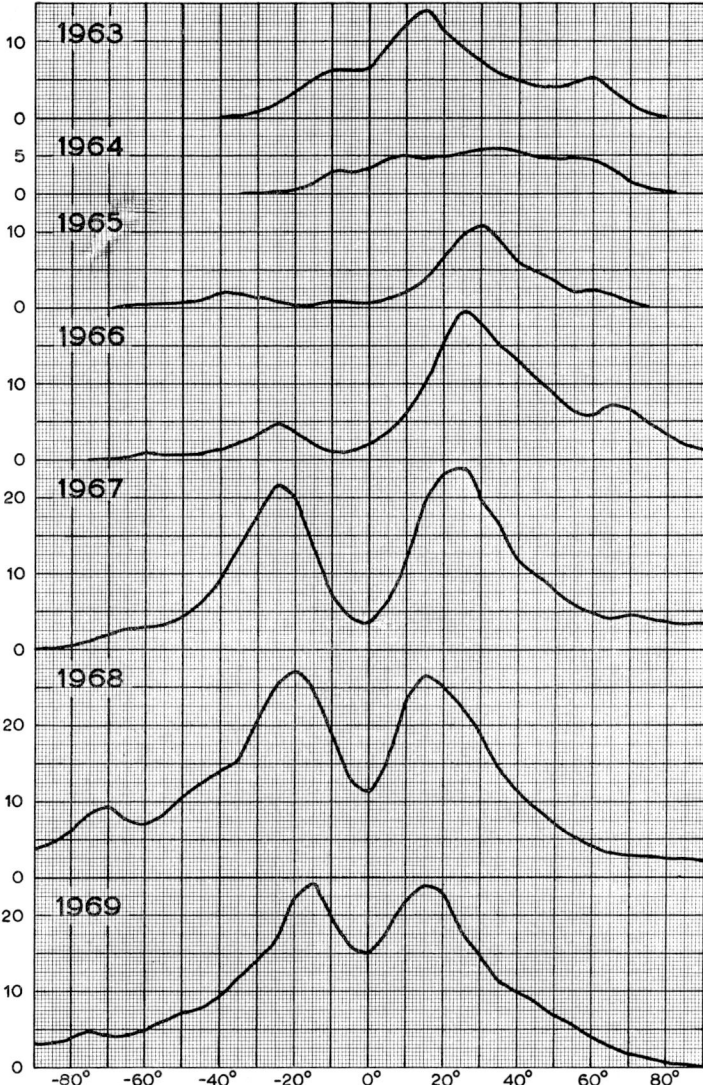

Fig. 3b. Latitude distribution of the line 5303 Å for 1963–1969.

All available measurements of N_e and T indicate that these two parameters are not independent of each other. The ratio r and therefore T have their lowest values at the poles at sunspot minimum. At the same time this is the place of lowest N_e-values. As we proceed to lower latitudes both N_e and T are increasing, reaching high values over plages. Finally the highest values of N_e (indicated by a strong continuum) and T (indicated by the appearance of the yellow line 5694 Å) are found over strong centers of activity. Any theory of the heating mechanism of the corona has to give an explanation of the density-temperature-relation that is observed in the corona (Observatory

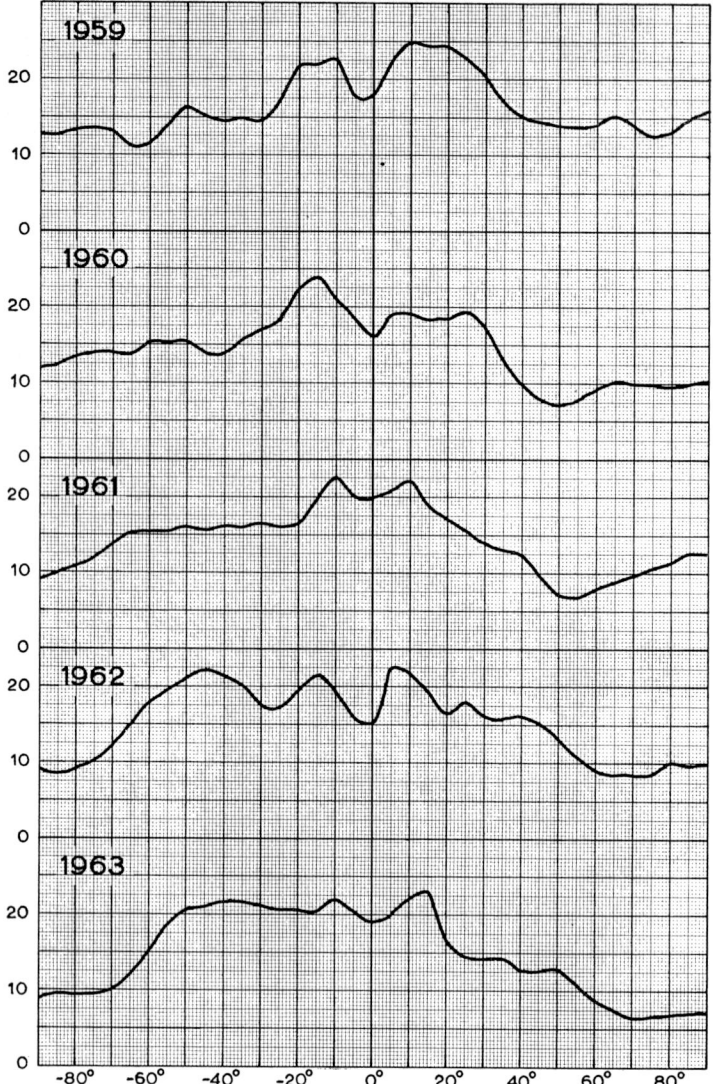

Fig. 4a. Latitude distribution of the line 6374 Å for 1959–1963.

87, 83, 1967). Table III gives corresponding values of N_e and T for some typical coronal regions. From these values one gets the following formula for the density-temperature-relation: $N_e = 10^{-10} \cdot T^3$. This relation means that actually the physical state of the corona can be described by one parameter only, N_e or T. Low temperature is always connected with low density, high temperature with high density. Combinations of low density (weak continuum) with high temperature (high r-value) or of high density (strong continuum) with low temperature (low r-value) have not been observed.

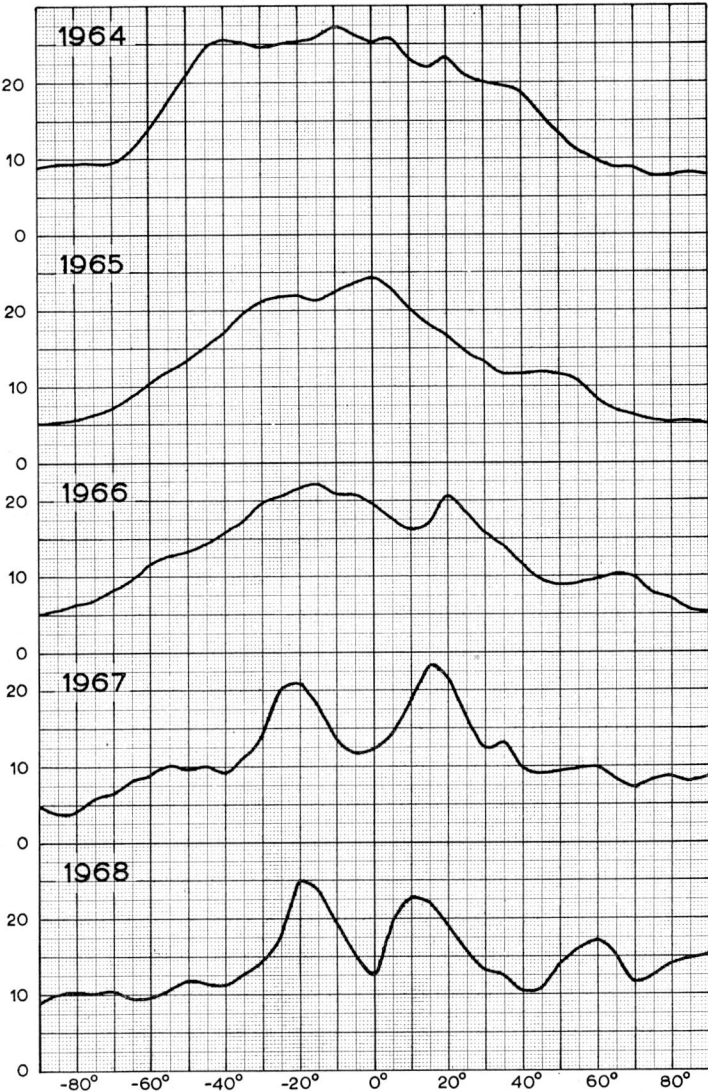

Fig. 4b. Latitude distribution of the line 6374 Å for 1964–1968.

TABLE III

The temperature-density-relation for the innermost corona

Region	$T \cdot 10^{-6}$	$N_e \cdot 10^{-8}$ cm^{-3}
pole at minimum	1	1
undisturbed corona	1.5	3
weak plage region	2	6
strong plage region	2.5	12
weak condensation	3	30
strong condensation	4	80

10. CORONAL ACTIVE REGIONS AND FLARE-ASSOCIATED EVENTS

J. B. ZIRKER

Hawaii Institute of Geophysics, University of Hawaii, Honolulu, Hawaii

1. Introduction

This paper has three parts. In the first, we review recent empirical models of coronal active regions that are based on observations outside the optical wavelength band. Next we consider models derived from optical observations, with special emphasis on the optical forbidden line spectrum. The next section discusses the formation of sporadic condensations and loop prominences following a flare.

Coronal active regions radiate strongly throughout the electromagnetic spectrum and observations of each region of the spectrum have been analyzed (often independently of all others) to yield a model of the electron temperature and density distributions. We attempt, in this review, to compare coronal models and to point out their common features. We discuss the models in order of complexity: Homogeneous models (one or two dimensional) first and inhomogeneous models next. The reader should refer to the articles in these Proceedings by Noyes, Dollfus, Dunn, Wilson and Neupert for detailed discussions of the properties of coronal active regions.

2. Models of Coronal Active Regions: Non-Optical Observations

Figure 1 shows the schematic evolution of a typical coronal active region (after Newkirk, 1967) as it might appear in white light photographs. The region evolves

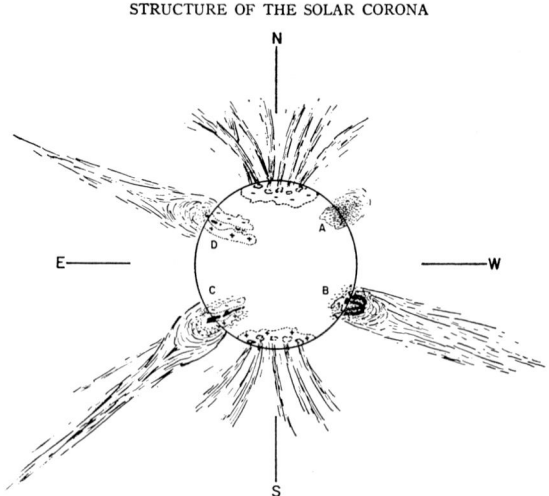

Fig. 1. Schematic time development for a typical coronal active region (Newkirk, 1967).

from a "diffuse enhancement" (A) into a mature active region (B) with a coronal condensation at its base and an active region streamer trailing outward as a hot dense plume of gas. The subsequent appearance of a "helmet" structure and eventual migration and dissolution of the region is shown in stages (C) and (D).

Photographs of the condensations and enhancements, made in the light of the optical forbidden lines (see the articles of Dunn and of Dollfus in these proceedings) show a fine-structure of loops, arches and rays with typical cross-sections of ~15 arc-sec. These fine-structures, as well as the larger-scale features, are thought to arise from the interaction of the gas with the region's magnetic field (see Newkirk's article in these Proceedings).

The main point of this section is the following: observations of coronal active regions (in any portion of the electromagnetic spectrum) with a spatial resolution insufficient to resolve these fine structures tend to imply a *homogeneous* distribution of electron temperature and density, while more highly resolved observations tend to lead to *inhomogeneous* models. We illustrate this point in the remainder of this section.

Noyes, Withbroe and Kirschner (1970) have derived a model for typical coronal regions which is based on EUV line intensities observed with the Harvard spectrometer aboard OSO IV. The spatial resolution of the EUV spectroheliograms is one arc-minute. Their model is mainly valid for the chromospheric-coronal "interface", i.e., they interpret the line intensities in terms of a one-dimensional, homogeneous model. The model rests on three assumptions: hydrostatic equilibrium, constant flux of conducted heat, and an isothermal corona, and introduces only three free parameters: the conducting flux (F_c), the base gas pressure (P_0) and the coronal temperature (T_c).

Figure 2 shows their results. The height-distribution of temperature is similar to that determinated for the quiet sun by Withbroe (1970), Dupree and Goldberg (1967)

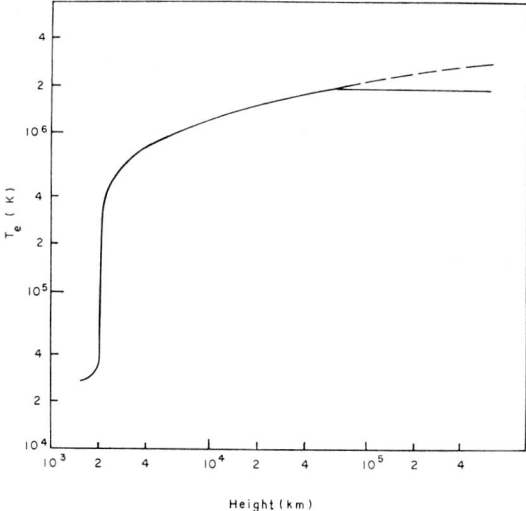

Fig. 2. Model for the Coronal Interface above active regions (Noyes *et al.*, 1970).

and by Athay (1966), except that the conductive flux and gas pressure are each five times larger than the quiet sun values. Thus, their EUV observations imply a steeper temperature gradient and higher gas pressure than exist in the quiet corona.

Another one-dimensional model is shown in Figure 3. It was derived from the thermal radio emission of a single active region, observed simultaneously at 7 frequencies by a group of investigators (Christiansen *et al.*, 1960). The spatial resolution

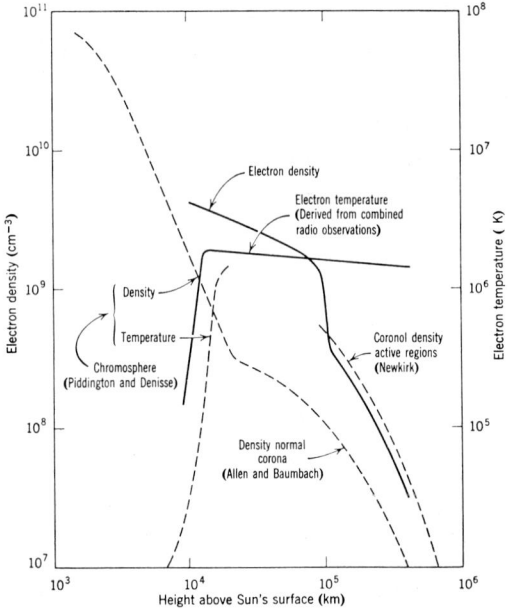

Fig. 3. Height distribution of electron density and temperature, determined from Multi-Frequency Radio Observations (Christiansen *et al.*, 1960).

of the observations varies with frequency but is typically a few arc-minutes. The radio observations, especially at decimetric and metric wavelengths give information about the temperature and density at heights $\sim 10^5$ km above the disk.

This radio model is similar in several respects to the EUV model of Noyes *et al.* It shows a sharp temperature gradient low in the corona, an isothermal coronal higher up, and electron densities an order of magnitude higher than quiet sun values.

It is interesting to compare these models to another, which is based upon EUV line observations. Boardman and Billings (1969) have analyzed spectroheliograms of a coronal enhancement at the limb obtained by the Navy Research Laboratory Group. The enhancement was recorded in 10 lines (in the range 170 to 630 Å) of Fe IX, XIV, XVI, and Mn IX and Si XII. These rockets spectra have a spatial resolution of about 10 arc-sec.

Billings and Boardman derived the three-dimensional temperature and density distribution in the enhancement. They assumed axial symmetry for the enhancement, and adopted Pottasch's chemical abundances (1967), but made no assumptions on the pressure or temperature gradient. In order to interpret the observations, they require an *inhomogeneous* temperature structure: in each volume element they find the emission measure $\int n_e^2 \, dV$ associated with each of three discrete temperatures (1×10^6, 2.5×10^6, 4.8×10^6 K). The average of the three temperatures (weighed according to the emission measure) and the average electron density is shown in Figure 4. Their model consists of a low, hot dense core surrounded by a cooler halo.

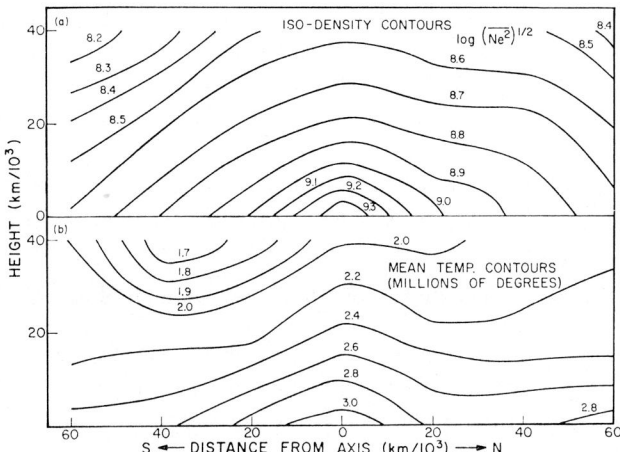

Fig. 4. Electron density and temperature distribution within a coronal enhancement, derived from EUV line intensities (Boardman and Billings, 1969).

Several comments are appropriate. First, this hot core, cool halo structure emerges also from the analysis of the optical observations, as we shall see below, and is consistent with other EUV and XUV observations. As a very recent example, we note the high-resolution X-ray images obtained by the American Science and Engineering group (Vaiana, 1970) in three bands: 3–8, 8–12, and 20–40 Å. A comparison of the same coronal active region in these images shows the presence of a hot central core and cooler shell.

The most striking feature of these X-ray images, is, of course, the presence of coronal arches and filamentary structure, similar to that observable in the forbidden Fe XIV line, $\lambda 5303$ (Dunn, 1970). Almost all EUV spectral observations published to date do not resolve these fine structures. When they are resolved, even incipiently, as in the NRL images analyzed by Billings and Boardman, an *inhomogeneous* temperature structure emerges.

The basic difficulty in deriving models from such EUV (and optical forbidden line) observations is that the gas is transparent in the line radiation. As a result, the distribution of radiating material along the line of sight is undetermined and must be specified by an assumption, such as that of symmetry, or by some additional observational constraint. A relation between the electron temperature and density, if it existed, would serve as such a constraint.

The work of Rugge and Walker (1970) suggested an explicit relation between T_e and n_e. They observed the intercombination (2^3P-1^1S) and forbidden (2^3S-1^1S) lines of ions in the helium isoelectronic sequence up to SXV. According to the theory of Gabriel and Jordan (1969), the intensity ratio of these two lines depends on electron density, so that values of n_e can be determined as a function of ionization temperature. A relation of the form $n_e \alpha T_e^5$ fits their results.

Unfortunately, this relation may be imposed by the Z-dependence of the radiative transition probability for the forbidden line (Batstone, 1970) rather the physics of the coronal region. A recomputation of the density variation of the ratio (Gabriel and Jordan, 1970), based upon corrected radiative transition probabilities, raises the inferred densities by an order of magnitude and suggests that the method does not apply to the available observations.

We turn now to models of coronal condensations derived from optical observations, for which the resolution is adequate to resolve fine structures.

3. Models of Coronal Active Regions: Optical Observations

Observations of the electron-scattered coronal continuum obtained during eclipse or (by means of K-coronameter) without an eclipse, lead to direct determinations of the electron density distribution in action regions. In Figure 5 we show the isophotes of

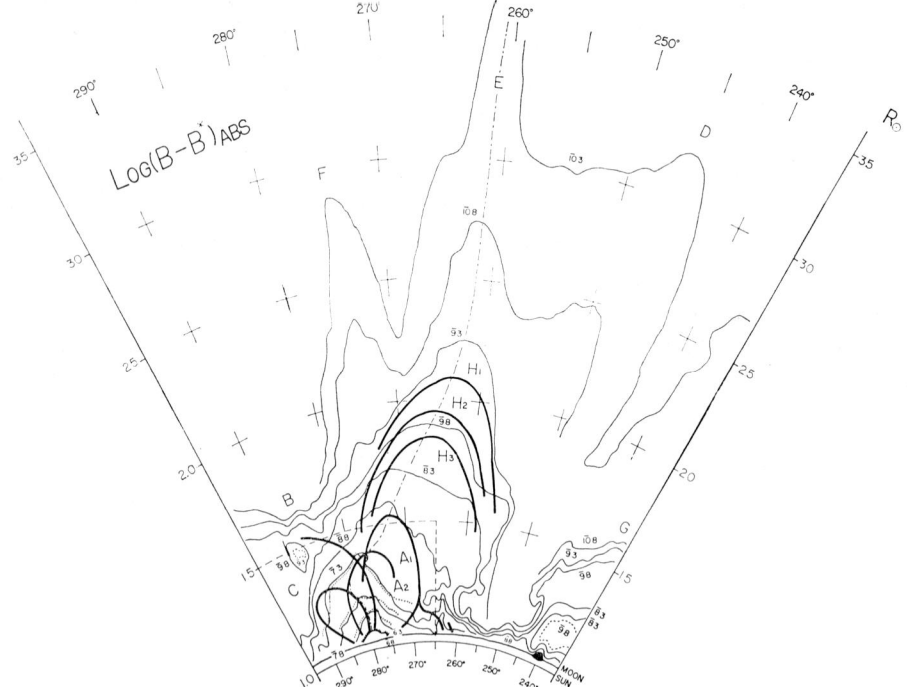

Fig. 5. White-light isophotes of a coronal active region streamer (Saito and Owaki, 1967).

an active region streamer, derived by Saito and Owaki, from polarimetric white-light photographs made at the February 5, 1962 eclipse. At the base of the streamer, a dense coronal condensation was visible. Figure 5 also shows the position of a number of *arches*, visible in white light within the condensation. Tsubaki (1966) discussed the form and contrast of these arches but was unable to extract quantitative information from the photographs. These arches probably coincide with those observable in the green coronal line (Dunn, 1970), and presumably outline the magnetic field.

Saito and Billings (1964) carried out a detailed study of the coronal condensation shown in Figure 5. They ignored the arches and determined the (smoothed) three-dimensional distribution of electrons by assuming circular cylindrical symmetry about a distorted axis. Figure 6 shows their results; the density in the condensation falls off

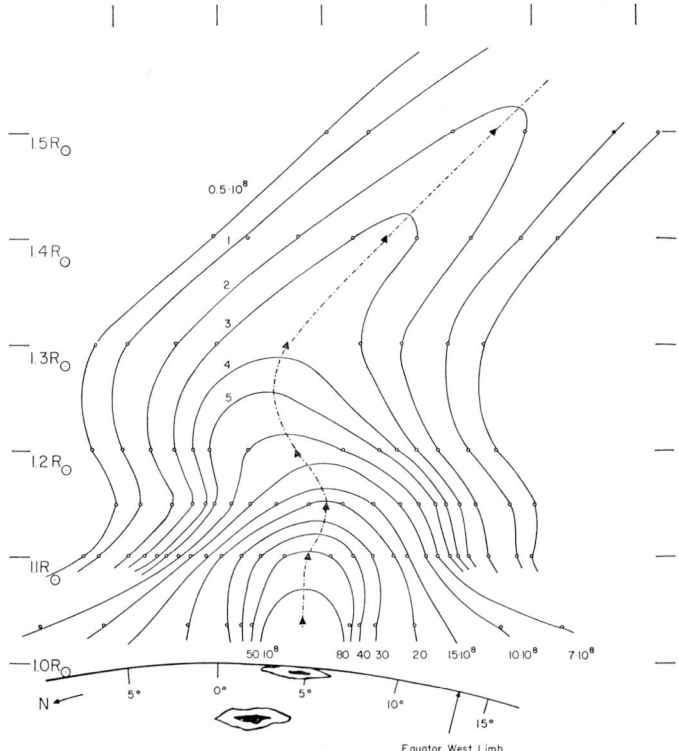

Fig. 6. The electron density distribution in the coronal condensation observed at the February 5, 1962 Eclipse (Saito and Billings, 1966).

with height and with distance from the axis. Note the resemblance of this density model to Figure 5, which is based on EUV line observations.

The change in appearance of coronal condensations, when viewed in optical forbidden lines that arise from different ions (Dollfus, 1970), suggest that these lines contain much information on the thermodynamic variables within the condensation.

Indeed, the study of these lines has formed the basis for much of the information we now have concerning the coronal electron temperature distribution and chemical composition. As we have mentioned earlier, however, the analysis of these line intensities is complicated by the fact that the radiation arises from transparent layers, so that additional information is required to find the distribution of radiating material along any given line of sight. Unfortunately, the most complete coronal spectra, obtained during solar eclipses, seldom resolve the fine-structures that are visible in green line filtergrams.

Recent efforts to observe the forbidden line spectrum as completely as possible (Jefferies, 1969) were stimulated by Lyot's success at the 1952 eclipse in observing 17 emission lines. Figure 7 shows the intensity distribution along the limb at a fixed height in that condensation for a number of lines (Aly *et al.*, 1962). The symmetry of the distributions, coupled with the dip in the red line intensity and the peak in the

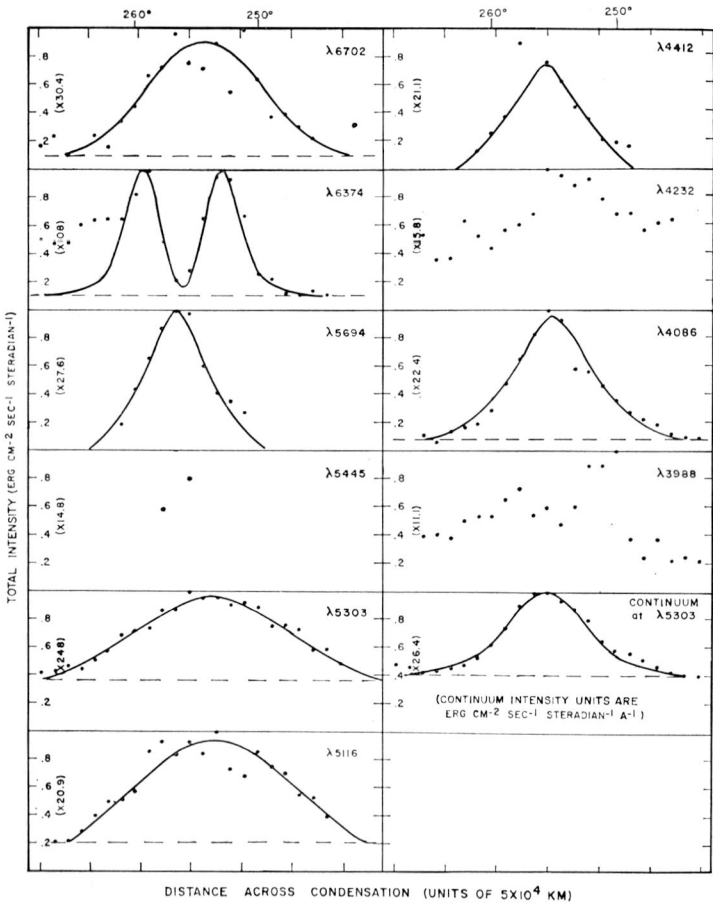

Fig. 7. Forbidden Line intensities in the condensation observed at the February 12, 1952 Eclipse (Aly *et al.*, 1962).

Caxv line $\lambda 5694$ at the center of the condensation, suggested that the condensation was axially symmetric, with a radial negative temperature gradient. Aly, Evans and Orrall, assuming axial symmetry and a homogeneous structure, determined the volume emissivity in each line from the observed intensity distribution. Their subsequent attempts to construct a *homogeneous* model for this condensation led to inconsistent results, however. (Aly *et al.*, 1963).

Suzuki and Hirayama (1964) constructed an inhomogeneous model for this same condensation in which four discrete temperature components were postulated in each volume element. Even this complicated model fails to account for all the line observations, however. Despite *apparent* symmetry and homogeneity, the condensation possesses unresolved inhomogeneities that seriously limit attempts to determine its physical properties.

Both the observational material and the theoretical data required to interpret it have improved markedly since this work in the early 1960s. Jefferies (1969) has reviewed recent eclipse photometry of the coronal forbidden lines. A series of three eclipses has now been observed, with the same instrument, by the University of Hawaii group, in collaboration with the Sacramento Peak Observatory. Some 45 lines are now known, 29 of which have suggested identifications. Intensity measurements for many of these lines have been made at the 1965, 1966 and 1970 eclipses (Jefferies *et al.*, 1969, 1971). The data is most complete for Fe (for which five and possibly six ionization stages are seen), Ni (four stages) and Ca (three stages).

In parallel with this improvement in the quality and quantity of spectroscopic data, has come improved calculations for the excitation and ionization equilibrium of coronal lions. In order to interpret the empirical volume emissivity for a given line, a prediction is needed of the number density of ions excited to the upper level, over a wide range of temperatures and densities. Following the discovery by Burgess in 1964 of the importance of dielectronic recombination (and the later suggestion, by Goldberg, of the influence of autoionization) new calculations for the ionization equilibrium have been carried out by Jordan (1969) and Allen and Dupree (1969).

Detailed calculations, based upon the best available atomic parameters have been carried out for the excitation of specific ions: Fexiii (Chevalier and Lambert, 1969), Fexv (Bely and Blaha, 1969) and Caxv (Chevalier and Lambert, 1970). Recently, Rozelot (1969) and Zirker (1970) have published extensive calculations for a large number of ions.

Figure 8 is taken from Zirker's calculations. It illustrates the dependence on electron density of two upper-state populations of Nixv. In general, each level of an ion is populated by a combination of three processes: direction collisional excitation from the ground state; collisional excitation of a high-lying state, followed by a permitted radiative transition to the level; or radiative excitation. The latter process is, of course, independent of the number of colliding electrons, while the other two processes vary linearly with n_e. As a result, the relative populations of levels in the ground term (where arise the observed forbidden optical transitions) vary with electron density.

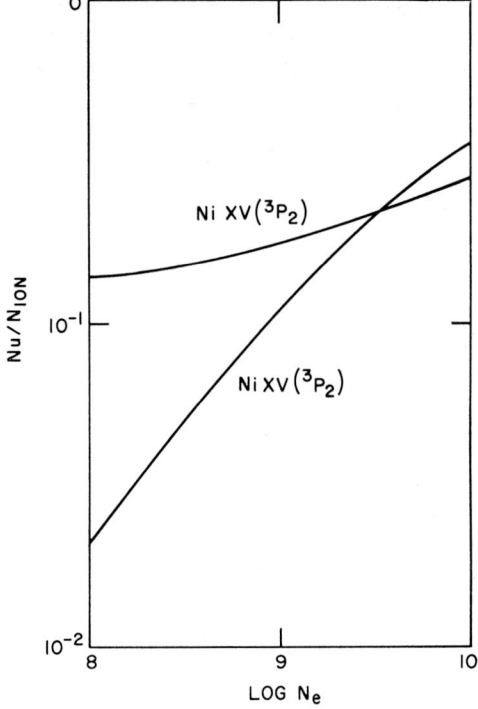

Fig. 8. The Density-variation of Ni xv levels (Zirker, 1970).

Thus, if the relative populations of levels in the ground term can be determined empirically from simultaneous observations of two or more lines from the same ion, they can be used to determine the local value of electron density over the temperature range where the ion is formed.

As an example of this method, we present a preliminary analysis of the line intensities emitted from the center of a coronal condensation at the 1965 eclipse. For each line arising from an upper state u, we can write the following equation for the number of ions in the line of sight excited to that level.:

$$N_u = \int \frac{n_u}{n_{ion}} \frac{n_{ion}}{n_{element}} \frac{n_{element}}{n_e} n_e \, dx$$

where the integration is carried out along the line of sight. The excitation ratio n_u/n_{ion} can be expressed as a power of the electron density $C_u(n_e^{\alpha_u})$ over the range of electron density between 10^8 and 10^{10}. The ionization ratio $n_{ion}/n_{element}$ is a function of electron temperature alone: $g(T)$. The ratio $n_{element}/n_e$ is the chemical abundance for that element. With these substitutions, we have

$$N_u = C_u A_{element} \int n_e^{1+\alpha_u} g(T) \, dx$$

As the first approximation, let us take an average value for α (numerically α varies between about 0.2 and 1.0 for different upper levels). Then, following Boardman and Billings, we may look for the fraction Φ of the total emission measure, $\int n_e^{1+\bar{\alpha}} dx$ that is associated with each of four discrete temperatures. We have, therefore

$$N_u = C_u A_{el} \int n_e^{1+\bar{\alpha}} dx \sum_i g(T_i) \phi(T_i)$$

Each observation of a line intensity yields a value of N_u and we may solve for the unknowns: A, the emission measure, and the four values of Φ. Six lines of iron, five lines of nickel and two lines of calcium were used to determine the fractional emission measures Φ shown in Table Ia.

Most of the emission arises from a temperature region between 1 and 1.5 million degrees and another one between 2.5 and 5 million degrees. The relative abundances of iron to nickel and nickel to calcium are found to be 30 and 14 respectively.

Table Ib shows the fit of the data to this model. In the fourth column, the relative population of each level as predicted by the model is shown. Compare this to the observed relative populations in column 5. The agreement is generally within a factor of 2 with two glaring exceptions (See the asterisks).

TABLE I

Model I solutions

(a) Fractional Emission Measure (Φ)

$T \times 10^{-6}$	Φ
1.0–1.5	0.48
1.5–2.0	0.02
2.0–2.5	0.02
2.5–5.0	0.48

(b) Fit of Data

			$\log N_u/N_0$	
ion	level	α	model	obs.
Fe x	$^2P_{1/2}$	0.60	0.00	0.00
Fe xi	$^1D_2$	0.91	−0.39	−0.64
	$^3P_1$	0.50	0.50	−0.64*
Fe xiii	$^1D_2$	1.0	−1.21	−1.24
Fe xiv	$^2P_{3/2}$	0.62	−0.21	−0.34
Fe xv	$^3P_2$	0.97	−1.34	−1.16
Ni xii	$^2P_{1/2}$	0.84	0.00	0.00
Ni xiii	$^1D_2$	0.86	0.06	1.05*
	$^3P_1$	0.66	−0.11	0.00
Ni xv	$^3P_2$	0.15	0.55	0.65
Ni xvi	$^3P_{3/2}$	0.88	−0.32	−0.25
Ca xii	$^2P_{1/2}$	0.69	0.00	0.00
Ca xiii	$^3P_1$	0.52	0.30	0.18

This preliminary model (Model I), in which a mean value has been taken for α, assigns each ion to the temperature range at which it contributes most of its emission to the observed line. We now proceed with a two temperature component model and use the differences in α for the individual lines to determine the electron density associated with each component. Thus we write

$$N_u = A_{el} C_u n_e^{1+\alpha_u} L_i g(T_i)$$

where the symbols are as before and L_i is the path length in the line of sight corresponding to a particular temperature component. Thus we have

$$\log \left[\frac{N_u}{C_u g(T_i)} \right] = \log (A_{el} L_i) + (1 + \alpha_u) \log n_e.$$

The left hand side is known from observation and from the excitation and ionization equilibrium calculations. If we plot this quantity against $1 + \alpha_u$, a linear relation

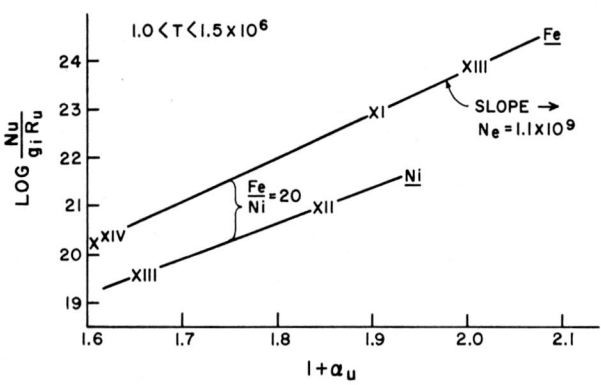

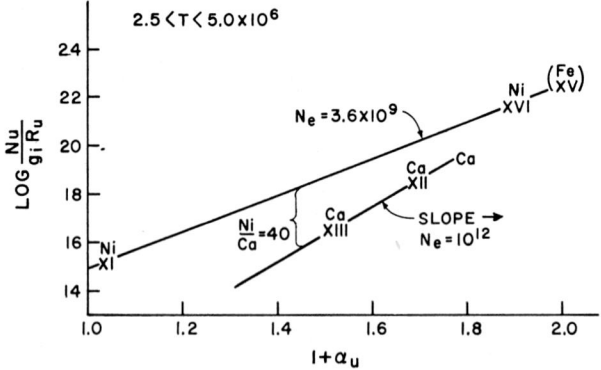

Fig. 9. Density determinations for a model with two temperature components, based upon forbidden line intensities observed at the 1965 Eclipse.

results as shown in Figure 9. The slope of this relation, according to the preceding equation, is a measure of $\log n_e$ in the component we are considering, while the intercept determines the product AL_i. Two halves of Figure 9 show the solutions for the two temperature intervals. Table II summarizes the electron densities and path lengths

TABLE II
Model II solutions

$T \times 10^{-6}$ (K)	n_e (CM^{-3})	L_i (CM)
1.0–1.5	1.1×10^9	4.4×10^9
2.5–5.0	3.6×10^9	0.4×10^9
(5.0)	(10^{12}?)	

Abundances ($\log N_H = 12.0$)

	Model II	Photosphere
Fe	7.1	6.7
Ni	5.8	5.8
Ca	4.2	6.2 ($n_e = 3.6 \times 10^6$)

associated with each temperature interval as well as the absolute, abundances. Note that the slope of the line for calcium, Figure 9, indicates a larger electron density (of the order of 10^{12}) than does the nickel and iron data. This result suggests that the data would be better represented by a three temperature component.

The trend is clear, however. Increasing density is associated with increasing electron temperature in accord with the findings of Billings and Boardman, and Walker and Rugge. In order to proceed beyond models such as this, we need auxiliary information on the spatial distribution of radiating material. This information might come from three types of observations: (a) velocity measurements within the condensation, (b) comparison of monochromatic filtergrams obtained at different times and showing, possibly, a stereo effect, and (c) from polarization measurements within the condensation.

4. Loop Prominences and Sporadic Condensations

Figure 10 shows a typical loop prominence system, photographed with the twin coronagraphs in Hawaii. These spectacular transient objects form 15 or 20 min after a major flare – particularly flares that produce high energy proton events. Bruzek (1969) gives a short discussion of these objects and a complete bibliography. As Figure 10 shows, the loops are equally visible in Hα, characteristic of a temperature of 10^4 K and in $\lambda 5303$ of Fe XIV, characteristic of a temperature of 2×10^6 K. The Hα loops are referred to as a loop prominence, the green line loops as a sporadic condensation but clearly they are different views of the same physical object.

Individual loops (when viewed in Hα) appear to materialize out of the background

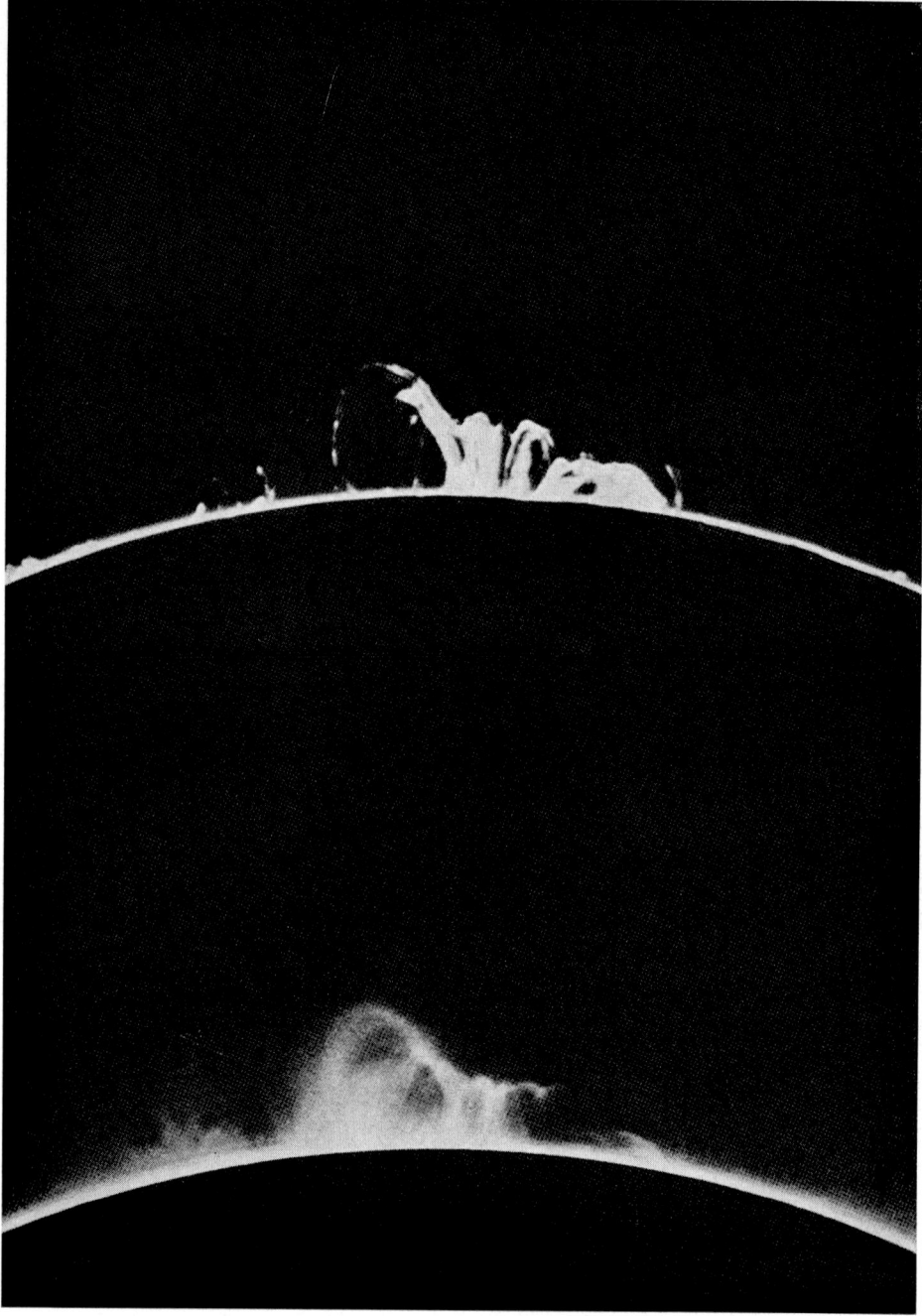

Fig. 10. A loop prominence, photographed in Hα (top) and $\lambda 5303$ (bottom) simultaneously (R. R. Fisher, University of Hawaii).

corona to grow in brightness and then fade away. Material within a given loop seems to stream down from the top of the loop and fall into the chromosphere with free-fall velocities of the order of 100 km/sec. An individual loop lasts about 20 min while the system of loops grows in height at a speed of order 10 km/sec to a maximum height of 40 to 150000 kilometers. The loop system (while initiated by a flare) may last as long as 10 or 12 h.

When seen near the limb or against the disk, the loops are found to be rooted in two parallel ribbons within the flare – sometimes on opposite sides of the magnetic field neutral line. Thus, the loops appear to connect plage opposite magnetic polarity and there seems no doubt that we are looking at individual tubes of force illuminated by the hot, streaming material within it. Field strengths of the order of 50 gauss have been measured by Hyder (1965) and by Harvey (1968).

The loops are visible in spectral lines covering a wide range of temperature. Orrall (1965) reported spectroscopic observations of the sporadic condensation and loop prominence observed at Sacramento Peak in February, 1962. The same structures were visible in the yellow, green and red coronal lines, $\lambda 5411$ of He II, $\lambda 6678$ of He I, the sodium D lines, the b group of magnesium I, and Hα. There is some suggestion on high resolution images – for example, Figure 10 – that the loops have a smaller diameter when seen in Hα than in the green line so that all the emissions from different temperatures are not exactly co-spatial. Orrall determined the diameters of the emitting region for different lines by comparing the observed line intensity with theoretical emissivities. He finds diameters of 2×10^8, 1.5×10^8, and 15×10^8 cm, for regions which emit Hα, the red line and the green line. These lengths suggest a model for the loops with several 10^4 K cores, surrounded by a hotter, more diffuse 2 million degree region, with a sharp transition zone at 1 million deg. The yellow line, formed at about 5 million deg, is confined to knots of high density (10^{10} cm^{-3}).

The formation of loop prominences following a large flare involves processes that are, as yet, incompletely understood. During the last few years, however, a gradual revision of earlier ideas has been forced by new observations.

The appearance of loop prominences in Hα movies strongly suggests that the cold material we see streaming down loops has condensed out of the surrounding hot coronal condensation. This condensation process was first suggested by Kiepenheuer, and explored theoretically by Parker (1955a). According to tehse ideas, if an initial condensation can be made to form by some external force, perhaps the surrounding magnetic field, and if the non-thermal energy supply to this condensation does not increase with density, then the condensation becomes radiatively unstable, i.e., it radiates more energy than it receives, and, therefore, cools catastrophically. To explain the very rapid formation of loop prominences, Lüst and Zirin (1960) suggested that the thermal energy of the condensation is carried away by heat conduction rather than by radiation.

These ideas were criticized by Jefferies and Orrall (1963) and by Kleczek (1963). They pointed out that very reasonable estimates of the mass (10^{16} g) and energy (10^{31} erg) dissipated in a loop prominence exceed by an order of magnitude the mass and thermal energy of the entire coronal condensation in which it is imbedded. Moreover, a macroscopic flow from a larger volume into the loop would be inhibited because the ionized gas is not free to cross the magnetic field lines. Finally, after the sporadic condensation and loop prominence has disappeared, the permanent condensation seems unaffected.

Jefferies and Orrall suggested an alternative source for the energy and mass of loop prominences. They noted that loop prominence systems appear very frequently following flares that produce solar proton events and type IV radiation. (See Bruzek, 1969.) They speculated that the flare accelerates some protons to energies of tens of keV and that these lower energy particles are trapped in magnetic fields high in the corona where the low collision rates enable them to survive for hours. Collisions with the ambient gas near the foot-points of the field lines may shift them into other field lines that outline the loop prominence. Thus, fast particles are gradually transferred from a storage area high in the corona into the loop prominence system.

Jefferies and Orrall showed that the radiative energy observed from a loop system (10^{31} erg) could be accounted for if the total mass supplied to the system (10^{16} g) were in the form of fast protons each having an energy of order 10 keV. In 1965, they presented observational evidence for the presence in loops of protons with energies of 10 keV. The Hα line in a loop prominence observed in 1962, showed Hα wings 30 Å wide. An electron density of 10^{15} cm^{-3} is required to account for such wings by Stark broadening. It is more reasonable, therefore, to account for the wings by postulating fast-moving hydrogen atoms. Jefferies and Orrall argued that these H atoms are formed by charge exchange between fast protons and thermal protons in the condensation. These fast H atoms thermalize in seconds, compared with a lifetime of the loop (20 min).

The rapid heating of the coronal gas produces *macroscopic* streaming from the tops of the loops with the observed speed of $\sim 10^2$ km/sec.

This model has several virtues. In particular, it interprets the strong association between loop prominences and proton flares. Evidence is being sought, with a coronagraph in Hawaii, for the presence of high velocity upward streaming at the foot of the loops. However, a complete model would answer the following questions: (1) Why the Hα emission first appears at the *tops* of the loops? What is the significance of the tops? (2) Why don't fast protons thermalize at the field foot points, where the density is highest? (3) Why does the loop system grow upward, i.e., why do successive foot points appear further from the neutral line?

In summary, we may say that although loop prominences represent a secondary phenomenon of solar flares, they incorporate many of the features of thermalization of super-thermal particles that are involved in more energetic processes – such as the generation of X-ray and centimeter wave bursts.

Acknowledgements

This research was supported by NASA Grant No. NGL 12-001-011.

References

Aly, M. K., Evans, J. W., and Orrall, F. Q.: 1962, *Astrophys. J.* **136**, 956.
Aly, M. K., Evans, J. W., and Orrall, F. Q.: 1963, *Astrophys. J.* **137**, 1313.
Allen, J. W. and Dupree, A. K.: 1969, *Astrophys. J.* **155**, 27.
Athay, R. G.: 1966, *Astrophys. J.* **145**, 784.
Batstone, R. M.: 1970, private communication.
Beley, O. and Blaha, M.: 1969, *Solar Phys.* **10**, 115.
Boardman, W. J. and Billings, D.: 1969, *Astrophys. J.* **156**, 731.
Bruzek, A.: 1969, 'Solar Flares and Space Research', in *COSPAR Symp.*, p. 61, C. DeJager and Z. Švestka, (eds.).
Chevalier, R. A. and Lambert, D. L.: 1969, *Solar Phys.* **10**, 115.
Chevalier, R. A. and Lambert, D. L.: 1970, *Solar Phys.* **11**, 243.
Christiansen, W. N., Mathewson, D. S., Pawsy, J. L., Smerd, S. F., Boishot, A., Denisse, J. F., Simon, P., Kakinuma, T., Dodson-Prince, H., and Firor, J.: 1969, *Ann. Astrophys.* **23**, 75.
Dollfus, A.: 1970, these proceedings, Paper 7, p. 97.
Dunn, R. B.: 1970, these proceedings, Paper 8, p. 114.
Dupree, A. K. and Goldberg, L.: 1967, *Solar Phys.* **1**, 729.
Gabriel, A., and Jordan, C.: 1970, *Monthly Notices Roy. Astron. Soc.*, in press.
Harvey, J.: 1968, *Astron. J.* **73**, 62.
Hyder, C. L.: 1965, *Astrophys. J.* **141**, 1374.
Jefferies, J. T.: 1969, *Mém. Soc. Roy. Sci. Liège*, 5th Series **XVII**, 213.
Jefferies, J. T. and Orrall, F. Q.: 1963, AAS-NASA Symposium on Physics of Solar Flares, October 28–30, 1963, p. 71.
Jefferies, J. T., Orrall, F. Q., and Zirker, J. B.: 1969, *BAAS* **1**, 3.
Jefferies, J. T., Orrall, F. Q., and Zirker, J. B.: 1969, *Mém. Soc. Roy. Sci. Liège*, 5th Series **XVII**, 235.
Jefferies, J. T., Orrall, F. Q., and Zirker, J. B.: 1971, *Solar Phys.*, **16**, 103.
Jordan, C.: 1969, *Monthly Notices Roy. Astron. Soc.* **142**, 501.
Kleczek, J.: 1963, AAS-NASA Symposium on Physics of Solar Flares, October 28–30, 1963, p. 71.
Lüst, R. and Zirin, H.: 1960, *Z. Astrophys.* **49**, 8.
Newkirk, G.: 1967, *Ann. Rev. Astron. Astrophys.* **5**, 213.
Noyes, R. W., Withbroe, G. L., and Kirschner, R. P.: 1970, *Solar Phys.* **11**, 388.
Orrall, F. Q.: 1965, in *Solar Spectrum*, (ed. by C. de Jager), Reidel Publ. Co., Dordrecht, p. 308.
Parker, E. N.: 1955a, *Astrophys. J.* **117**, 431.
Parker, E. N.: 1955b, *Astrophys. J.* **121**, 491.
Pottasch, S.: 1967, *Bull. Astron. Inst. Neth.* **19**, 113.
Rozelot, J. P.: 1969, Doctoral Thesis, University of Paris.
Rugge, H. R. and Walker, A. B. C.: 1970, *Astron. Astrophys.* **5**, 4.
Saito, K. and Billings, D.: 1964, *Astrophys. J.* **140**, 760.
Saito, K. and Owaki, N.: 1967, *Publ. Astron. Soc. Japan* **19** (No. 4), 535.
Suzuki, T. and Hirayama, T.: 1964, *Publ. Astron. Soc. Japan* **16**, 58.
Tsubaki, T.: 1966, *Publ. Astron. Soc. Japan* **18**, 1.
Vaiana, G. S.: 1970, paper presented at IAU Meeting in Brighton, England.
Withbroe, G. L.: 1970, *Solar Phys.* **11**, 42.
Zirker, J. B.: 1970, *Solar Phys.* **11**, 68.

11. OBSERVATIONAL EFFECTS OF FLARE-ASSOCIATED WAVES

SARA F. SMITH and KAREN L. HARVEY
Lockheed Solar Observatory

1. Introduction

Flare-associated waves have been referred to in the literature as "shock-waves", "magnetohydrodynamic waves", "solar blast waves", "Moreton waves", "high speed disturbances", and "rapidly expanding features". To minimize prejudicing our discussion of the observational effects of this phenomenon, however, we prefer to simply refer to this kind of event as a "wave" or "flare-associated wave", postponing further labelling or interpretation until the observational facts are presented. In most of the observations, we are only indirectly inferring the existence of a so-called "wave" from the consistency and repeated occurrence of four types of observational effects associated with flares. These effects are (1) an abruptly initiated activation or oscillation of a filament outside the boundaries of a flare, (2) a front, interpreted as a depression and relaxation of the chromospheric fine structure, rapidly moving away from the active region of a flare, observed in the wings of Hα, (3) a very fast, bright, diffuse front of emission propagating away from a flare, usually only seen in the center of Hα (4) progressive short-lived brightenings of small points of the chromosphere outside the active region of a flare.

Definite evidence of the possible existence of waves was first presented by Dodson (1949) in a detailed study of the activation of a filament during a flare on 10 May 1949. Other work relating to the possible existence of waves were studies of sympathetic flares by Richardson (1936, 1951) and Becker (1958). Becker studied 45 pairs of flares and suggested that flares and some filament disturbances were triggered by long distance traveling disturbances with velocities of on the order of 2000 km/sec. He also described a flare occurring on 22 August 1957 which was shortly followed by two filament activations and another flare. Becker estimated the velocity of the inferred disturbance to be 1500–2000 km/sec.

Valnicek (1961, 1964) discussed disturbances of velocities greater than 1000 km/sec. In 1961, as a result of the work of Richardson and Becker, he found impulses traveling at velocities greater than 1000 km/sec from a flare which produced another flare. Valnicek did not observe the wave but inferred it from the occurrence of sympathetic flares. In 1960 evidence of these disturbances on the chromosphere were finally observed (Moreton and Ramsey, 1960; Moreton, 1960) and have been further studied by Moreton (1961, 1964), Athay and Moreton (1961), Dodson (1964), Ramsey and Smith (1965), and Dodson and Hedeman (1968). Recent evidence for relating waves and Type II radio bursts has been reported by Kai (1970).

In 1964, Valnicek divided flare ejecta into three classes: Group I – vel. 1000 km/sec, detached part of flares, initial high velocity, then deceleration (Sprays); Group II –

blow-off prominences, vel. 250 km/sec; Group III – waves, vel. 330–4200 km/sec. Dodson and Hedeman (1968) suggest a relationship between eruptive prominences (Group II) and waves (Group III) phenomena since eruptive prominences may exhibit velocities comparable to some wave-effects. However, possible relationships between all three of Valnicek's classes need further investigation since sprays also exhibit velocities comparable to waves. There are also numerous suggestions in the literature relating Type II radio bursts to both sprays and waves. A limited discussion of the association of Type II bursts to flare-associated waves is included in the last section. However, our primary intention in this paper is to contribute additional information on the optically observed characteristics of phenomena which suggest the existence of "waves". This study is based on time-lapse photography of the solar chromosphere taken through narrow band Hα filters.

2. Velocities Derived From Wave Effects

Velocity is a key parameter in the interpretation of waves and the association of the different wave effects. However, the velocities which can be inferred from wave effects offer many sources of appreciable error because of the difficulty in determining the physical nature of the events, the exact time and spatial origin of the source of the waves, and the length of time after the passage of a wave that each of the wave effects become visible. Our most accurate measurements of velocity come from those events in which an absorption or emission front appears to propagate across the chromosphere outside the active region of the flare. Fifteen such cases existed in our sample of wave-effects associated with 60 flares.

Relevant data for each one of these fifteen directly observed wave fronts are listed in Table I. In the last two columns in Table I, two velocities are given: (1) the transverse component of velocity assuming that the wave effect is moving through the corona, and (2) the velocity calculated assuming that the wave-effect is propagating along or just above the surface of the chromosphere. Velocity curves of the transverse component of the 15 events were made. Each of these curves was constructed by measuring successive positions of the wave front on different frames of film along a straight line from an assumed starting point in the flare. As an example, the velocity curves of the bright wave front observed on 23 May 1967 is shown in Figure 1. The plotted points in this illustration have been corrected for foreshortening. Of the events studied this one shows the most pronounced difference in velocity in different parts of the wave front. In this case the measured transverse component of velocity varied in a continuous manner from 340 to 1060 km/sec from east to west along the wave front.

The chromospheric or surface velocity in Table I was similarly measured except that drawings of the apparent positions of the wave front were transferred to a globe and the distances were measured along great circles on the globe surface from the same assumed starting point at the flare. Both velocities are presented because we feel that there are still several possible interpretations of the data which could lead

TABLE I

Flare data				Filter	Wave data							
Date	Start (UT)	Explosive phase (UT)	Error in expl. phase (sec)	Wavelength	Emission wave	Dark wave	Fil. Osc.	Chromospheric brightening	Cen. Dir. prop. (deg)	Angular ext. (deg)	Mean transverse components of velocity (km/sec)	Mean chromospheric velocity (km/sec)
25 Jun. 1960	2037:00	2041:45	±60	Hα−0.5 Å					250	165	810	810
7 Sep. 1960	2309:00	2309:00	±15	Hα+0.5 Å		×	×		230	35	710	710
12 Oct. 1960	1740:00	1745:45	±15	Hα−0.5 Å		×	×	×	190	115	680	680
18 Mar. 1961	1908:30	1908:30	±30	Hα−0.5 Å		×	×		300	70	630	630
3 Sep. 1961	2039:00	2041:00[a]	±60	Hα, Hα+0.5 Å		×		×	270	60	440	440
20 Sep. 1963	2050:00	2056:15	±15	Hα+0.5 Å, Hα−0.5 Å		×	×		230	70	920	940
30 Dec. 1966	2231:00	2233:00	±20	Hα	×		×		70	75	440	440
4 Feb. 1967	1640:00	1645:15[a]	±30	Hα−0.5 Å		×	×		185	125	650	750
27 Feb. 1967	1637:00	1643:15[a]	±15	Hα	×	×	×		180	115	350	650
27 Mar. 1967	2107:00	2111:00	±15	Hα−0.5 Å					340	80	460	620
23 May 1967	1835:30	1837:10	±10	Hα	×		×	× ×	340	85	340–1060	750–1125
30 Jul. 1967	1554:45	1559:45	±15	Hα	×		×		195	60	500	500
31 Jul. 1967	1715:00	1721:00	±15	Hα	×				160	65	560	560
1 Aug. 1967	0056:00	0117:00	±15	Hα	×		×		160	65	710	710
26 Aug. 1967	0006:00	0011:00[a]	±30	Hα	×		×		355	70	490	490

[a] Rapid rise − not explosive

to different models of the wave phenomena. Also, the wave effects vary so much from event to event that it is not yet certain that any single interpretation or model yet published (Anderson, 1966; Meyer, 1966, 1968; Dodson and Hedeman, 1968, Uchida, 1969) adequately accounts for all of the observed effects.

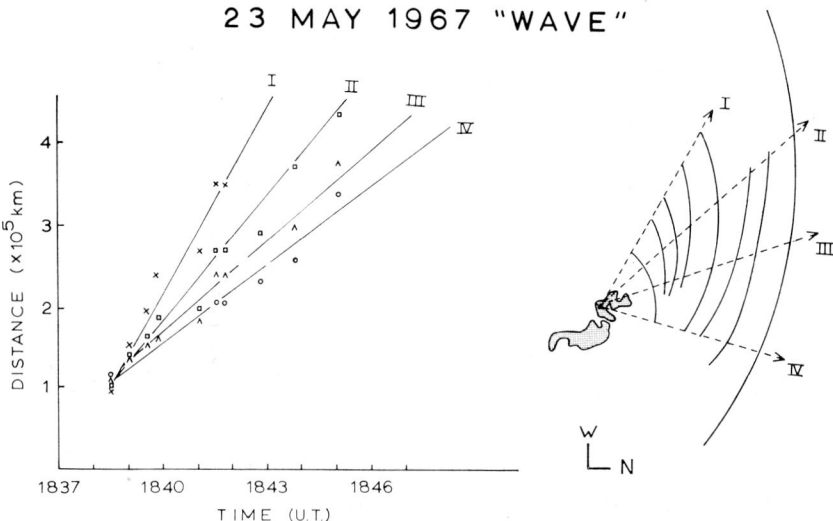

Fig. 1. The measured transverse component of velocity varied from 340 to 1060 km/sec from east to west along the arcs depicting the observed wave effect. The actual emission front is shown in the upper right corner of Figure 3.

Several sources of error are possible for each measurement. The visibility of the wave fronts varies greatly from event to event. The wave front is faint and seldom a definite clearly outlined feature on any single frame of the film. The error due to the lack of well defined boundaries is further compounded by variations in the image quality from frame to frame. Additional error is probably introduced by incorrectly assumed initiation points for each wave.

In *many* waves, excluding the one on 23 May 1967, the effects are only seen well outside the active region of the flare. Thus, the time of origin of the waves must be determined by extrapolating the velocity curve back to the time coordinate. The time origin of waves determined in this manner are compared in Figure 2 with the starting time and the explosive phase of the associated flares. It is seen that for these 15 cases, the waves originated during the first 7 minutes of the flares. The extrapolated origin of these waves, however, is shown in the lower half of Figure 2 to agree much better with the start of the explosive phase than the start of the flares. In all except three of these events, the start of the explosive phase was quantitatively determined using the definition suggested by Harvey (1970). The start of the explosive phase is the time at which the increase in a photometric measure of a whole flare exceeds in 1 min 25% of the total increase in light from flare start to flare maximum. According to this definition not all flares will exhibit an explosive phase. For the few events in our sample which did not fulfil the definition proposed by Harvey (1970), the explosive phase was defined as the time at which the rate of increase of a photometric measure of a whole flare first reached a maximum. For 3 of the events, the start of the explosive phase was determined visually. As shown by Harvey (1970) the visual estimates on an

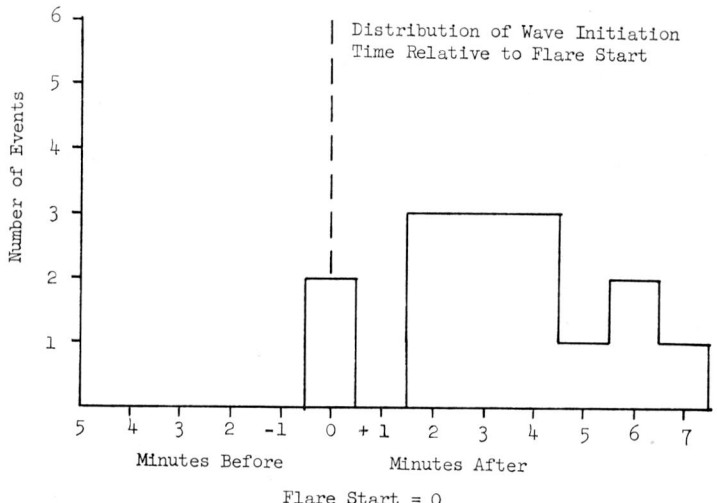

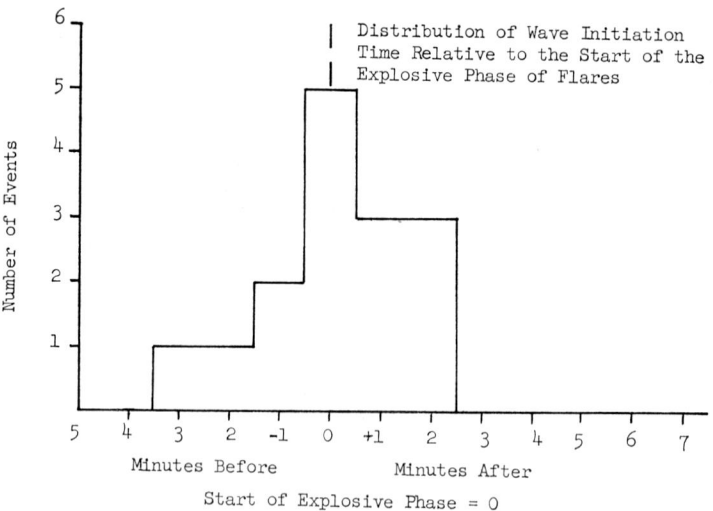

Fig. 2. It is shown that the initiation time of waves is more closely related to the start of the explosive phase in flares than to the start of the flares. In two cases, the start of the flares and the start of the explosive phase were nearly coincident.

average, show the start of the explosive phase as beginning 30 sec later than the quantitative definition.

Since the extrapolated origin of these waves agrees with these definitions of the explosive phase we are justified in using this phase as the assumed time of initiation of waves for those cases in which the only visible wave effects are oscillating filaments (Ramsey and Smith, 1966) or chromospheric brightenings. In our sample of 45 events with one or more oscillating or disrupted filaments, the range of velocities for the invisible waves ranged between 410 to 2000 km/sec. The mean velocity of waves producing these filament oscillations is 880 km/sec while the mean velocity of the 15 visible wave fronts is 600 km/sec. The higher mean velocity for filament oscillations was substantiated for a few cases in which both a wave front and one or more oscillating filaments are associated with the same flare. It is likely that this difference is due to the more readily visible change in a filament as opposed to the poor visibility of the leading edge of either dark or bright wave fronts observed against the chromosphere.

3. Possible Interpretations of Dark and Bright Wave Fronts

In Figure 3 is an illustration of a wave observed on 20 September 1963 with a filter of 0.5 Å bandpass cyclically tuned to Hα+0.5 Å, Hα−0.5 Å, and the center of Hα line. The top row of pictures in the red wing (Hα+0.5 Å) shows a dark leading edge and a bright wake whereas the second row of pictures in the blue wing (Hα−0.5 Å) show a bright leading edge and a dark wake. This type of observation is readily interpreted as a depression and relaxation of the fine structure of the chromosphere (Dodson and Hedeman, 1968). We concur with this interpretation for the event described by Dodson and Hedeman (1968) and for the event shown in Figure 3, and for five other events listed in Table I. However, we contend that the remaining events in Table I can be more readily interpreted as material ejected into the corona than as a Doppler shift of the chromospheric structure. Those events in Table I which are similar to the 20 September 63 event (Figure 5) are 25 June 1960, 12 October 1960, 18 March 1961, 3 September 1961, and 27 March 1967. All of these events were observed in the blue wing (Hα−0.5 Å), with the exception of 3 September 1961 which was observed in the red wing (Hα+0.5 Å). They are like 20 September 63 wing observations in that successive small chromospheric structures appear to be enhanced and reduced in visibility.

Figure 4 (23 May 1967) is an excellent example of a wave observed in emission in the center of Hα with a filter having a bandpass of 0.5 Å. There are several noteworthy differences between this type of event and the event on 20 September 63 observed also in the wings of Hα as shown in Figure 3: (1) The emission wave of 23 May 1967 can be seen while still within the active region while the wave front of 20 September 63 cannot be seen until it is well outside the active region of the flare. (2) The emission front of 23 May 1967 is semi-transparent while the effects on the 20 September 63 event are not distinguishable from an enhancement of the background. (3) The emission front of 23 May 67 is not unlike the ejected bright knot visible in the center line

Hα picture of 20 September 63, except that in the 23 May event the emission is more diffuse and covers a considerably larger area. (4) The emission front of the 23 May 1967 event exhibits a wide range of velocities; 340–1060 km/sec as shown in Figure 1. These differences indicate that the bright emission fronts, although extensive, are probably material ejected from the flare region which is emitting in Hα while traveling through the corona.

23 MAY 1967

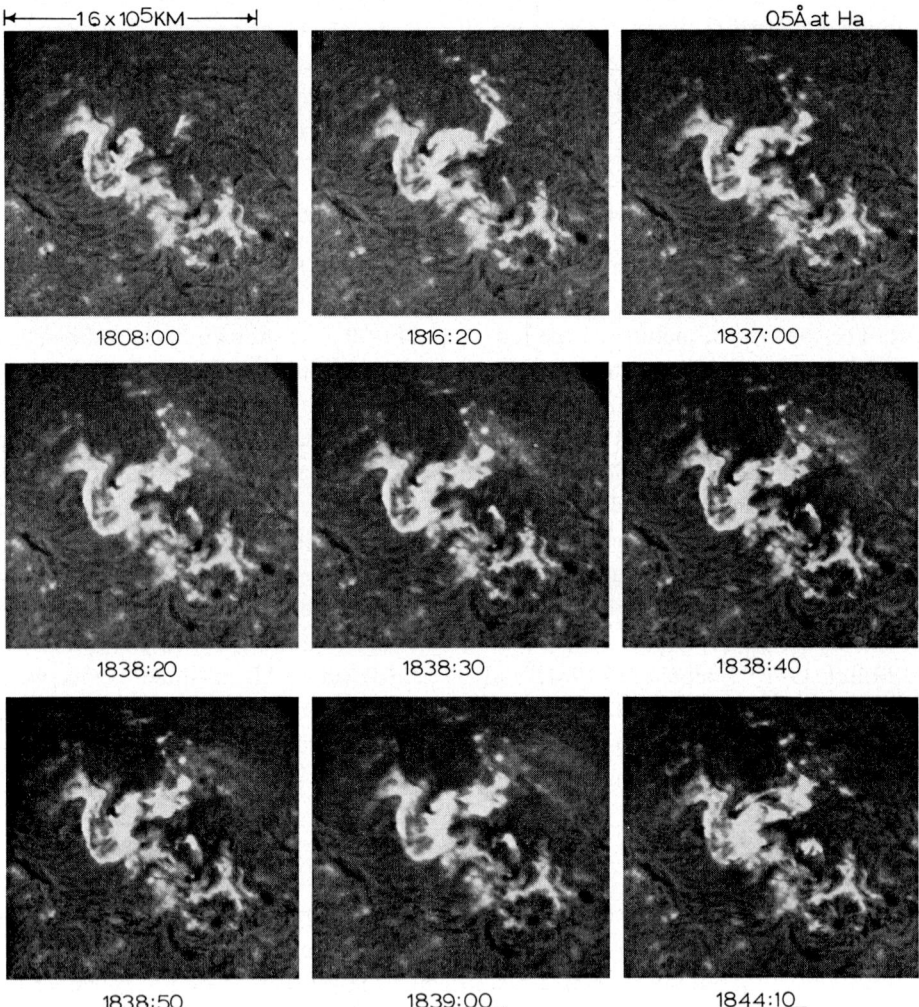

Fig. 3. The Bright "wave" of emission leaves the flare at approximately 1837 UT travelling in a north westerly direction (upper right corners). After the emission passes, small fixed points of the chromosphere remain bright like the flare. We refer to these secondary small emission points as "chromospheric brightenings".

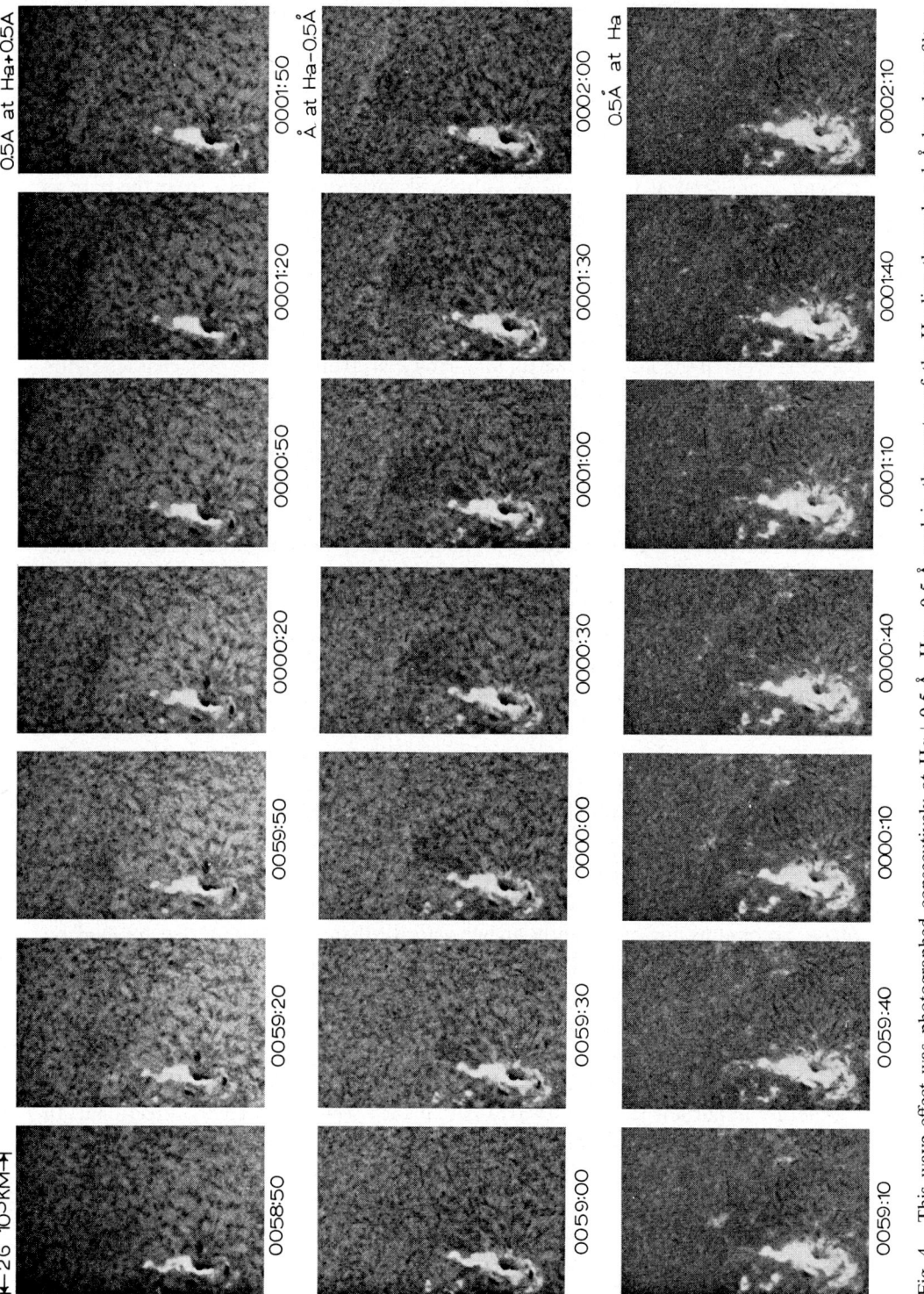

Fig. 4. This wave effect was photographed consecutively at $H\alpha+0.5$ Å, $H\alpha-0.5$ Å and in the center of the $H\alpha$ line through a $\frac{1}{2}$ Å bandpass filter. Note that in the last row of pictures the only visible effect of the wave front is a rapidly moving knot of emission.

Additional supporting evidence for the interpretation of mass ejection is provided by other events such as 7 September 1960 and 4 February 1967 which appear to partially obscure the chromospheric network in the wings of the Hα line. Similarly, most of the bright wave fronts observed in the center of Hα also have the characteristic of appearing semi-transparent in varying degrees as in the described case of the bright wavefront observed on 23 May 1967 (Figure 4).

Some events appear to expand in width with increasing distance from the flare as in 20 September 1963 (Figure 3) (red and blue wings only) and as in 23 May 1967 (Figure 4) in the center of Hα. However, the dark wave event of 7 September 1960, and the rapidly moving bright ejecta observed in the center line Hα (bottom row of Figure 4) do not appreciably expand. Notably, the bright ejecta in the lower row of Figure 3 is not a broad bright front having dimensions similar to the wave front of the same event in the wings of the Hα line. However, there are common characteristics between the events in Figures 3 and 4. The bright ejecta in Figure 3 has approximately the same brightness and transparency as the broad "front" in Figure 4. Also, in both events an oscillation of a distant filament is initiated. In both cases the start of the oscillation is in agreement with the arrival of the wave front at the location of the filament.

Concerning the event on 25 June 1960, there was a large discrepancy in the velocities of the presumed wave reported by Dodson (1964) and Moreton (1964). Dodson reported that a single velocity of 600 km/sec tied all observable filament oscillations to a disturbance initiated at the flash phase of the flare. Moreton maintained that direct measures of the wave front as observed $\frac{1}{2}$ Å into the blue wing of Hα gave a velocity of approximately 1000 km/sec. As given in Table I, we find another result: a velocity of 810 km/sec for the disturbance propagating along the chromosphere and a velocity of 1040 km/sec for the filament oscillations as observed at Hα-0.5 Å. However, we also find that all of these discrepancies can be resolved by considering additional factors. It has been shown that the initiation time of the waves related more closely to the start of the explosive phase of the flare than to flare start as illustrated in Figure 3. When extrapolating our measurements back to the flare, in this particular case, however, we arrive at a wave initiation time $1^m 20^s$ after the start of the explosive phase, accounting for our slightly higher velocity than the one reported by Dodson (1964). The even higher velocity of 1040 km/sec we believe to be of realistic lower limit for the velocity of the wave in the corona which initiates the filament oscillations and chromospheric wave effect. The exposures were made at a rate of one per 10 seconds and the initial fading of the filaments as seen in the blue wing (Hα-0.5 Å) is generally detectable within 30 sec or 3 frames.

Another important feature of waves is that they seem to be highly directional. In most cases, the angular width of the observed effects is less than 90°. However, in several cases, effects were observed with a separation angle of nearly 180 deg, as in the 25 June 1960 event.

A diagram of the central direction of all the various wave effects is shown in Figure 5. The length of each arrow represents the mean velocity of each propagation.

The solid lines are those effects for which velocity error is less than 30%. The dashed lines represent those cases in which the velocity could not be ascertained or the probable error was greater than 30%. The first thing to notice is that wave effects may occur in any direction from a flare. There is, however, in this sample a tendency for more wave effects to be observed to the south and west of the flare's position.

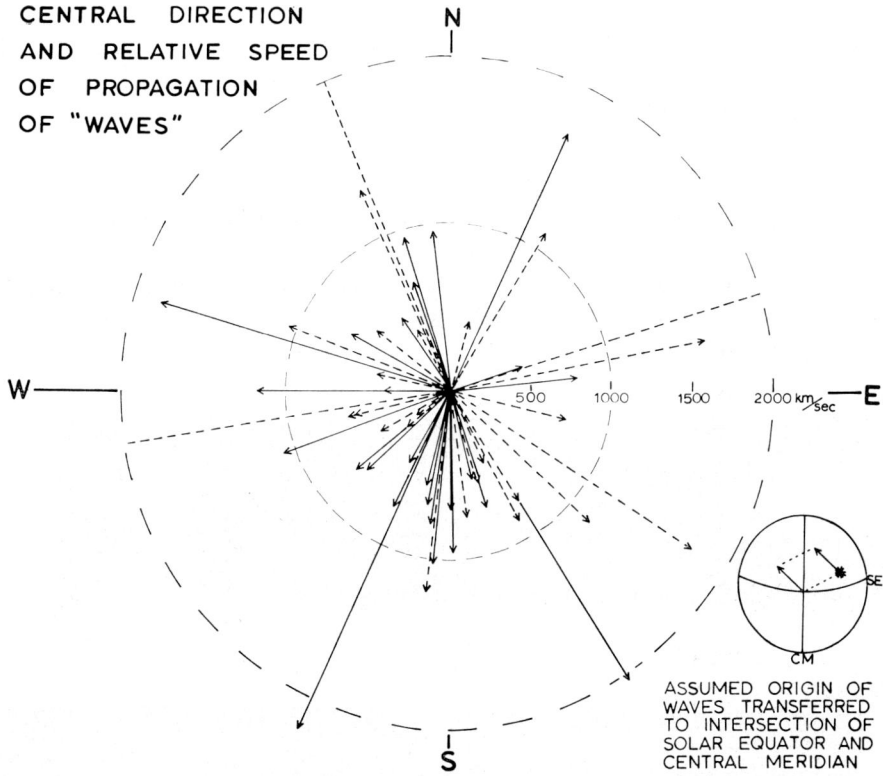

Fig. 5. The arrows represent the central direction of the observed wave effects shown as if all of the flares occurred at the same point on the sun. Wave effects can appear in any direction relative to the flare although, to date, more wave effects have been observed to the south and west of the flares than to the north or east.

Since the configuration of flares is related to the configuration of the local magnetic field (Smith and Ramsey, 1966), it is possible that the visible direction of propagation of the wave could be related to the geometry of the coronal fields. This possible relationship needs investigation.

4. Waves and Type II Radio Events

Another subject which needs further investigation is the relationship of Type II bursts to flare-associated waves. It has been hypothesized that flare-waves are generated by the same cloud of ions which produce Type II bursts (Wild, 1963; Kundu, 1964; Lin and Anderson, 1967). However, to date very little observational material has been studied to substantiate this hypothesis.

The list of waves discussed in this paper were checked for the occurrence of Type II events during the associated flares from reports in *Solar-Geophysical Data*. Only eighteen of the fifty (36%) events for which spectral radio observations were available, were found also to have associated Type II bursts. It is interesting to note that prior to 1963, seven of twenty-nine wave events ($<25\%$) were associated with Type II bursts while since 1963, eleven of twenty-one wave events ($>50\%$) were associated with Type II bursts. This increase in the percent of concurrent waves and Type II bursts is attributed to improved radio spectrum observations.

To determine whether or not Type II events were really absent when no Type II is reported, it is necessary to carefully scrutinize the actual radio spectra. Fourteen radio spectra were available from the Harvard Radio Astronomy Observatory at Fort Davis, Texas during flares which also produced waves. Nine of these radio spectra showed strong Type II events; two were such weak Type II's as to be almost invisible; three, at best, could be considered questionable Type II events.

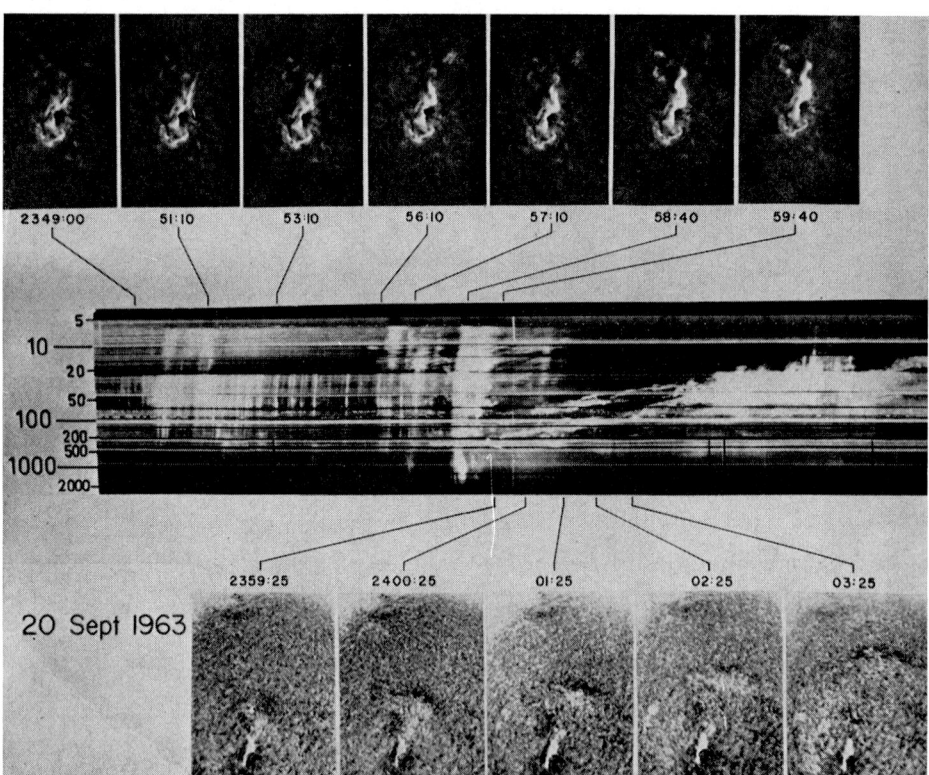

Fig. 6. The wave effect, also illustrated in Figure 4, is shown here relative to the radio spectrum from Fort Davis, Texas. The Type II Burst is seen to occur simultaneously with observed wave front.

Further evidence for associating waves and Type II events can be sought by comparing the timing and velocities of individual events as exemplified by the Type II events studied by Kai (1970).

The clearest example of a wave that has been observed at Lockheed occurred on 20 September 1963. Photographically subtracted observations at H$\alpha \pm 0.5$ Å of the event are shown in Figure 6 below the corresponding radio spectrum. It is obvious that the timing of the wave with the Type II is very precise. The start of the explosive phase of the flare occurred at 2056:15 UT $\pm$ 15 sec. At exactly this same time a very broad Type III occurs throughout the radio spectrum, similar to an event described by Dodson and Hedeman (1968). As shown, the beginning of the explosive phase is also the time at which the wave events are typically initiated during flares. The occurrence of a Type III marking the start of the explosive phase was an occasional but not frequent characteristic in the radio records surveyed. The Type II event began at approximately 2059:15 UT at 200 Mc/sec, three minutes after the start of the explosive phase.

Although Type II events can be coincident in time with a wave, it is not clearly established from this limited study that the generation of this type of flare-associated wave is necessarily related to Type II bursts. Additional Type II records should be studied for the correspondence in time and velocity between individual wave events and Type II bursts. Also such a study should not neglect the existence of sprays of material from a flare. Sprays may have velocities comparable to waves and Type II events and they may be observed at the same time during flares. Consequently, a more complete analysis should compare or relate observed or inferred heights for sprays, waves, and Type II events.

References

Anderson, G.: 1966, Dissertation, University of Colorado, Boulder, Colorado.
Athay, R. G. and Moreton, G. E.: 1961, *Astrophys. J.* **133**, 935.
Becker, U.: 1958, *Z. Astrophys.* **44**, 243.
Dodson, H. W.: 1949, *Astrophys J.* **110**, 382.
Dodson-Prince, H., 1963, ASA-NASA Symp. Phys. of Solar Flares, p. 41.
Dodson, H. W. and Hedeman, E. R.: 1968, *Solar Phys.* **4**, 229.
Harvey, K. L.: 1971, *Solar Phys.*, in press.
Kai, K.: 1970, *Solar Phys.* **11**, 310.
Meyer, F.: 1966, Abstracts of Papers for 123rd American Astronomical Society Meeting.
Meyer, F.: 1968, in K. O. Kiepenheuer (ed.), 'Structure and Development of Solar Active Regions', *IAU Symp.* 35, p. 458–489.
Moreton, G. E.: 1960, *Astrophys. J.* **65**, 494.
Moreton, G. E.: 1961, *Sky and Telescope* **21**, 145.
Moreton, G. E.: 1964, *Astron, J.* **69**, 145.
Moreton, G. E. and Ramsey, H. E.: 1960, *Publ. Astron. Soc. Pacific* **72**, 357.
Ramsey, H. E. and S. F. Smith: 1966, *Astron. J.* **71**, 197.
Richardson, R. S.: 1936, Annual Report of the Director, Mt. Wilson Obs., No. 35, p. 871.
Richardson, R. S.: 1951, *Astrophys. J.* **114**, 356.
Smith, E. v. P.: 1968, *Ninth Nobel Symp.* Anacapri, Italy, p. 137.
Uchida, Y.: 1968, *Solar Phys.* **4**, 30.
Valnicek, B.: 1964, *Bull. Astron. Inst. Czech.* **15**, 207.

12. THE SURGES

CONSTANTIN J. MACRIS

Research Center for Astronomy and Applied Mathematics, Academy of Athens, Athens, Greece

The surge is a fast phenomenon of the solar Chromosphere and the Corona; it is some kind of solar prominence. The prominences extend in the Corona much higher than the top of the Chromosphere. Their behaviour constitutes one of the unsolved problems of solar physics. Questions about their origin, the mechanism of support and excitation, the transfer of energy to the high layers, and their complicated motions, have not been answered in a satisfactory way.

In this paper we deal only with the prominences which are strongly related to the solar flares, have a hot spectrum and reveal violent activity. According to the Menzel-Evans classification they belong to the BSs group. (Menzel and Evans, 1953).

The surges accompany some flares and active sunspots (Zirin, 1966) and seem to be closely associated with the Type II radio bursts. A complete model of a flare with the general characteristics observed in optical, X-ray and corpuscular radiation has been given by Wild (1963).

The appearance of the surge is a phenomenon which has not been studied and explained well. Flares of great importance very frequently do not give rise to surges, while many times some small chromospheric brightenings have as a result the appearance of surges. We do not know why some of the largest flares produce surges, while others do not. The lifetimes of the surges range from 2 min for the smallest, to about $2\frac{1}{2}$ hours for the largest (mean duration 25 min). They are characterised as short-lived prominences, a name which also applies to the sprays, explosions, puffs, loops and coronal rain. All those are prominences associated with active regions, but the loops and coronal rain are produced in the Corona, while the others are produced in the Chromosphere.

The spray is a violent explosion of material from the main mass of the flare. The material moves upwards with velocities between 200 and 2000 km/sec and the great Doppler shifts make the observation of this phenomenon on the disk extremely difficult. They are usually observed at the limb and the ejected matter frequently attains escape velocities. At any rate, the distinction among sprays, surges and ejected prominences is difficult. The mechanism appears to be the same for these events, the difference among them being only in the time of appearance relatively to the evolution of the flare and sometimes in the velocities of the ejected material. Sometimes the term "surge" is restricted to the cases where the material returns to the same point of the solar surface. From another viewpoint, sprays are the events which have velocities higher than 500 km/sec. However, all agree that the surge is a less violent phenomenon; it is related to the end of the flash phase of the flare, but this is not always the case.

When they are seen at the limb, the surges appear bright. Viewed on the disk, they may be brighter or darker than the surrounding Chromosphere; they are often bright in their initial phase, becoming dark in the next phase as their material, moving upwards, expands and becomes colder.

The limb surge consists of streamers or knots of luminous material ejected from active areas of the Chromosphere, usually flares. The opinion is that the ejected knots return to the Sun almost through the same paths, at least in certain surges. Some surges become almost invisible because their matter fades out, either when they ascend or when they descend. This happens because a great part of the material may expand in the Corona or escape in the outer space. This is the case of the surge of September 14, 1966 which is described below. The velocity of ascend of a limb surge may be as high as 500 km/sec or even higher. Line of sight velocities may be detected as Doppler shifts; large line of sight velocities may move part of the surge outside the pass-band of the filter, this being another reason for the fading of the surge.

On the disk surges are usually observed at the wings of the Hα because of the predominately upward motions. The ascending part of the surge is visible in the blue wing, the descending at the red wing, while the stationary parts are visible at the center of the line. Sometimes surges are observed as dark structures in white light. These white light surges are visible only for some minutes during the first phases of the surge.

We observed a remarkable surge on November 2, 1968, which was produced from a 2b flare (Figure 1), within the active region $\varphi = 15°S$, $L = 173°$ (Macris and Alissandrakis, 1970). The flare consisted of two parts (Figure 1a); from one of them a rapidly evolving surge appeared at 1002 UT. The surge began as an emitting feature, but gradually it evolved to an absorbing feature; the change began at first in the lower parts of the surge and propagated to the higher parts (Figures 1b, 1c, 1d, 1e). It showed a clear filamentary structure, parallel to the direction of the motion. The width of the filaments was of the order of 2" of arc; they showed significant changes during the evolution of the event. An interesting remark is that after the flare maximum and while the surge was developing, the area of the underlying part of the flare decreased and that part disappeared 40 min earlier than the other part, indicating that the surge material comes from the flare.

As it is seen on the photographs at the center and the wings of Hα the material of the surge rose near the main spot umbra (North polarity, Figure 1f) followed the magnetic lines of force and descended in the vicinity of a spot of opposite polarity. The maximum height the material reached in its curved trajectory increases gradually. At 1125 UT the maximum height was about 50000 km.

It has been suggested that the event of November 2, 1968, might be a flare loop, not a surge. The criterion is whether the material moved from the flare to the Corona or from the Corona downwards. The fact that the one branch is blue shifted and the other red shifted supports the opinion that the two branches moved in opposite directions, which does not happen in the loops. However, it is possible that these Doppler-shifts are due to the relative orientation of the branches. Since the event was observed in projection on the disk, at a heliocentric angle of 66 deg, it is difficult to

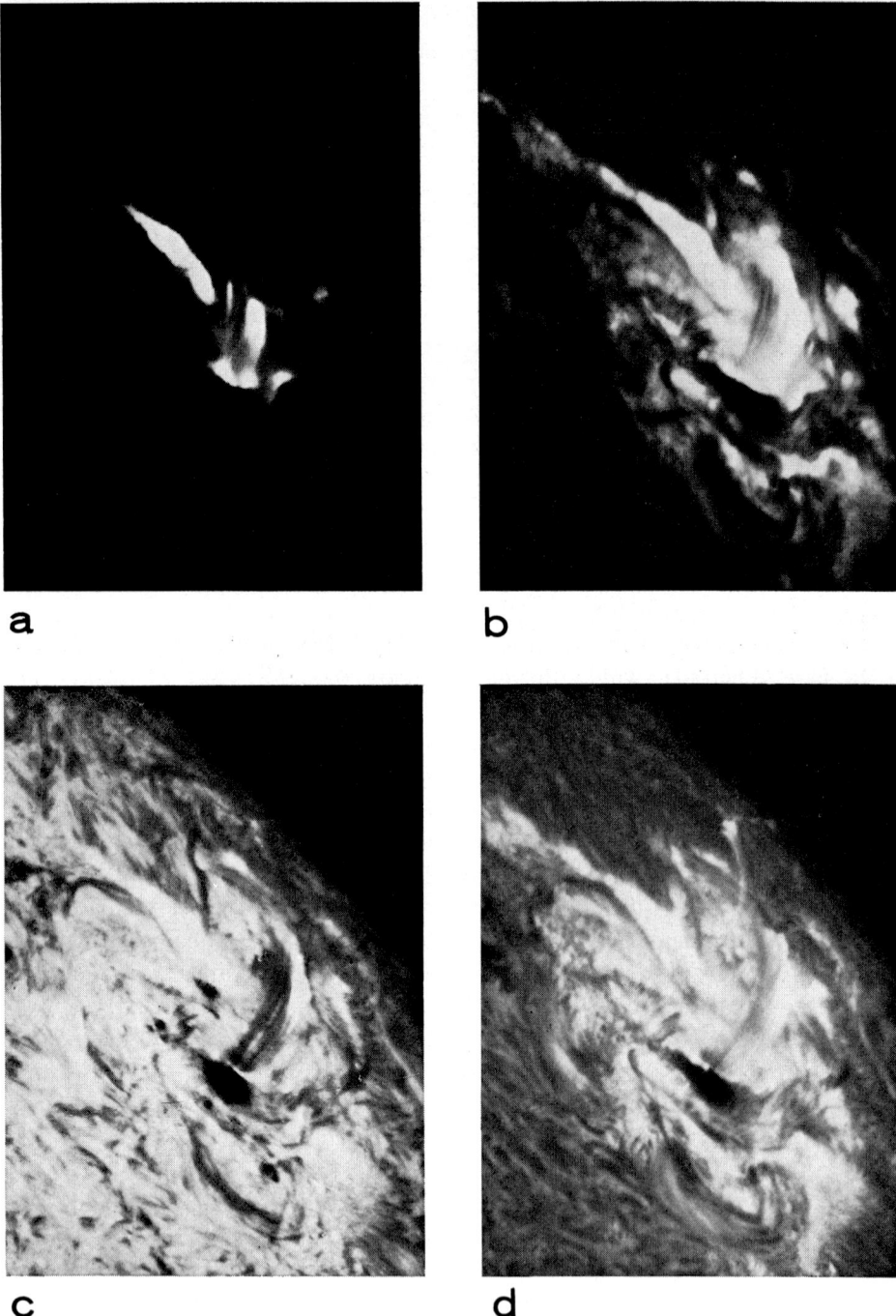

Fig. 1. The development of the flare surge of November 2, 1968. (a) $10^h19^m05^s$ UT, Hα + 0.0 Å; (b) $10^h38^m12^s$ UT, Hα + 0.0 Å; (c) $10^h53^m42^s$ UT, Hα − 0.5 Å; (d) $11^h01^m43^s$ UT, Hα + 0.0 Å; (e) $11^h20^m37^s$ UT, Hα − 0.5 Å; (f) $11^h24^m44^s$ UT, Hα − 1.0 Å.

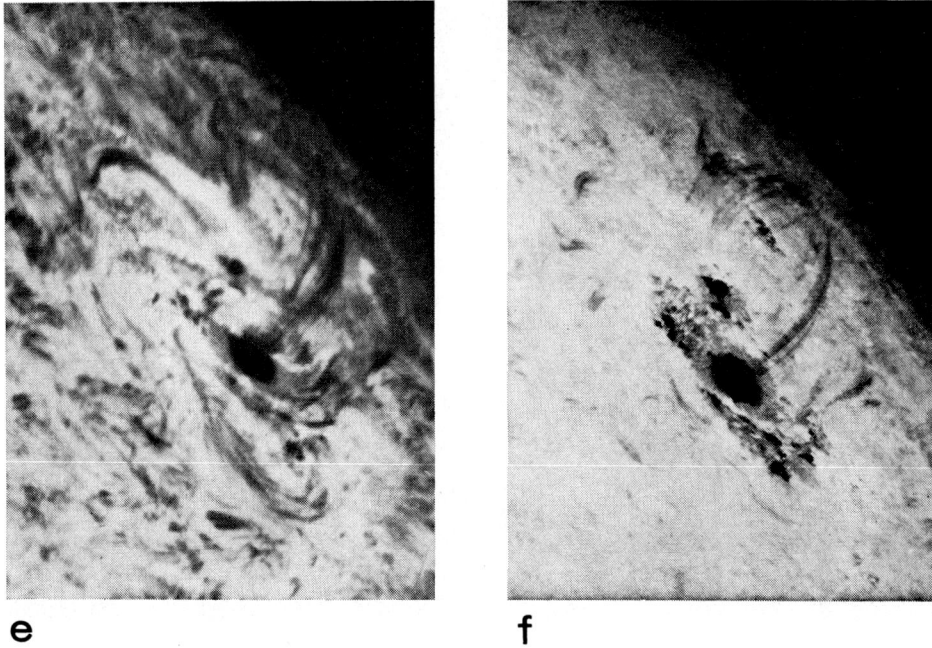

e f

Fig. 1.

be sure about the orientation. On the other hand the event has not any condensations along the filaments, that would help to detect the true direction of the motion. It is not impossible for a surge to have a curved trajectory. This is shown clearly in Figure 4 (see also Kleczek, 1964). In our opinion the event is a surge similar to that of November 4, 1968.

It was mentioned above that there is the opinion that the surge material returns to the same point of the surface from which it was produced. This may happen in very few surges, mainly in the ones with great velocities, but in many cases the material moves and returns in regions of opposite magnetic polarity.

The material of a surge on the top of its trajectory may break up into two branches having opposite direction. A similar and very interesting surge was observed on the 20th March 1966 (Macris, 1967). An Importance 2b flare (Figure 2a) evolved within the active region $\varphi = 21°$ N, $L = 144°$ at 0934 UT. About 15 minutes later, a dark surge with filamentary structure appeared (Figure 2b), which within a few minutes evolved rapidly and at the top broke up into two branches (Figures 2c, 2d). The trunk of the surge, as well as each branch consisted of smaller elements. The influence of the magnetic field on the trajectories of the small elements of the surge is clear. In many cases we observed spiral motions which may also be explained by the presence of magnetic fields.

The velocity of the material, in the two surges we described above, was not too

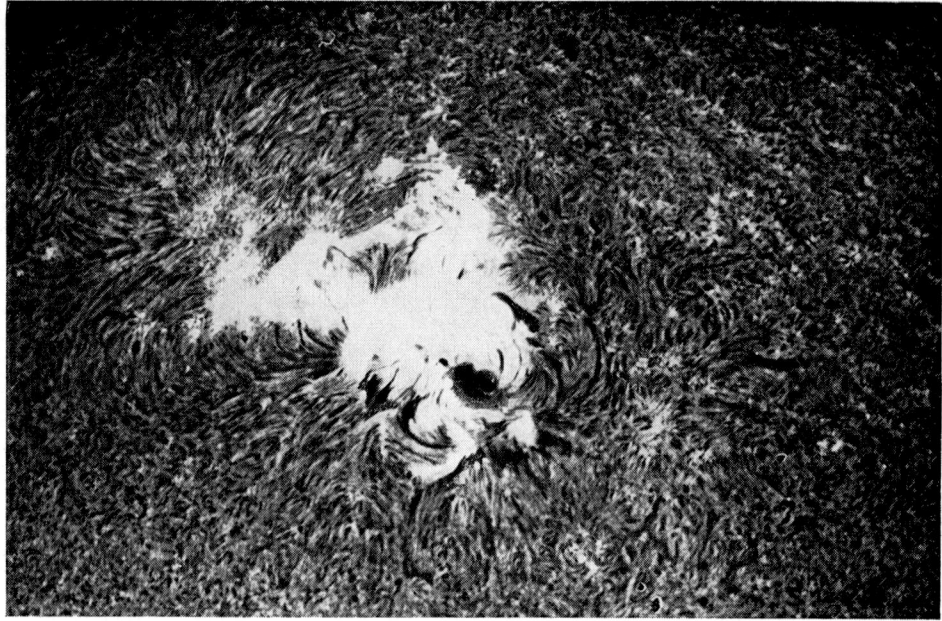

a

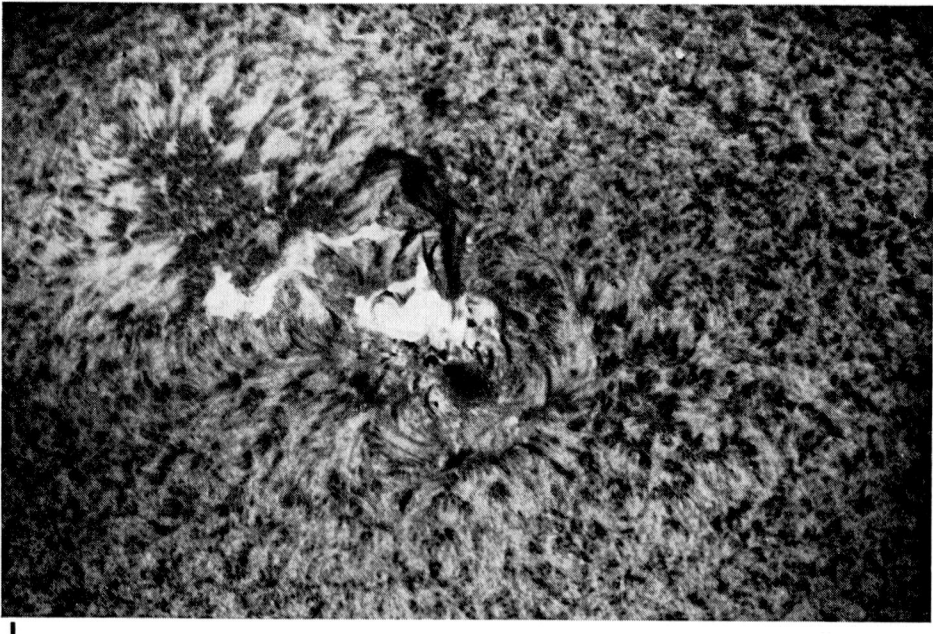

b

Fig. 2. The 2b flare and dark surge of March 20, 1966. (a) The flare of $10^h00^m46^s$ UT at the center of Hα; (b) Ejection of dark material at $10^h02^m19^s$ UT, H$\alpha-0.5$ Å; (c) The upper part of the surge brakes up to two branches moving in opposite direction; their composition of smaller elements is evident. Photographed at $10^h15^m45^s$ at H$\alpha-0.5$ Å; (d) The surge at H$\alpha-1.0$ Å a few minutes later. $10^h17^m30^s$ UT.

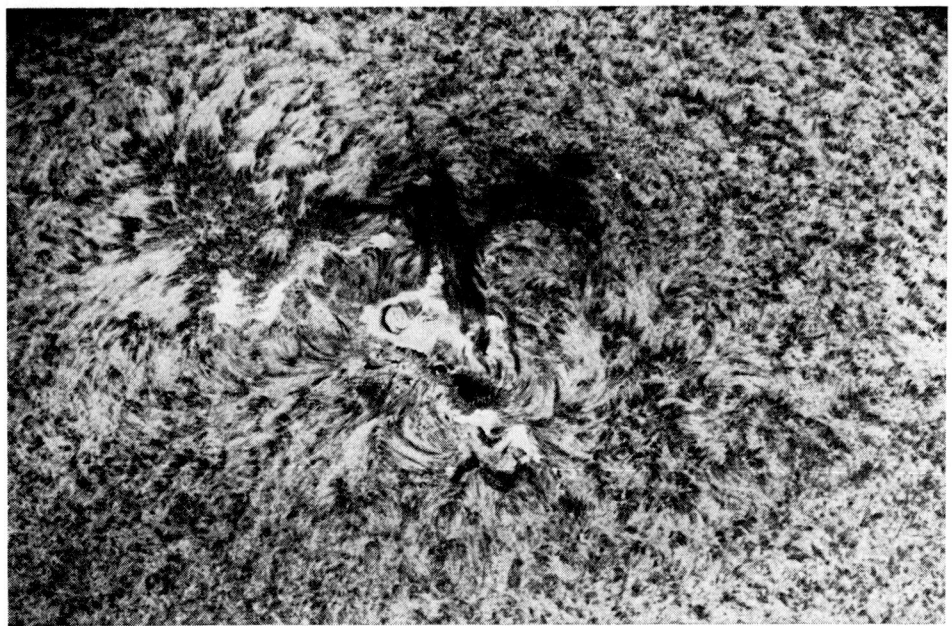

c

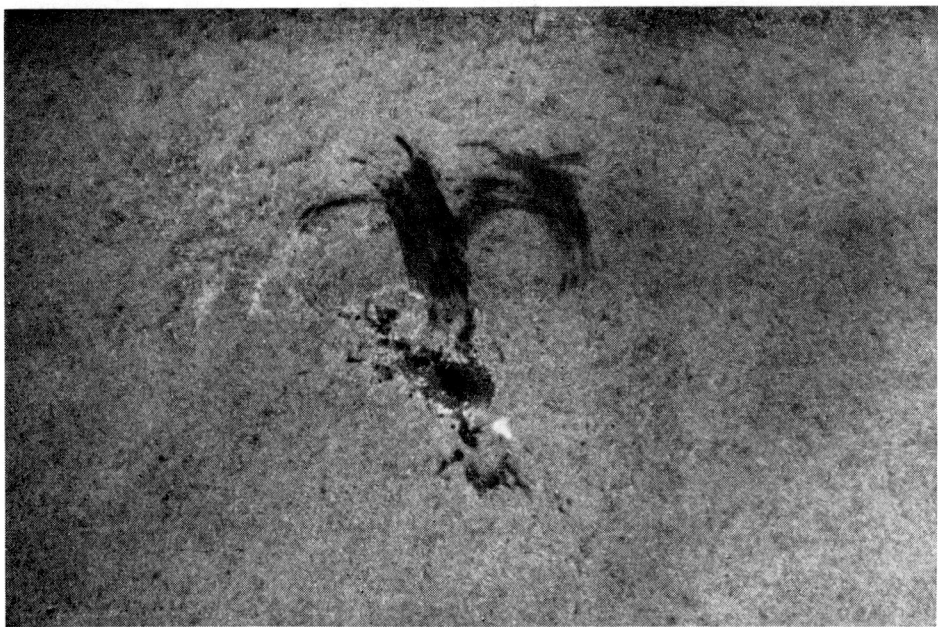

d

Fig. 2.

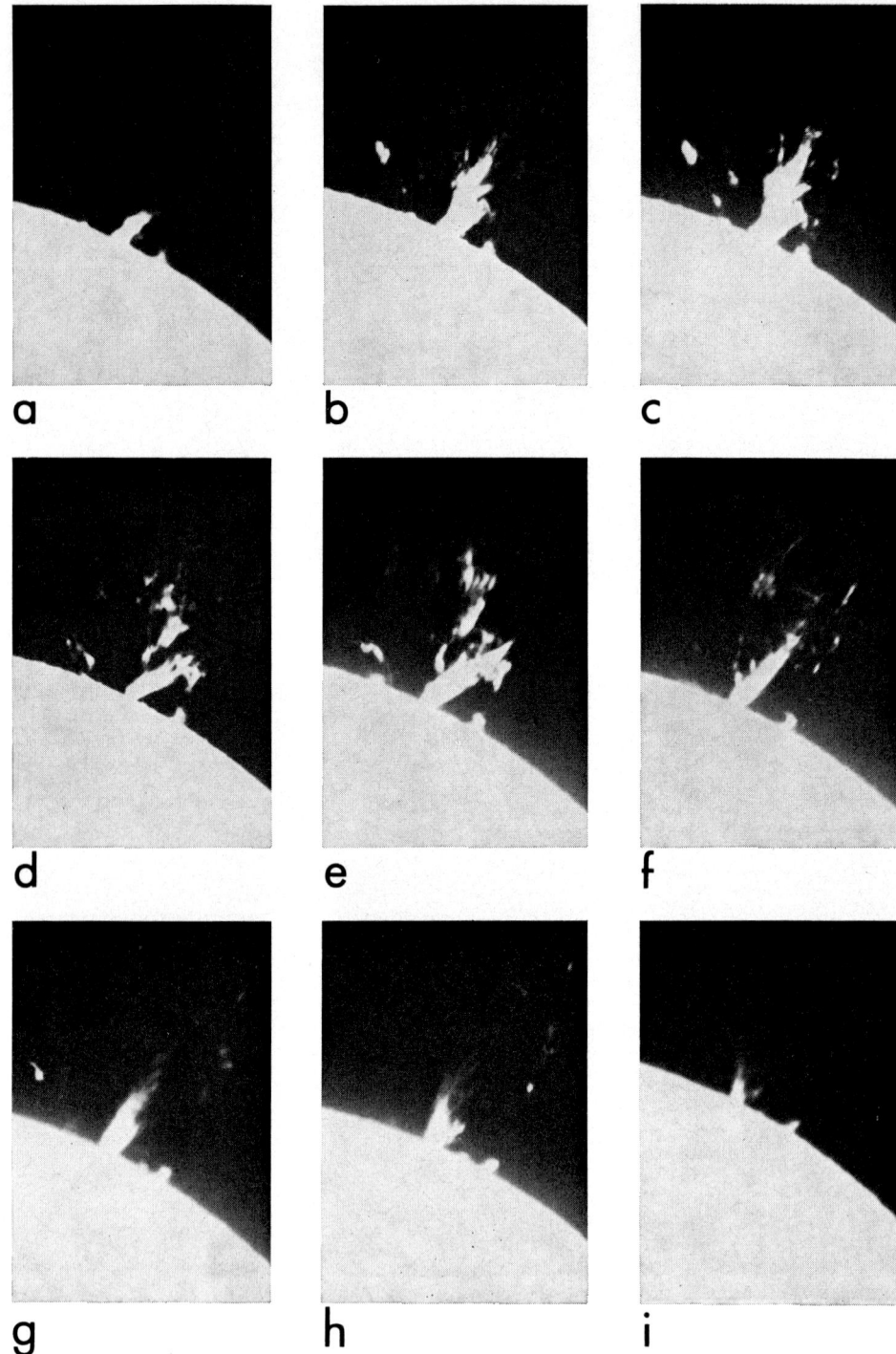

Fig. 3. The flare surge of September 14, 1968. All photographs are at the center of Hα. (a) $10^h16^m20^s$ UT; (b) $10^h18^m45^s$ UT; (c) $10^h19^m10^s$ UT; (d) $10^h21^m35^s$ UT; (e) $10^h24^m30^s$ UT; (f) $10^h30^m06^s$ UT; (g) $10^h35^m10^s$ UT; (h) $10^h37^m44^s$ UT; (i) $10^h39^m40^s$ UT.

high; they both were slow moving surges, making evident the curving of the trajectory of their matter. In case the velocity of the matter of the surge is great the shape of the trajectory of the material is different. The material moves almost radially out in a trajectory of little or no curvature.

As an example we will discuss the flare surge of September 14, 1968 which occurred in the active region 8484 (McMath Observatory) which was already 10° behind the western solar limb (Figure 3); the event was not visible at chromospheric levels, but it was at greater heights. The ejection was very powerful and the velocities of certain parts reached the velocity of escape. Part of the material disappeared into the corona. Figure 3 gives the main phases of the surge; its duration was about 40 min, the main mass reached a projected height of more than 100 000 km. and many radio and X-ray events accompanied the surge (Boischot and Claverier, 1968).

The surge of September 14, 1968 was a fast moving surge with a trajectory almost vertical to the disk and most of the surge material returned through the same path to the region where it originated after the end of the cause that had produced it.

The surge of November 4, 1968 was a slow moving surge. The active region, $\varphi = 15°$ S, $L = 173°$, which produced the event of November 2 was on the western limb when a flare gave rise to a surge; its evolution is shown in Figure 4. The ejected material moved from South to North and fell on the solar surface 8° to the North of the initial point of ejection. Because of the small velocity of the gases and the influence of the magnetic field, the material followed the magnetic lines of force in the Corona. At the point where the gases fell back there was no center of activity, but it seems very probable that it had a strong magnetic field.

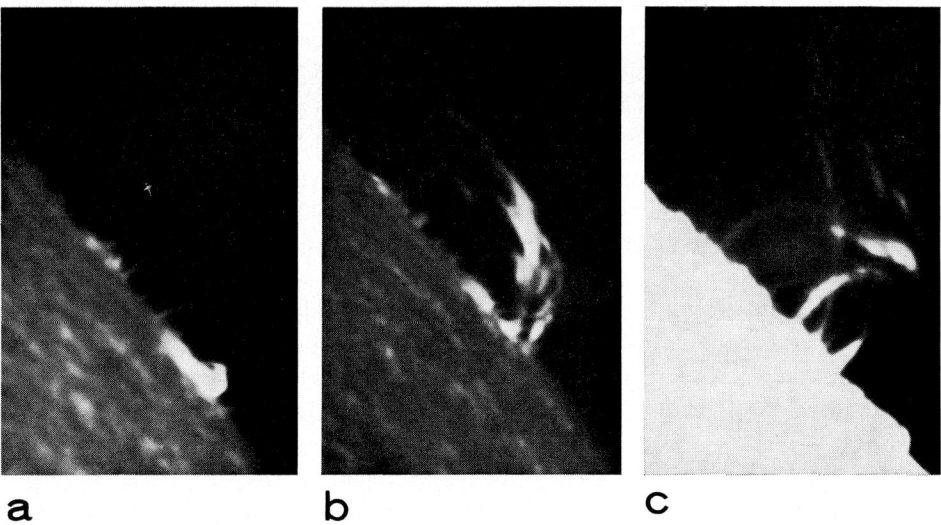

Fig. 4. The surge of the November 4, 1968. (a) $09^h35^m14^s$ UT, Hα + 0.0 Å; (b) $09^h42^m59^s$ UT, Hα + 0.0 Å; (c) $10^h58^m32^s$ UT, Hα + 0.0 Å.

In slow moving surges the elements of the material move along the lines of force of a bipolar magnetic field. The whole structure and geometry of the active regions are controlled by the configuration and strength of their magnetic fields.

The surges we studied above show motions only inside one active region. However, there are also some surges in which the material follows the magnetic lines of force to great heights in the Corona, thus connecting active regions separated by large distances on the solar disk. Such an event was observed on January 29, 1968 (Macris, 1968). In the eastern extreme part of the active region $\varphi = 13°N$, $L = 169°$ (center I, Figure 5a) a flare of Importance 1b appeared at 0806 UT (Figure 5b). Almost simultaneously a dark surge occurred (Figure 5c). The characteristic point was that within a few minutes the main mass of the surge, which was moving eastwards, diverged 45° to the North-East. At the same time, well defined individual streamers became clearly visible (Figure 5d). The two primary branches of the streamers moved independently; the first towards the active center 2 ($\varphi = 21°N$, $L = 128°$) and the other towards the active center 3 ($\varphi = 24°N$, $L = 116°$) (Figure 5e). The ejected gases extended to great distances from the point of ejection. The primary lower branch reached a distance of about 20° (center 2, Figure 5d, 5e) while the higher branch extended beyond a distance of 35° of heliographic longitude (center 3, Figure 5d, 5e). Figure 5f shows the conditions near the end of the event. The streamers extend up to the active centers 2 and 3, not beyond them; this provides evidence that the material moved in fact towards those centers and it was not just a projection effect.

The 45° change in direction of the surge material and the orientation of the dark filaments may be attributed to an interconnection of the magnetic fields of the active centers 1, 2 and 3. The ejected material is compressed along the lines of force under the pressure of the colliding fields. The streamers, in which the dark matter slits, follow the lines of force connecting the three active centers through the Corona. The shape and the way of development of the above event are in good agreement with Babcock's theory; large magnetic loops extend into the Corona and join the Centers 1–2 and 1–3 (Figure 5e). The pronounced curvature of the trajectories reveals the rather remarkable extent, to which the magnetic fields of the active regions are interconnected. This can not be deduced from the morphology of the static regions.

After the description of the above surges it is important to examine whether, in the course of such events, chromospheric matter is ejected, or the matter comes from coronal condensations. The material of the Corona, under the influence of the magnetic field, may cool locally so as to give Hα emission. According to Giovanelli (1959), sometimes the surges appear at first in emission outside the solar limb, while on the disk they become visible in absorption only after 10 min; that is, the ejection of a surge may begin before the temperature is low enough to give rise to Hα emission or absorption. This mechanism is accepted for the formation of loop prominences, i.e. the sparse concentration above an active region. There is, however, much objection concerning the formation of the surges. The observations, mainly those at the limb, reveal clearly the ejection of material from the Chromosphere to the Corona.

When we observe a surge with the line-shifter of an Hα filter, we find essential

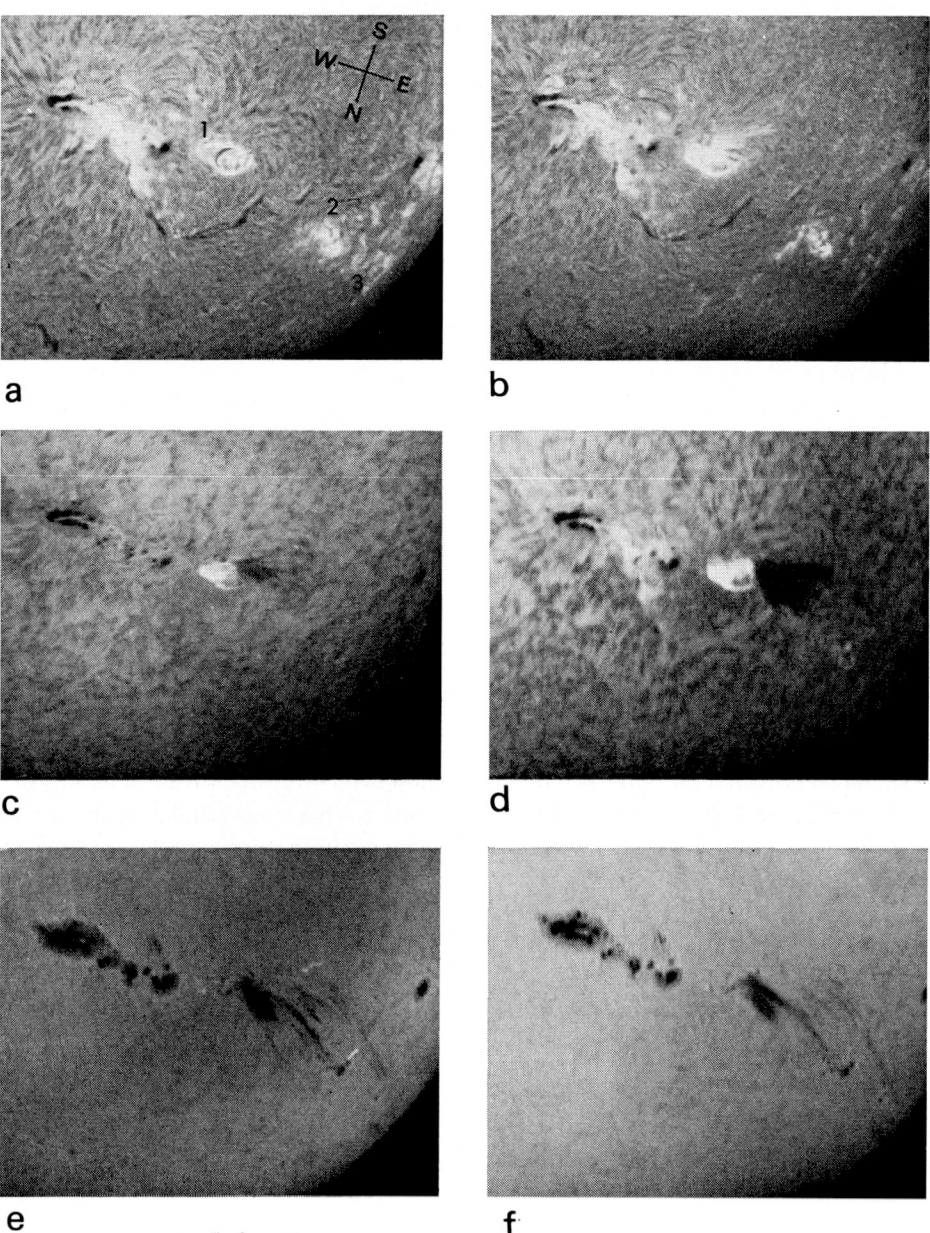

Fig. 5. The flare surge of January 29, 1968. The weather conditions were not favorable on that day. The quality of the pictures is not good because most of them have been taken through clouds. (a) The active regions $\varphi = 13°$N, $L = 169°$ (1), $\varphi = 21°$N, $L = 128°$ (2), $\varphi = 24°$N, $L = 116°$ (3) at $06^h52^m10^s$ at the center of Hα; (b) The flare near the beginning at $08^h09^m10^s$ UT; (c) The dark surge at $08^h13^m40^s$ UT. The picture was taken at $+0.5$ Å; (d) The main mass of the surge at $08^h20^m00^s$ UT, H$\alpha - 0.5$ Å; (e) The 45° change of direction of the main mass of the surge. Filamentary structures extend as far as region 2. H$\alpha + 1.0$ Å, at $08^h25^m58^s$ UT; (f) Picture at H$\alpha + 1.0$ Å, $08^h27^m45^s$. Part of the dark matter arrives at center 3, 35° heliographic longitude away from center I.

differences in the form of the surge when we compare observations at the center and the wings of Hα. If we have a real movement of material upwards, the photographs taken at the beginning of the event at the violet wing of Hα(-0.5 Å), should have significant differences with photographs taken at the center and the red wing of the Hα($+0.5$ Å). This actually happens (see Figures 2a, 2b) and therefore we have a substantial evidence that the material moves up from chromospheric layers.

The disturbances in the Corona may be explained as produced by the energy which is released in the form of corpuscular radiation, produced by the ejection of material during the fast phenomena (surges, sprays, puffs). These disturbances have been observed optically by the direct emission or absorption in the Hα, due to a change in the physical conditions.

The spray and puff events indicate clearly, that matter from the lower layers of the solar atmosphere is thrown up and if the velocity of the ejection is high enough, the matter either escapes or replenishes the Corona. The surges, that are in fact the same phenomena as the sprays and puffs, but with smaller ejection velocities, follow the same mechanism of production. Consequently, appears most likely that during these events the chromospheric material ascends in the Corona, guided by the lines of force of the local magnetic field.

Acknowledgements

I wish to express my thanks to the Scientific Committee of NATO which, with its Grants, provided me with the necessary equipment with which I was able to work and make observations of the Sun. The author is also deeply indebted to Mr C. Alissandrakis for his valuable assistance and cooperation in this work.

References

Boischot, A. and Clavelier, B.: 1968, *Ann. Astrophys.* **32**, 445.
Giovanelli, R. G.: 1959, *Suppl. Nuovo Cimento* **13**, 290.
Kleczek, Z.: 1964, *Bull. Astr. Inst. Czech.* **15**, No 4, 123.
Macris, C. J.: 1968, *Solar Phys.* **5**, 361.
Macris, C. J. and Alissandrakis, C.: 1970, World Data Center A, Upper Atmosphere Geophysics, Report UAG-8, Part I (ed. by J. V. Lincoln), p. 54.
Menzel, D. H. and Evans, J. W.: 1953, *Atti del'Convegno Volta*, Roma Academia Lincei, p. 119.
Wild, J. P.: 1963, *IAU Symp.* **16** (ed. by J. W. Evans), p. 122–125.
Zirin, H.: 1966, *The Solar Atmosphere*, p. 326.

13. RELATIONS BETWEEN THE AREAS INDEX AND DIFFERENT PHENOMENA IN THE CHROMOSPHERE, THE CORONA AND THE INTERPLANETARY SPACE

JOHN XANTHAKIS

Research Center for Astronomy and Applied Mathematics, Academy of Athens, Athens, Greece

1. Introduction

It is well known that the physical conditions prevailing in the solar corona depend to a great extent on solar activity. The shape, the dimensions and the brightness of the corona, for example, are different during the maximum, the minimum or the intermediate phases of solar activity. Furthermore, the evolution of different other phenomena in the chromosphere and the Corona, as it is, for example, the emission of the green coronal line at 5303 Å or the emission of the radio waves follows more or less closely the evolution of solar activity within the eleven-year sunspot cycle.

For the comparison of solar activity with the different phenomena of the chromosphere, the corona, the interplanetary space and the ionosphere solar activity has been expressed, as a rule, with the help of the relative sunspot numbers. In some cases, as for example, in the case of the solar radio emission in the frequency range $1000 \leqslant f \leqslant 10000$ MHz, this comparison leads indeed to satisfactory results, i.e high correlation coefficients. In many other cases, on the contrary, the corresponding correlation coefficients are more or less low.

This result led us to the conclusion that another, more representative index of solar activity in the photosphere than the relative sunspot numbers should be used. To this end the areas index I_a has been introduced which is defined with the help of the relation

$$I_a = \tfrac{1}{2}[\sqrt{A} + \sqrt{f}] \tag{1}$$

where A and f represent respectively, the areas of the sunspots and faculae corrected for foreshortening. From relation (1) one can see that the areas index I_a represents in some way the importance of the activity centers of the solar photosphere. Doubtless the computation of the areas index I_a is more complicated than that of the relative sunspot numbers. This index however, presents many advantages as compared with the relative sunspot numbers. The areas index I_a, for example, has a more obvious physical significance than the relative sunspot, because it depends directly on the areas of the sunspots and faculae i.e. on two phenomena which depend to a great extent on the solar magnetic field strength. On the other hand the comparison of the variation of the areas index I_a with the variation of different phenomena in the chromosphere and the corona during the sunspot cycle leads to higher values of the correlation coefficient than that of the relative sunspot numbers. Moreover, in some

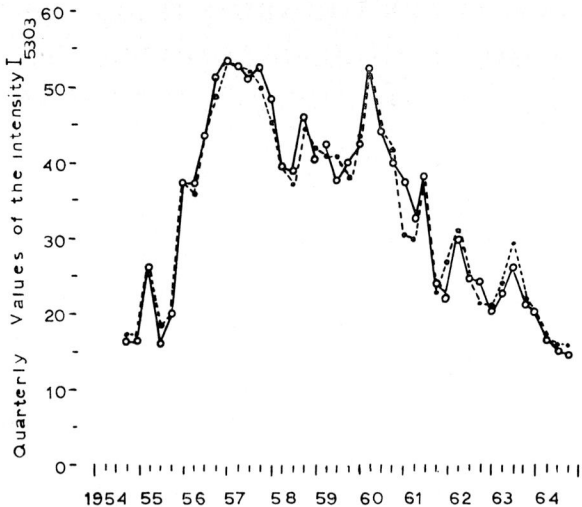

Fig. 1. The continuous line represents the quarterly values of the intensity I_{5303} of the green coronal line at 5303 Å observed at Pic-du-Midi and the dashed line gives the values of the same quantity computed with the help of Relation (2).

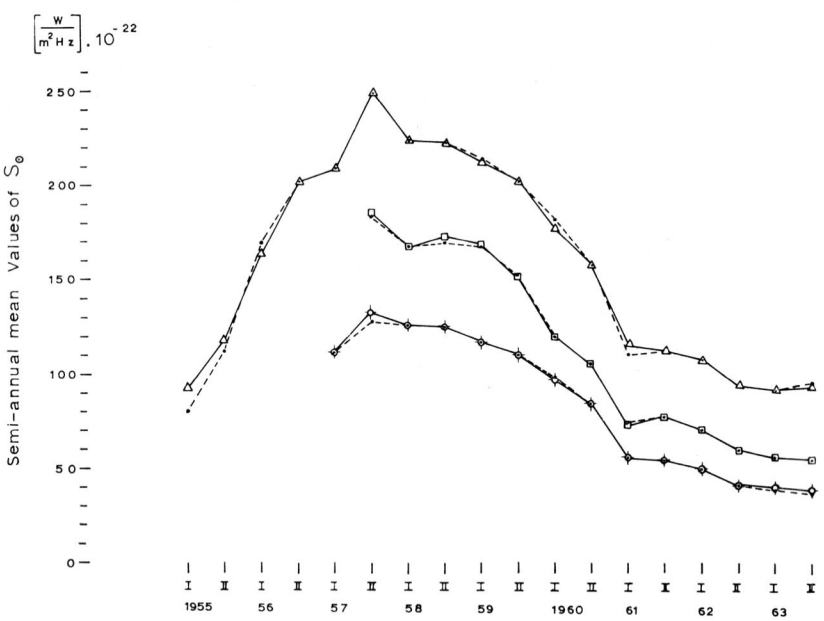

Fig. 2a. The continuous lines represent the semi-annual values of the intensity $S_\odot$ of the solar radio emission in the frequencies 2800 MHz ($\triangle$), 2000 MHz ($\square$) and 1000 MHz ($\diamondsuit$) as given by the observations at different stations (Xanthakis, 1969), and the dashed lines give the values of the same quantities computed with the help of Relation (3).

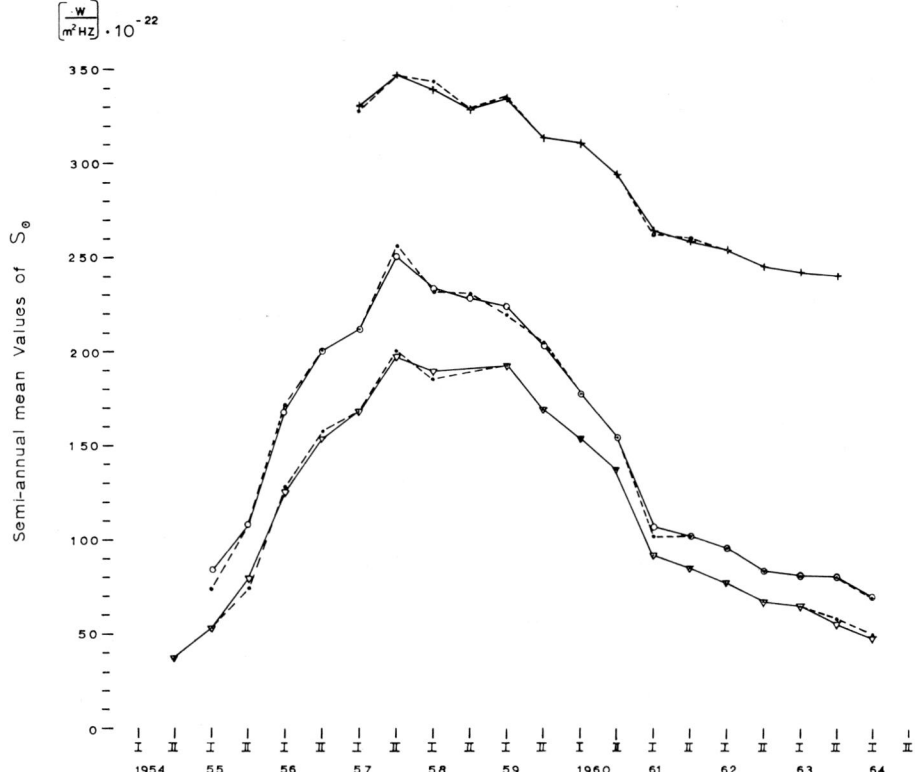

Fig. 2b. The continuous lines represent the semi-annual values of the intensity $S_\odot$ of the solar radio emission in the frequencies 9400 MHz (+) 3730 MHz (○) and 1500 MHz (△) as given by the observations at different stations (Xanthakis, 1969) and the dashed lines give the values of the same quantities computed with the help of Relation (3).

cases such comparisons lead to simple analytical relations between these phenomena, the areas index I_a and a number of supplementary factors, like the flare index I_f and the number of proton flares N_{PF} corresponding to different phases of the sunspot cycle.

In the present paper a study will be made of some correlations or analytical relations between the areas index I_a of solar activity and different phenomena in the chromosphere, the corona and the interplanetary space.

In a previous paper (Xanthakis, 1969) has been shown that the values of the intensity I_{5303} of the green coronal line at 5303 Å observed at Pic-du-Midi during the time interval 1957–1964 can be satisfactorily expressed as a function of the areas index I_a and the number of proton flares N_{PF} with the help of the relation

$$I_{5303} = I_0 + K(1 + \sqrt{N_{PF}}) I_a, \qquad (2)$$

where $K=0.165$ is a constant coefficient and I_0 represents the value of I_{5303} corresponding to the epoch of the quiet Sun i.e. to a period of time during which $I_a=0$.

In Figure 1 the continuous line represents the quarterly values of I_{5303} observed at Pic-du-Midi while the dashed line gives the values of the same quantity computed with the help of relation (2).

In Figures 2a and 2b the continuous lines represent the semi-annual values of the intensity $S_\odot$ of the solar radio emission in the frequency range $1000 \leqslant f \leqslant 10000$ MHz as given by the observations at different stations while the dashed lines give the values of the same quantity computed with the help of relation

$$S_\odot = S_0 + K[1 + 10^{-1}\sqrt{I_f}]\, I_a. \tag{3}$$

In this relation S_0 represents the value of $S_\odot$ corresponding to the value $I_a = 0$, I_f is the flare index and k is a coefficient whose value depends on the frequency of the radioemission considered.

In the following it is shown that two more phenomena related to solar activity can be analytically expressed in a more or less satisfactory way as functions of the areas index I_a and the number of proton flares N_{PF}.

2. The Coronal Radio Diameter During the Solar Cycle

Leblanc (1969) has compared the mean quarterly values of the Coronal radio diameter during the time interval 1957 II to 1964 IV with the corresponding values of the intensity I_{5303} of the green coronal line averaged around the solar disc and found that these two quantities are well correlated during the time interval considered and

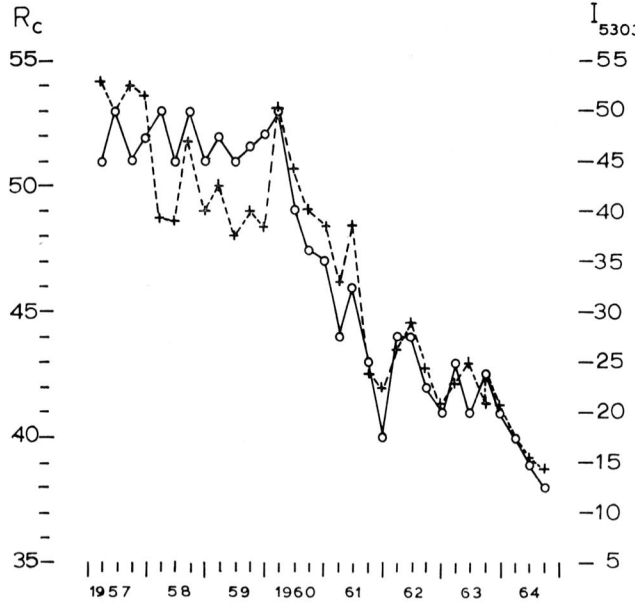

Fig. 3. The continuous line represents the values of the equatorial diameter R_c of the radio corona and the dashed line gives the corresponding values of the intensity I_{5303} of the green coronal line at 5303 Å (Leblanc, 1969).

especially during the decay and minimum of the solar cycle. This is shown in Figure 3 where the continuous line represents the values of the equatorial diameter R_c of the radio corona while the dashed line gives the corresponding values of I_{5303}. From this Figure we see that the correlation between these two quantities is very high during the descending branch of sunspot cycle No. 19 i.e. during the time interval 1960 I–1964 IV.

During the years of maximum solar activity i.e. during the time interval 1957 II–1959 IV on the contrary, the variation of these two quantities presents noticeable differences. During this time interval, in fact the equatorial diameter R_c of the radio corona presents a smoothed maximum while the intensity I_{5303} of the green coronal line shows two distinct maxima, the first in 1957 II and the second in 1960 II.

If instead of the intensity I_{5303} of the green coronal line we compare the equatorial diameter R_c the radio corona with the areas index I_a, which also presents a smoother maximum during the time interval 1957 II–1959 IV, then the correlation becomes more satisfactory. On the basis of the data given in Table I, one can find, in fact, that the correlation coefficients between the equatorial diameter R_c of the radio corona and the intensity I_{5303} of the green coronal line on the one hand and the areas index I_a on the other hand are respectively equal to $r_{R_c, I_{5303}} = 0.925$ and $r_{R_c, I_a} = 0.96$.

At the time of minimum solar activity the equatorial diameter of the radio corona was equal to $38' \pm 1'$. (Leblanc, 1969). If we accept that in an ideal epoch during which $I_a = 0$ i.e. no sunspots or faculae are present on the Sun the equatorial diameter of the radio corona assumes the value $R_0 = 34'.5 \pm 1'$ then the following relation (4) can be easily found on the basis of the data given in Table I,

$$R_c - R_0 = KI_a, \qquad (4)$$

TABLE I

Mean quarterly values of the equatorial diameter (R_c) of the radio corona, the intensity (I_{5303}) of the grean coronal line at 5303 Å and the areas index (I_a).

	R_c	I_a	I_{5303}		R_c	I_a	I_{5303}
1957 II	51	51.3	53.2	1961 I	47	33.0	38.4
III	53	52.4	50.2	II	44	34.7	32.8
IV	51	56.7	52.4	III	46	36.0	38.4
1958 I	52	52.0	52.0	IV	43	26.9	24.7
II	53	51.6	39.6	1962 I	40	30.3	22.5
III	51	51.9	39.0	II	44	32.4	26.0
IV	53	47.7	47.2	III	44	26.9	28.8
1959 I	51	52.6	40.7	IV	42	23.0	24.5
II	52	47.5	42.6	1963 I	41	20.6	20.4
III	51	51.5	37.5	II	43	28.2	23.0
IV	51.5	48.4	39.9	III	41	25.5	25.1
1960 I	52	48.3	38.7	IV	42.5	20.3	21.2
II	53	45.6	52.5	1964 I	41	18.6	20.6
III	49	45.9	44.2	II	40	15.2	17.1
IV	47.5	39.2	40.2	III	39	11.6	15.3
				IV	38	11.5	14.8

where $R_0 = 34'.5$ and $K = \frac{1}{3}$. This relation represents the variation of the equatorial diameter of the radio corona as a function of the areas index I_a with an accuracy $(1 - \sigma/\overline{(R_c - R_0)}) \, 100\% = 90\%$.

In Figure 4 the continuous line represents the 'variable component' $R_c - R_0$ (quarterly values) of the equatorial diameter R_c of the radio corona during the sunspot cycle No. 19 (1957 II–1964 IV) while the dashed line represents the values of the quantity $KI_a (K = \frac{1}{3})$.

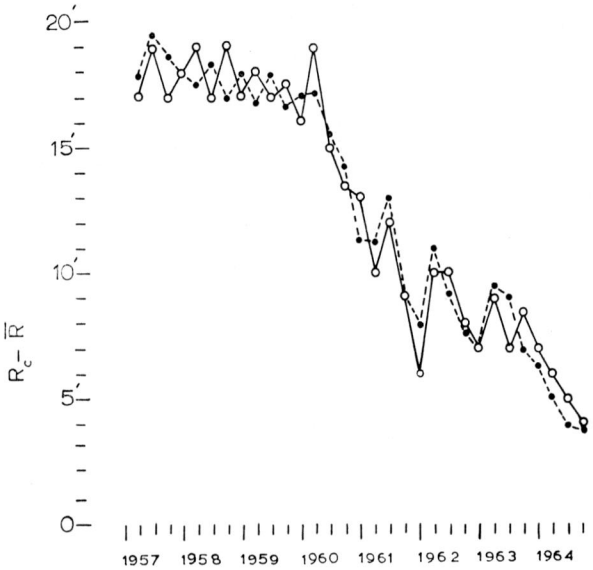

Fig. 4. The continuous line represents the 'variable component' $R_c - R_0$ (quarterly values) of the equatorial diameter R_c of the radio corona and the dashed line gives the values of the quantity $KI_a \, (K = \frac{1}{3})$.

A comparison of Relation (2) and (4) can easily explain the different behavior of the quantities $R_c - R_0$ and $I_{5303} - I_0$ during the time interval 1957–1960. The 'variable component' $R_c - R_0$ of the equatorial diameter of the radio corona seems to depend, according to Relation (4), only on the areas index I_a and, therefore, presents the smoothed maximum shown by I_a during the same interval time. The 'variable component' $I_{5303} - I_0$ of the intensity of the green coronal line, on the contrary, seems to depend, according to Relation (2) both on the areas index I_a and the proton flare number N_{PF} and, therefore, presents the two distinct maxima shown by N_{PF} during the years 1957 and 1960.

3. Cosmic Rays

It is well known (Forbush, 1958) that between the sunspot numbers and the cosmic-rays intensity there is an inverse correlation. A comparison of the curves representing the variation of these quantities shows moreover that there is also an hysteresis phenomenon between these two curves (Balasubrahmanyan, 1969).

As this hysteresis phenomenon is not easy to understand on the basis of the theory of cosmic-ray modulation by the solar wind the question arises whether this hysteresis appears because of the use of the sunspot numbers as an index of solar activity.

Balasubrahmanyan (1969) using the ground neutron monitor data N_e for the sunspot cycle No. 19 showed that Bartel's index A_p correlates with the cosmic-ray intensity without pronounced phase lag. This is shown in Figure 5 where the dashed line represents the Mt. Washington neutron monitor data N_e during the sunspot cycle No. 19 while the continuous line gives the corresponding values of the Bartel's index A_p.

In the following a comparison will be made between the cosmic-ray data during the sunspot cycle No. 19 and the areas index I_a. As shown in Figure 6.

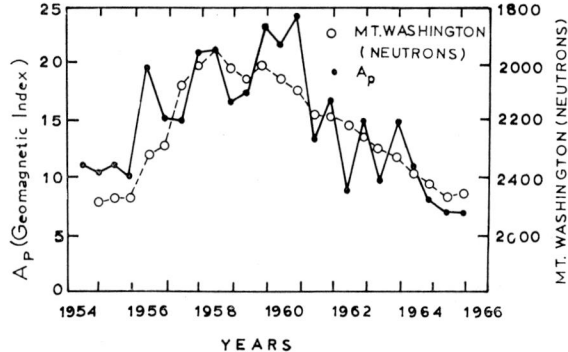

Fig. 5. The dashed line represents the Mt. Washington neutron monitor data N_e and the continuous line gives the corresponding values of the Bartel's index A_P. (Balasubrahmanyan, 1969).

Fig. 6. The continuous line represents the Mt. Washington neutron monitor data N_e and the dashed line gives the corresponding values of the areas index I_a.

The hysteresis phenomenon is also present in the case of the areas index I_a (dashed line) like in the case of the sunspot numbers with the only difference that in this case the phase lag is approximately equal to one year. For this reason if we compare the values of the neutron monitor data N_e in the year t with the values of the areas index I_a corresponding to the year $t-1$ the phase lag almost disappears and the correlation coefficient assumes satisfactory values. This is shown in Figure 7 where the continuous line represents the values of the Mt. Washington neutron monitor data N_e corre-

Fig. 7. The continuous line represents the Mt. Washington neutron monitor data N_e and the dashed line gives the values of the areas index I_a. The time scale corresponding to I_a is shifted toward the left of the time scale corresponding to N_e by one year.

sponding to the time interval 1955–1965 while the dashed line gives the values of the areas index I_a corresponding to the time interval 1954–1964. The time scale corresponding to the areas index I_a is shifted toward the left of the time scale corresponding to the cosmic ray data N_e by one year. The correlation coefficient between the cosmic ray data N_e and the areas index I_a corresponding to the same year t is equal to 0.84. If, on the contrary, we correlate the value of N_e corresponding to the year t with the values of I_a corresponding to the year $t-1$ the correlation coefficient becomes equal to 0.98.

It is worth mentioning that also between the Bartel's index A_p and the areas index I_a exists a phase lag equal to one year. This is shown in Figure 8 where the continuous line represents the semi-annual values of the Bartel's index A_p during the sunspot cycles Nos. 18 and 19 (1944–1965) while the dashed line gives the values of the areas index I_a. In this figure too, like in Figure 7, the time-scale corresponding to I_a is shifted toward the left of the time scale corresponding to A_p by one year. In this way one can explain why no noticeable phase lag between the curves representing the values of N_e and A_p exists.

In Figure 9 the values of the cosmic-ray data N_e corresponding to the year t are correlated with the values of the areas index I_a corresponding to the year $t-1$. Symbols ✧ correspond to the years of the ascending branch of sunspot cycle No. 19 (1955 I–1957 II) while the small circles correspond to the years of the descending branch of the same cycle.

The small deviation from the mean line shown during the year of the ascending branch are very probably due to the influence of the proton flares the number of which during the years 1955 and 1956 is smaller than the mean value corresponding to the sunspot cycle No. 19 while during the years of maximum (1957 I–1957 II) is higher than the same mean value. In fact, if we consider the differences $C-[11(I_a)_{t-1}+(N_e)_t]$ we find out that these differences vary during the sunspot cycle No. 19 almost in

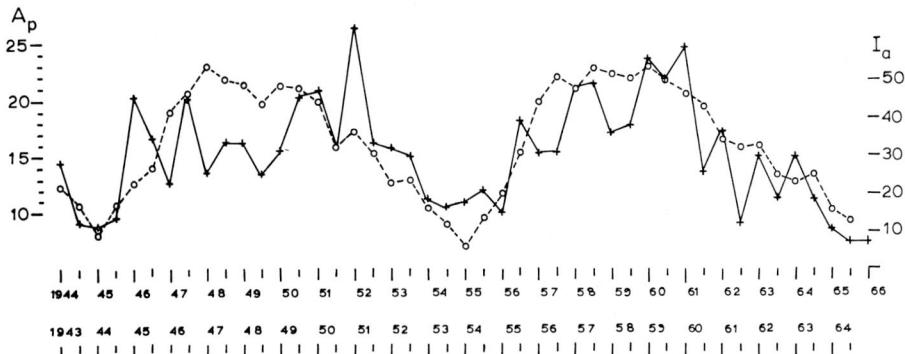

Fig. 8. The continuous line represents the semi-annual values of the Bartel's index A_P and the dashed line gives the values of the areas index I_a. The time scale corresponding to I_a is shifted toward the left of the time scale corresponding to A_P by one year.

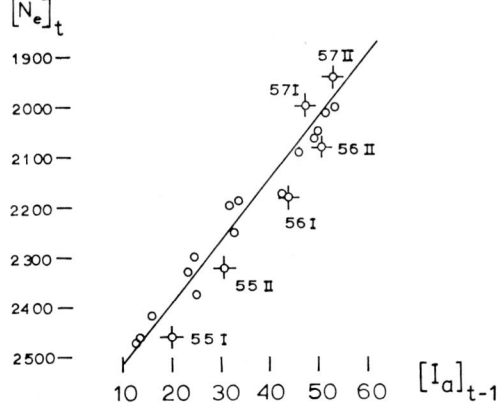

Fig. 9. Correlation between the Mt. Washington neutron monitor data N_e corresponding to the year t and the areas index I_a corresponding to the year $t-1$. Symbols ✧ correspond to the years of the ascending branch of sunspot cycle No. 19 (1955 I–1957 II) while open circles correspond to the years of the descending branch of the same cycle.

the same way with the number of the proton flares N_{PF}. This is clearly shown by Figure 10 where the dashed line represents the number N_{PF} of the proton flares (semi-annual values) while the continuous line gives the corresponding values of the difference $C-[11(I_a)_{t-1}+(N_e)_t]$ smoothed with the help of the formula $(a+2b+c/4)$. The values of N_{PF} and N_e are given in Table II.

This remark led us to investigate whether a correlation between the quantities N_e, I_a and N_{PF} of the form

$$[N_e]_t + a[I_a]_{t-1} + b[N_{PF}]_{t-1} = C, \tag{5}$$

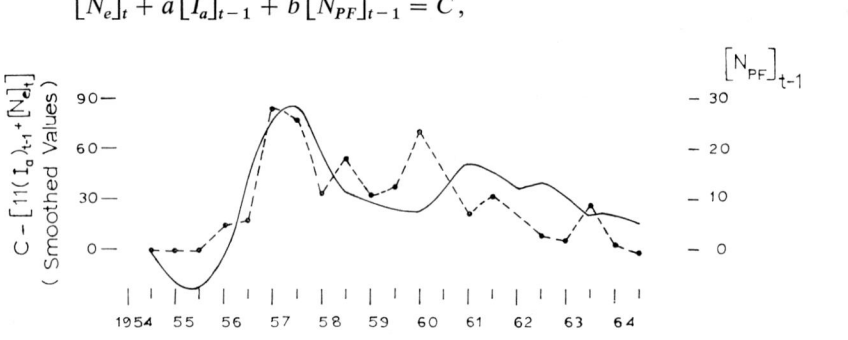

Fig. 10. The dashed line represents the number of the proton flares $(N_{PF})_t$ and the continuous line gives the corresponding values of the difference $C-[11(I_a)_{t-1}+(N_e)_t]$ smoothed with the help of the formula $(a+2b+c/4)$.

TABLE II

Values of the Mt. Washington neutron monitor data N_e the proton flare numbers (N_{PF}) and the areas index (I_a)

	N_e		I_a	N_{PF}	$11 I_a + 3 N_{PF}$	N_e^{comp}
1955 II	2460	1954 II	14.0	0	154	2476
6 I	2460	5 I	19.8	0	208	2422
II	2320	II	30.8	0	339	2291
7 I	2140	6 I	44.4	5	503	2133
II	2080	II	51.1	6	580	2050
8 I	2000	7 I	48.0	28	612	2018
II	1940	II	53.7	26	669	1961
9 I	2010	8 I	51.9	11	604	2026
II	2050	II	50.7	19	615	2015
60 I	2000	9 I	54.1	11	628	2002
II	2060	II	50.0	13	589	2041
61 I	2090	60 I	46.5	24	584	2046
II	2170	II	43.3	15	521	2109
62 I	2185	61 I	34.2	7	397	2233
II	2220	II	32.1	11	386	2244
63 I	2250	62 I	32.9	7	383	2247
II	2300	II	25.0	3	284	2346
64 I	2330	63 I	23.7	2	267	2363
II	2375	II	22.8	9	278	2352
65 I	2415	64 I	16.2	1	176	2454
II	2470	II	13.0	0	143	2487

where C is a constant representing the value of N_e for $I_a=0$ and $N_{PF}=0$ could be established. The data given in Table II show, in fact, that such a relation exists and has the form

$$C - [N_e]_t = [11 I_a + 3 N_{PF}]_{t-1},\qquad(6)$$

where $C=2634$. Relation (6) represents the data given in Table II with an accuracy $(1-\sigma/\overline{C-N_e})\,100\%=92.7\%$.

The same results are also shown in Figure 11 where the continuous line represents the values of the 'variable component' $C-[N_e]_t$ given by the observations while the dashed line gives the values of the same quantity computed with the help of Relation (6).

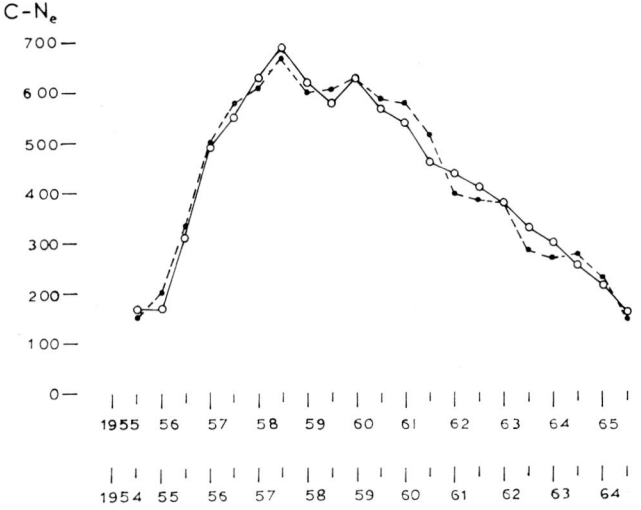

Fig. 11. The continuous line represents the values of the difference $C-N_e$ given by the observations while the dashed line gives the values of the same quantity computed with the help of Relation (6). The time scale corresponding to the values of I_a and N_{PF} is shifted toward the left of the time scale corresponding to N_e by one year.

The time scale corresponding to the values of I_a and N_{PF} is shifted toward the left of the time scale corresponding to N_e by one year.

Another interesting question would be to make an attempt to explain the observed phase lag between the values of the areas index I_a and the number of proton flares N_{PF} on the one hand and the cosmic-ray intensity and the geomagnetic index A_p on the other hand. Before any physical explanation of this phenomenon could be given, however an extension of the present investigation to the current sunspot cycle No. 20 should be made. If the same phase lag is found in this cycle too, then this phenomenon will give us the possibility of predicting the values of the cosmic-ray intensity in the year t on the basis of the already known values of the areas index I_a and the proton flares number N_{PF} during the preceding year $t-1$.

Finally it should be noted that the areas index I_a is very probably very closely correlated with the sunspot magnetic field strength. Figure 12 shows the relationship between the quantities $\sqrt{I_a}$ and $\log \Sigma H$, where ΣH represents the annual sums of the maximum sunspot magnetic field strengths observed for each group of sunspots

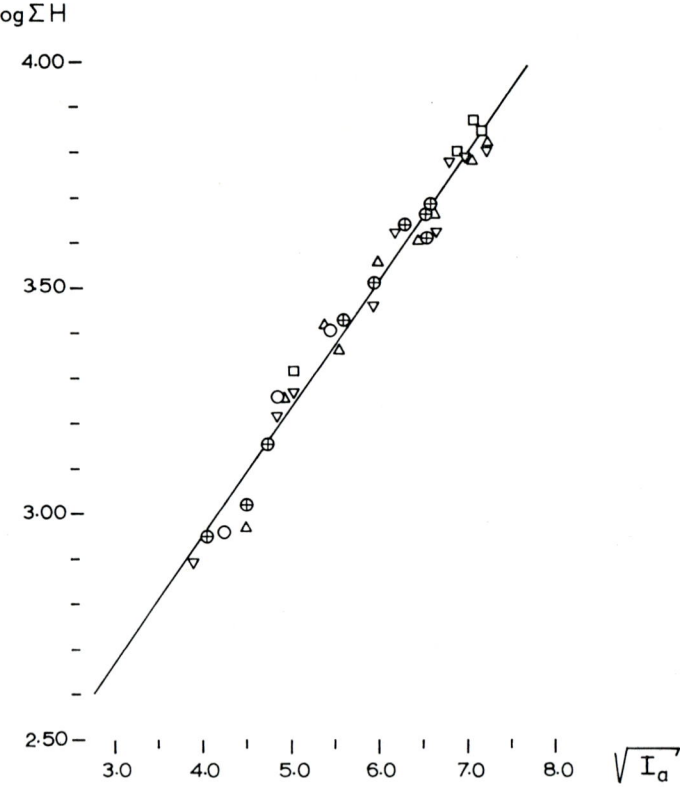

Fig. 12. Correlation between the quantities $\sqrt{I_a}$ and $\log \Sigma H$ where ΣH is the annual sum of the maximum values of the sunspot magnetic field strength observed for each group of sunspots during each passage of it over the visible solar hemisphere, as published by the Mt. Wilson Observatory. The years of minimum solar activity i.e. the years 1923, 1933–1934, 1944 and 1954 have been ommitted from this figure. The different symbols represent respectively the years of the sunspot cycles Nos. 15 (○), 16 (⊕), 17 (△), 18 (▽), and 19 (□).

during each passage of it over the visible solar hemisphere, as given by the observations carried out at the Mt. Wilson observatory during the years 1920–1959. The years of minimum solar activity i.e. the years 1923, 1933–1934, 1944 and 1954 have been omitted from this figure. If we take into account that the coronal phenomenon studied above are closely correlated with the magnetic fields in the solar photosphere we can easily explain why the areas index I_a is also closely correlated with the same phenomena.

Acknowledgements

Because of an illness the author has been prevented from presenting the present paper during the corresponding session of the NATO Advanced Study Institute. Therefore, he wishes to express his thanks to Professor L. N. Mavridis who took over the task of replacing him in this presentation.

References

Balasubrahmanyan, V. K.: 1969, *Solar Phys.* **7**, 39–45.
Leblanc, Y.: 1969, *Astron. Astrophys.* **1**, 467–472.
Svestka, Z. and Olmr, J.: 1966, *Bull. Astron. Inst. Czech.* **17**, 4–16.
Xanthakis, J.: 1969, *Solar Phys.* **10**, 168–177.
Xanthakis, J.: 1970, *Compt. Rend.* (Ser. A and B) **271**, in press.

14. MODELS OF THE QUIET AND ACTIVE SOLAR ATMOSPHERE FROM HARVARD OSO DATA

ROBERT W. NOYES

Smithsonian Astrophysical Observatory and Harvard College Observatory, Cambridge, Mass., U.S.A.

1. Introduction

In this paper we shall review some of the programs at Harvard aimed at defining the physical conditions in quiet and active solar regions. By way of introduction, we shall briefly describe the data obtained by the Harvard College Observatory from the spacecrafts OSO-IV and OSO-VI. The details of the experiment and its mode of operation are described elsewhere (Goldberg *et al.*, 1968), so we simply summarize them here, with reference to Figure 1. The spectral range covered is from 300 Å to 1400 Å, as shown in the lower curve in Figure 1. This spectral range consists of

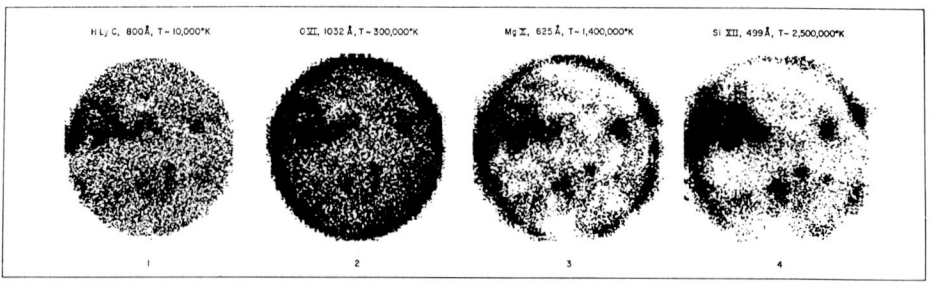

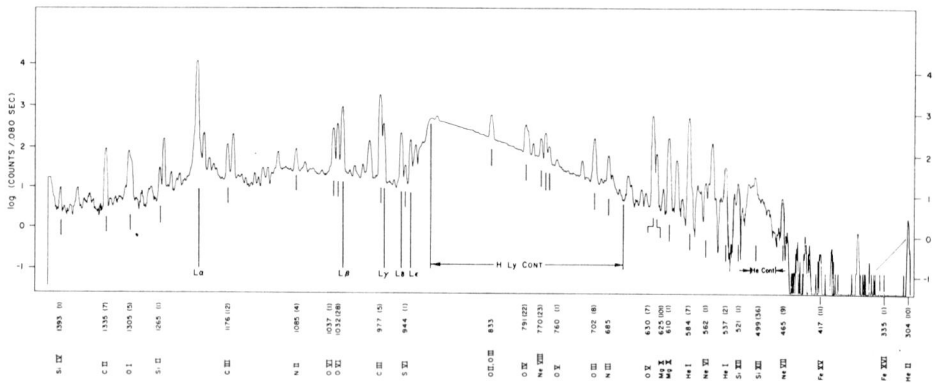

Fig. 1. (Bottom) OSO-IV Spectrum of the quiet sun observed at disk center with a one arc min square entrance aperture. Prominent features include the Lyman lines and continuum and the helium continuum. Other strong emission lines are labeled with their wavelength and, in parentheses, the number of orbits of observation in the spectroheliographic mode. (Top) Typical spectroheliograms in lines whose emission originates in the chromosphere, transition zone, and corona.

Macris (ed.), Physics of the Solar Corona, 192–218. All Rights Reserved.
Copyright © 1971 by D. Reidel Publishing Company, Dordrecht-Holland.

emission lines and continua from abundant elements such as hydrogen, helium, carbon, nitrogen, oxygen, silicon, magnesium, aluminum, neon, iron, and calcium in various ionization states ranging from neutral to 15 times ionized. Especially prominent are the hydrogen lines and Lyman continuum with the head at 912 Å, and the helium resonance lines and continuum with its head at 504 Å.

Spectra such as those illustrated could be obtained from OSO-IV only at the center of the disk, which during the time of operation of OSO-IV (October and November, 1967) generally reflected quiet solar conditions. The OSO-VI spacecraft, however, allowed us to position the slit at any point on or off the solar disk. Thus, we were able to obtain spectra of active regions, prominences and filaments, the quiet and active sun above the limb, and other features of interest.

A second operational mode of the spacecraft allowed us to make spectroheliograms of part or all of the solar disk at any chosen wavelength in the spectral region. The instrument was commanded by ground control to position the grating at any desired wavelength within the spectral range. Then the spacecraft was commanded to execute a raster motion which scanned the entrance slit of the spectrometer in two dimensions across the solar image. The raster array for OSO-IV covered a square 36 min of arc on a side, thus including the entire sun and the inner corona. Five minutes of time were required to acquire the data for a complete spectroheliogram. The OSO-VI spacecraft had two raster modes: a large raster covering an area 46 arc min square and requiring about 8 minutes to complete, and a small raster which covered an area about 7 arc min square once every 30 sec of time. The spatial resolution of these raster patterns, determined by the size of the entrance slit of the spectrometer, was 1 arc min for the OSO-IV experiment and 35 arc sec for the OSO-VI experiment.

The upper part of Figure 1 illustrates 4 typical spectroheliograms obtained by OSO-IV. The one arc minute resolution allows us to distinguish active regions from quiet regions and to measure the quiet sun intensity variation from center to limb. The left-hand image is of the chromospheric Lyman continuum, and shows an increased intensity in the active regions of about a factor of 5 and a relatively flat variation of the quiet sun intensity from center to limb. The second spectroheliogram is of O VI, a line formed principally in the chromosphere-corona transition zone; it exhibits considerable limb brightening in both the equatorial and polar regions. The third spectroheliogram, in the coronal line Mg X, also exhibits limb brightening but with a noticeable gap at the south pole. The contrast between the active and the quiet sun in Mg X is considerably greater than that in the Lyman continuum or in O VI. Finally the last spectroheliogram, in Si XII, a coronal line at a somewhat higher temperature than Mg X, shows that the emission is essentially absent at the poles and that the contrast between active and quiet regions is even larger than that for Mg X. We shall discuss these and other aspects of the OSO-IV and OSO-VI data in the remainder of this paper.

We divide our discussion into two parts. In Section 2 we shall discuss the structure of the quiet solar atmosphere as deduced from center-to-limb behavior of spectral lines and continua formed in the chromosphere and corona. In Section 3 we shall discuss our investigations of solar active regions using data from the OSO-IV and more recently the OSO-VI missions.

2. Structure of the Quiet Solar Atmosphere

Figure 2 is a schematic graph of the temperature structure of the quiet solar atmosphere which one may deduce from its ultraviolet emission. This graph forms the basis of the discussion in Part I. As Figure 2 shows, the temperature passes through a minimum of about 4300 K at a height of several hundred km above the limb. The

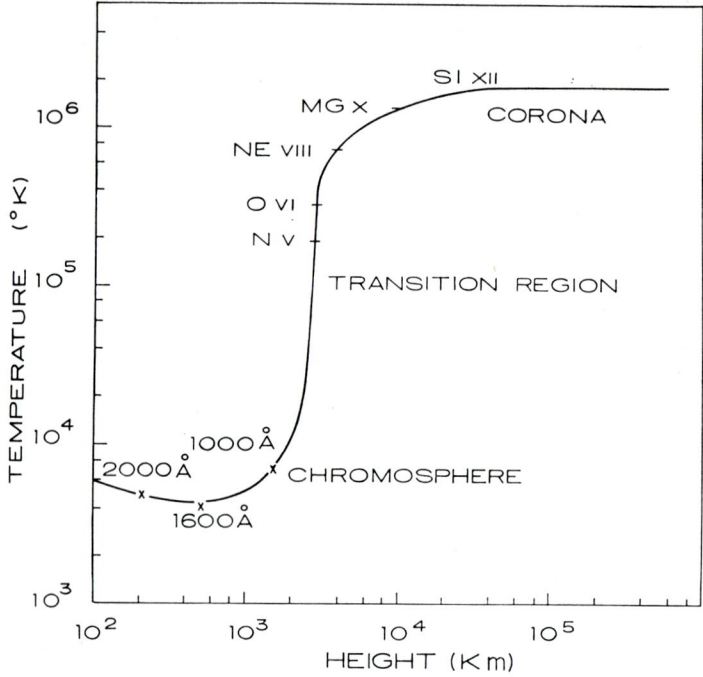

Fig. 2. Temperature structure of the mean quiet equatorial solar chromosphere, transition zone, and corona, as deduced from OSO-IV observations.

temperature rises rather slowly above the minimum until a height of about 2000 km is reached at which point a rapid rise occurs – the so-called transition region into the corona. Near the top of the transition region the temperature gradient again decreases until at a height of about 10000 km the temperature reaches its maximum value of approximately 2 million degrees. In the ensuing paragraphs we shall discuss how such a temperature structure is inferred treating first the transition region, then the corona, and finally the chromosphere.

A. UPPER TRANSITION ZONE AND CORONA, $T > 10^5$ K

In another paper in this conference Athay has described how one may deduce the structure of the transition zone from fluxes in the ultraviolet, i.e., from the total emission integrated over the entire sun. Considerable extra information is available from observations of the center-to-limb behavior of emission lines and of their intensities above the limb. A good example of this is provided by lines of the Li-like isoelectronic sequence. Li-like lines observable by the Harvard OSO spectrometers include the resonance doublets of N v, O vi, Ne viii, Mg x, and Si xii. N v, with maximum emission at $T_{max} = 2.0 \times 10^5$ K, is formed purely in the transition zone. O vi and Ne viii, with T_{max} of 3.3×10^5 K and 7.1×10^5 K respectively, are formed mainly in the transition zone but have a slight contribution from the corona. Mg x, with $T_{max} = 1.3 \times 10^6$ K is formed mainly in the corona, but with a slight contribution from the transition zone. Finally, Si xii, with $T_{max} = 2.2 \times 10^6$ K, is emitted entirelly from the corona. The regular shift of T_{max} along the Li-like isoelectronic sequence causes the center-to-limb variation and the intensity above the limb to change in the regular way from N v to Si xii. Withbroe (1970a) has calculated the expected behavior of the above lines for various solar models. If the optical thickness of the lines is much less than unity, one may write the intensity at any distance ϱ from the center of the disk as

$$I(\varrho) = \text{const} \frac{h\nu}{4\pi} A \int_{r_0}^{\infty} R_i(T) n_e C_{12}(T) \frac{dr}{\mu},$$

where ϱ is the distance of the point of observation from disk center in units of the solar radius, A is the abundance of the element relative to hydrogen, $R_i(T)$ is the abundance of the ionization state in question relative to the total abundance of the element, $n_e C_{12}(T)$ is the collisional excitation rate, r is the distance from the center of the sun to a point along the line of sight, $\mu = \sqrt{1 - \varrho^2/r^2}$, and r_0 is the larger of $(\varrho, 1)$. Above the limb, the intensity doubles. (For lines with optical thickness not much less than unity, the above equation must be modified to take into account the buildup of the radiation field in the line-forming region and self absorption due to the line. However, all of the lithium-like lines mentioned above are found to be optically thin.)

Optically thin lines which are formed almost entirely in the very thin transition zone only a few thousand km above the limb may be expected to show an intensity that varies from center to limb as $1/\mu_0$ where $\mu_0 = \mu(\varrho, r=1) = \cos\theta$, θ being defined in the usual way. On the other hand, a line such as Si xii which is formed over a rather extensive height range in the corona should show somewhat less limb brightening than a line formed in a thin shell. The OSO-IV data show an approximate $1/\cos\theta$ intensity variation for all Li-like lines; however, inhomogeneities cause intensity fluctuations sufficiently large to mask any model-dependent departures from a $1/\cos\theta$ law. Thus, observations from center to limb give little information on such important parameters as the height of the transition zone or its thickness.

Information on the thickness of the transition zone may be obtained, however, from measurements of the intensity above the limb relative to that on the disk. Observations made above the limb record emission only from the corona itself; the emission from the transition zone originates within about 2000 km of the limb (Figure 2) and therefore does not contribute to the intensity measured at heights of 1 or 2 arc min above the limb. On the other hand, measurements on the disk record the total emission both from the transition zone and from the corona. Thus the ratio of the intensity above the limb to that on the disk reflects the contribution of the coronal emission relative to that from the corona plus the transition zone.

Withbroe (1970a) has assumed that the structure of the transition zone and corona can be described by a simple model in hydrostatic equilibrium in which there are three free parameters, namely:

(1) The pressure P_0 in the transition zone. Under the assumption of hydrostatic equilibrium the total emission of any line formed in the transition zone or in the corona varies as the square of P_0.

(2) The 'thickness" of the transition zone, expressed in terms of a parameter denoting the conductive flux downward through the transition zone. Withbroe assumes that the conductive flux $F_c = \alpha T^{5/2} \, dT/dh$ is constant in the transition zone, following Athay (1966a), but that the value of the conductive flux F_c is a free parameter. Thus, he sets $T^{5/2} \, dT/dh = 1/C$ and uses the observations to determine the parameter C.

(3) The value of the coronal temperature T_{cor}, assumed to be constant. The temperature throughout Withbroe's model varies according to the relation given in (2) above until the value T_{cor} is reached, beyond which the corona is assumed isothermal. This description of the temperature structure, plus the assumption of hydrostatic equilibrium and specification of the pressure P_0 at the base completely determine the model.

It is easy to see the effect that a change of the conductive flux parameter C would have on the relative intensity above the limb and on the disk for a line that is formed partly in the transition zone and partly in the corona (for example, O VI or Ne VIII). An increase in C, corresponding to a decrease in the conductive flux and an increase in the thickness of the transition zone, would cause the relative contribution of the transition zone to the total emission of O VI or Ne VIII to be increased. Therefore, the relative intensity above the limb to that at disk center would be decreased. Withbroe has carried out detailed calculations for the model described above assuming various values of the conductive flux parameter C, and compared them with the observed data reduced from equatorial areas of OSO-IV spectroheliograms. Assuming that the corona is isothermal at a temperature of 2×10^6 K, the best match of the observations and theory appears to be for a value $C = 10^{-12}$, with an uncertainty of about a factor of 2. This leads to a conductive flux of $F_c = 6 \times 10^5$ erg/cm^2/sec, which is very close to the value obtained by other investigators from fluxes integrated over the entire sun.

However, factors other than the thickness of the transition zone could influence the ratio of intensities off the limb to those on the disk. In particular, a change in the temperature of the corona, assumed isothermal, would change the amount of emission from the corona without affecting the emission from the transition zone. For example, increasing coronal temperature would cause the coronal contribution from O VI to decrease, and therefore the intensity off the limb relative to the total intensity on the disk would also decrease. Withbroe (1970a) found that, within the observational uncertainties, an increase or decrease of the coronal temperature by 5×10^5 K from its assumed value of 2×10^6 K produces the same change in the intensity of O VI above the limb relative to disk center as does a decrease or increase of the thickness of the transition zone by a factor of two. In order to distinguish between the possible combinations of the two parameters C and T_{cor} then, it is necessary to obtain an independent estimate of one of them.

This evidence can be derived from the intensity ratios of lines observed above the limb and therefore emitted entirely in the corona. The intensity ratio of Mg X to Si XII for instance is very temperature sensitive, since at 2×10^6 K the abundance of

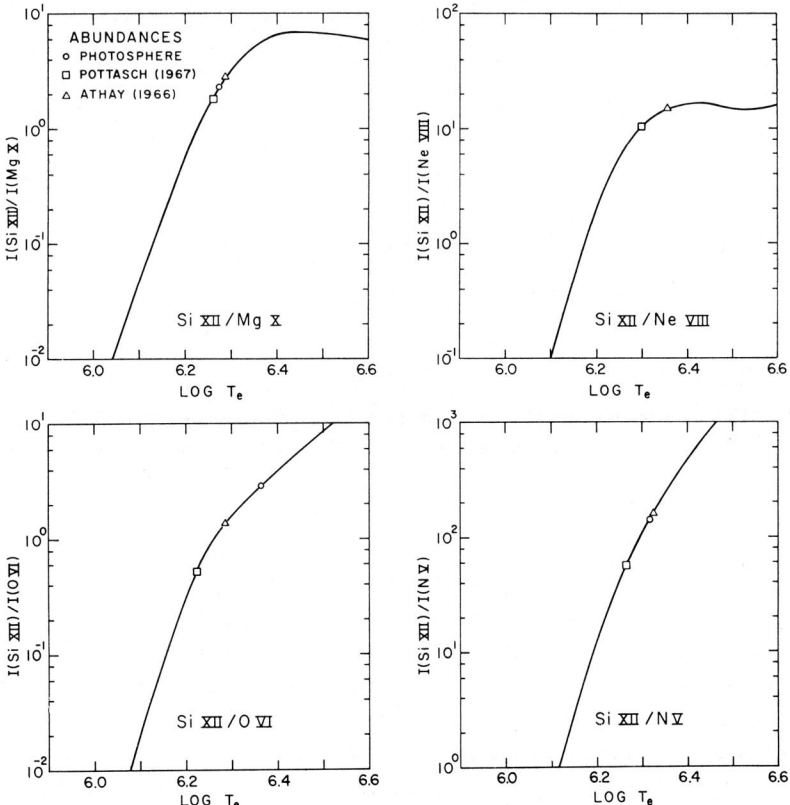

Fig. 3. Calculated ratio of lithium-like lines for various temperatures, assuming photospheric abundances. Also plotted are the observed ratios (circles) and the observed ratios scaled by the amount required if two other abundance distributions are used. (From Withbroe, 1970a.)

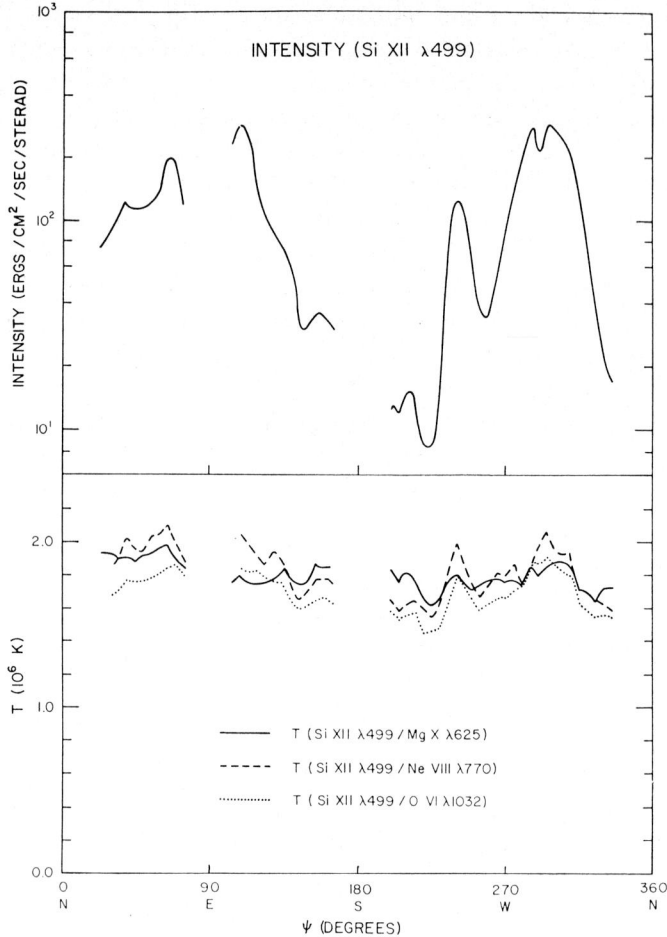

Fig. 4. (Bottom) Coronal temperature deduced from Li-like line ratios, versus limb position. (Top) Intensity of Si XII versus limb position. All curves were obtained from the same set of data. (From Withbroe 1971.)

Mg X decreases with increasing temperature while that of Si XII increases with increasing temperature. If we assume the corona is isothermal, then the emission from each of the two lines varies as the integral of the squared electron density along the line of sight multiplied by a temperature-dependent term. The ratio of the two intensities is thus independent of the electron density and depends only on the coronal temperature and the atomic abundances of the elements in question. Figure 3 plots calculations by Withbroe (1970a) of the expected intensity ratios of various lithium-like ions in the corona, assuming photospheric abundances. Also entered on the graph are the observed ratios (circles) and the observed ratios scaled by the amount required if two other abundance distributions are used. We see that the coronal temperatures

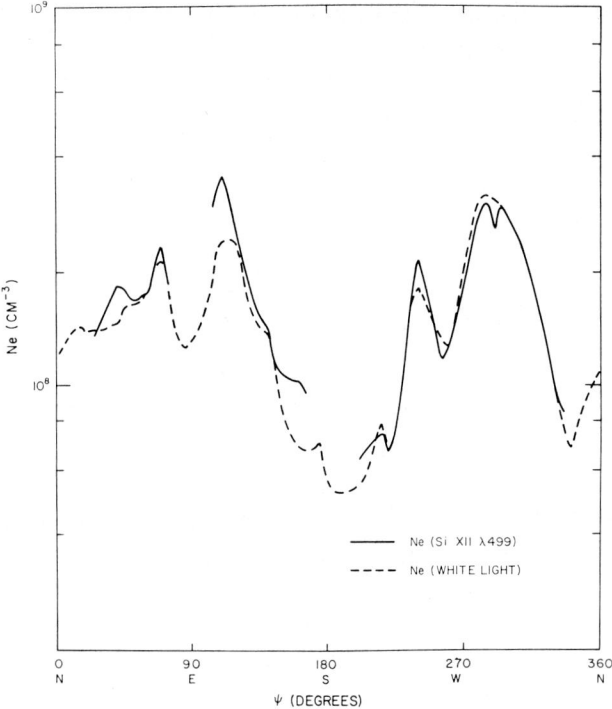

Fig. 5. Root-mean-square electron density from Si XII 499 intensity, and mean electron density from K-coronameter data, at 2 arc min above the limb, plotted versus position angle (From Withbroe, 1971.)

deduced from the ratios cluster around 2×10^6 K. It is this estimate of T_{cor} which permits the conclusion mentioned above, that in the transition zone

$$F_c = 6 \times 10^5 \text{ erg/cm}^2/\text{sec}.$$

In a related investigation Withbroe (1971) examined the variations of temperature in the corona with limb position. Figure 4 shows the temperature deduced from line ratios as described above, plotted against position angle. We see that the temperature is about 1.8×10^6 K and is rather constant with position angle, having a peak to peak fluctuation of perhaps 400000 K. The temperatures deduced from individual line ratios agree surprisingly well, in view of the very different temperatures of maximum emission for the various lines. It is interesting to note that the intensity of Si XII shown in the top part of Figure 4 varies by a factor of more than 10 even though the temperature varies only slightly. This suggests that the main factor causing variations in the emission of Si XII above the limb is the fluctuation in density, rather than in temperature, so that the upper curve very nearly maps the variation of the integrated electron density along the line of sight. Withbroe has used the individual temperatures deduced from line ratios along with intensities of observed line to calculate the actual electron density at various position angles, and compared the results with densities inferred from K-coronameter observations. The latter, obtained at the Hawaii station of the High Altitude Observatory of the University of Colorado on the same day as the EUV observations, yield the product of polarization and brightness of the white light corona pB. These observations were used to scale the electron densities of Van de Hulst's (1950) model of the corona by the ratio of the measured pB to that predicted by Van de Hulst's model. Figure 5 (Withbroe, 1971) compares electron density as a function of position angle derived from the EUV and white light measurement. The agreement is seen to be remarkably good, and in fact is somewhat surprising, in view of the different physical processes involved. The intensity of Si XII emission is proportional to the integral of the square of the electron density along the line of sight, whereas the white-light coronal emission is proportional to the integral of the electron density itself along the line of sight. Thus, one measurement gives us the root mean square electron density and the other, the mean electron density.

We see that the electron density inferred from either type of observation gives a mean of about 10^8 cm^{-3}, but with a variation of a factor of about 5 from equator to poles. The inferred electron density along with the assumption of hydrostatic equilibrium allows us to calculate what the electron pressure must be at the base of the corona and in the transition zone. This number may be compared with the number determined from the intensity of lines formed principally in the transition zone itself. Both numbers agree remarkably well and result in a pressure $P_0 = n_e T = 7 \times 10^{14}$ in the equatorial transition zone. This value supplies the third parameter for Withbroe's simple model of the quiet transition zone and corona, which may be summarized as: (1) $C = 10^{-12}$, $F_c = 6 \times 10^5$ erg/cm^2/sec, (2) $T_{cor} = 2 \times 10^6$ K, and (3) $P_0 = n_e T = 7 \times 10^{14}$.

B. LOW TRANSITION ZONE

We now turn to a discussion of the low transition zone where the temperature is less than 10^5 K. Here we expect serious difficulties with the simple model described above. For instance, if the conductive flux $F_c \sim T^{5/2} \, dT/dh$ is constant, the gradient must become enormous at low temperatures. For the conductive flux to be maintained at 6×10^5 K, the temperature must drop from 10^5 K to chromospheric temperatures in a distance of less than 1 km. Such a rapid temperature drop raises important theoretical questions as to whether a state of ionization equilibrium can be obtained, as has been discussed by Athay in another paper during this conference. Even disregarding these questions, such a temperature gradient is in disagreement with the observations, for lines formed below a temperature of 10^5 K would have a much lower intensity than has been observed, due to the exceedingly thin transition zone. Some information about the true thickness of the low transition zone may be obtained from the center-to-limb variation of optically thick lines in this region. The most outstanding candidate for an optically thick line is that of C III 977, which is the strongest line in the ultraviolet spectrum except for Ly-α. The limb brightening of this line is less than that predicted by a $1/\cos\theta$ law and may reflect the presence of self absorption in the line as viewed toward the limb (Withbroe, 1970b). If we could determine the optical thickness of this line, it might be of some help in deciding the physical thickness of the low transition region. The center-to-limb data are consistent with an optical thickness of about unity for the C III 977 line, but they are not precise enough to allow a unique determination. However, as Withbroe (1970b) has shown, an independent estimate of the optical thickness in the line may be obtained from the observed flux, provided one knows the electron pressure. The optical depth in the line may be written as

$$\tau = \text{const } AfP_e \int \frac{R_i(T)}{\Delta v_D T} (dT/dh)^{-1} \, dT.$$

It is difficult to calculate this integral because it is model dependent; in particular, it contains the temperature gradient which we desire to determine. The equation for the observed flux in the line also contains the temperature gradient in the integrand:

$$F = \text{const } AfgP_e^2 \int \frac{R_i(T)}{T^{5/2}} e^{-E_{21}/kT} (dT/dh)^{-1} \, dT.$$

By combining the two equations above, we find

$$\tau = \text{const } \frac{F}{gP_e} \frac{\int \frac{R_i}{T \, \Delta v_D} (dT/dh)^{-1} \, dT}{\int \frac{R_i}{T^{5/2}} e^{-E_{21}/kT} (dT/dh)^{-1} \, dT}.$$

The ratio of the two integrands in this equation is much less model dependent than either one alone, because both are proportional to the mean inverse temperature gradient. Using the admittedly incorrect model with constant conductive flux, Withbroe has evaluated the above equation. Assuming that the electron pressure is $P_0 = n_e T = 6 \times 10^{14}$ and that Δv_D is the thermal Doppler width, he finds an optical depth in the C III 977 line equal to about unity. This result of course is subject to considerable uncertainty, but it is encouraging to find that it is in good agreement with the observed center-to-limb variation. To produce optical depth unity in C III 977, the temperature gradient must be about 300 K/km at $T = 6 \times 10^4$ K. Although this is only one-fortieth the gradient predicted by the model with constant conductive flux, emission is confined to a region no more than 20 km thick. This result seems to rule out the possibility of an extended plateau of nearly uniform temperature near 50000 K as suggested earlier (Thomas and Athay, 1961). Athay (1966b) drew a similar conclusion for the low emission of He II 304. [Very recent work (Vernazza and Avrett, 1971; Vernazza and Noyes, 1971) suggests that a plateau may exist at a lower temperature near 20000 K.]

The final piece of information about the structure of the transition zone may be obtained from center-to-limb observations of lines such as O V 630, O IV 555, O III 702, and N IV 765. These lines all have optical depths much less than unity but nevertheless show limb brightening that is less than $1/\cos\theta$. They have in common that their wavelengths are shorter than 912 Å, the wavelength of the head of the hydrogen Lyman continuum. Withbroe (1970b) has suggested that the reduced limb brightening of these lines may be due to absorption by neutral hydrogen in spicules near the limb. Based on such an effect the transition zone would have to occur below the tops of the spicules, and in principle it would therefore be possible to deduce the height of the transition zone from knowledge of the geometry of spicules and the detailed center-to-limb variation of lines exhibiting the spicular absorption.

Withbroe has treated the problem of spicular absorption using a simple model in which spicules are characterized by a width W_s and height H_s that is greater than the height H_t of the transition zone. The spicules are distributed randomly over the solar surface with a mean number per unit area of N_s. If the spicules are opaque, the average intensity from the transition zone observed at position angle θ is related to the emitted intensity I_0 by

$$I = I_0(\theta) e^{-\sigma \tan \theta},$$

where $\sigma = N_s W_s (H_s - H_t)$, and $I_0(\theta) = I_0(0)/\cos\theta$ for an optically thin line.

By comparing center-to-limb observations from OSO-IV with the above relation, Withbroe found $\sigma \sim 0.1$ to 0.2 for lines of O III, N IV, O IV, and O V; $\sigma \sim 0.05$ for Ne VIII, and $\sigma \sim 0$ (no apparent absorption) for Mg X. These results were then compared with values of σ expected from the known numbers and dimensions of spicules. Using data of Beckers (1968) on both spicules and bright and dark mottles on the disk, Withbroe found $\sigma \sim 0.05$ to 0.2. Since the number density of spicules extending above a given height decreases with height, σ should also be found to decrease with height by a factor of 2 to 5 between 2000 and 3500 km. Thus, we may conclude that the transition zone itself must lie at a height not too different from 2000 km. If it were lower, the Ne VIII line would exhibit more spicular absorption than it does; if it were higher, there would be less absorption of O V and other lines near the limb. Unfortunately the data are insufficiently precise to determine the transition zone height to better than 1000 or 2000 km.

One other point emerges from this study: Because the spicules appear to absorb radiation from lines in the transition zone without emitting those same lines themselves, we conclude that the transition zone around the spicules is very thin or absent. If the transition zone at the edge of the spicules were the same as that in the quiet sun, we would not expect any effects of spicular absorption. It may be possible to explain the apparent difference in the transition zone structure around spicules and in the interspicular region in terms of the magnetic field which almost certainly lies along the axis of spicules and may prevent thermal conduction from smoothing out the radial temperature gradient in the spicules.

C. THE UPPER CHROMOSPHERE

We shall now turn our attention to the upper chromosphere which lies immediately below the transition zone and discuss how one may gain information on its structure from the hydrogen Lyman continuum. Figure 6 shows a typical Lyman continuum spectrum of the center of the solar disk obtained by the OSO-IV spectrometer. Long dashes represent the emission of a black body at 6500 K; note that the intensity at the head of the continuum corresponds to that brightness temperature. However, the intensity drops off towards shorter wavelengths at a slower rate than that of a 6500 K black body: it is matched much better by the solid line, the emission of a black body at 8300 K, but decreased by a factor of about 200. In other words, the brightness

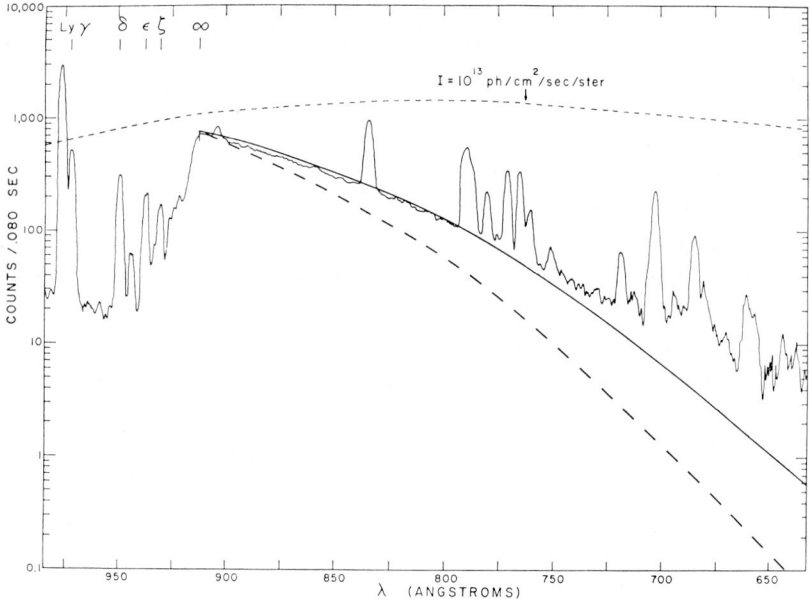

Fig. 6. OSO-IV spectrum of the quiet sun in the region of the Lyman continuum. Heavy dashed line is the expected curve for a blackbody of temperature $T_B = 6450$ K. Solid line, which is prediction of model of Noyes and Kalkofen (1970), also matches curve for a blackbody of $T_e = 8300$ K, decreased by a factor 200. Fine dashed line is calibration curve. (From Noyes and Kalkofen, 1970.)

temperature of the Lyman continuum is 6500 K but the color temperature is about 8300 K. This situation has been studied by Noyes and Kalkofen (1970), who have attributed the discrepancy between brightness temperature and color temperature to a non-LTE effect in the upper chromosphere, associating the color temperature with the true electron temperature at optical depth unity in the Lyman continuum. The decrease in the emergent intensity by a factor of 200 is due to the overpopulation of the ground state of hydrogen. This overpopulation has the effect of decreasing the source function (and therefore the emergent intensity) by the overpopulation factor b_1, since we may write the source function (neglecting stimulated mission) as

$$S(\tau) = \frac{1}{b_1} B\nu \left[T(\tau) \right].$$

An overpopulation of the ground state of hydrogen is to be expected in the upper chromosphere, where the electron density is too low for electron collisions between the ground and ionized states to maintain these states in a Boltzmann relation to each other. (The radiation field is of course far from thermodynamic equilibrium, since the brightness temperature of the continuum is 2000 K cooler than the true electron temperature.) Statistical equilibrium between the ground and ionized states of hydrogen is maintained by balancing the rates of photorecombinations in the Lyman continuum with ionization from the ground state. The principal ionization route is by collision from the ground state to the $n=2$ level followed by photoionization in the Balmer continuum. The collisional rate from the ground state to the $n=2$ level varies with the electron density as a result, that the overpopulation of the ground state varies roughly as the inverse square root of the electron density.

In addition to determining $T_e(\tau=1)$ and $b_1(\tau=1)$ from spectra at disk center, we can gain information about the variations of T_e and b_1 with τ from the variation of the intensity observed in the Lyman continuum from center to limb. Figure 7 shows schematically the spectrum near the limb of the quiet sun as compared with a similar spectrum at the center. We see that the Lyman continuum near the limb is flatter; in other words, it has a higher color temperature. If we associate the color temperature with the true electron temperature at $\tau=\cos\theta$, we immediately come to the not surprising conclusion that the temperature is rising outward. The precise

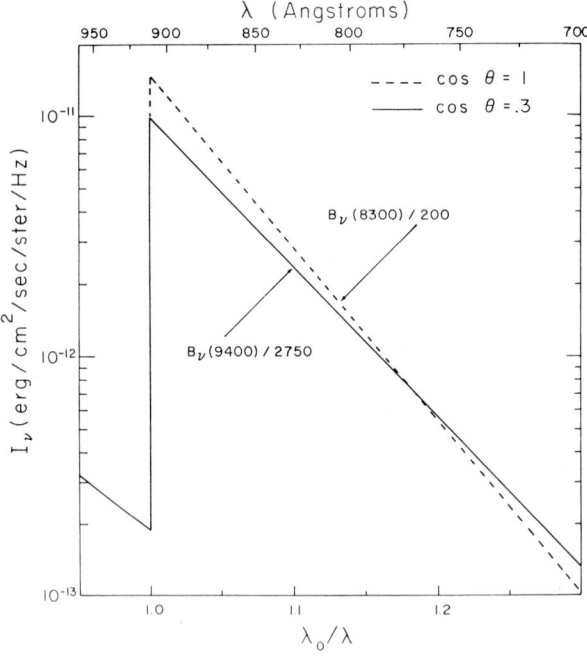

Fig. 7. Schematic diagram of the Lyman continuum spectrum at the center of the disk (dashed line) compared with that at the limb (solid line) observed with OSO-VI instrument.

variation of the color temperature with $\cos\theta$ allows us to deduce the temperature gradient $dT_e/d\tau$, where τ is optical depth in the Lyman continuum.

We also notice that the intensity, i.e., the brightness temperature, at the head of the Lyman continuum is less near the limb than at the center, even though the color temperature is higher at the limb. This can be explained by a greater overpopulation factor b_1 at the limb position than at the center. In fact, the spectrum at the limb corresponds to a black body of temperature $T_e = 9400$ K and $b_1 = 2750$. From the observations we can empirically determine the gradient $db_1/d\tau$ of the departure coefficient.

We note in passing the interesting point that the Lyman continuum exhibits limb darkening at its head and limb brightening at wavelengths below about 750 Å. This curious situation is not fundamental; it simply results from the differences in wavelength dependence of the Planck function at temperatures characteristic of the center and the limb. It is interesting, however, that the departure coefficient adjusts in such a way as to make the source function vary only slowly near optical depth unity (thus creating a very flat center-to-limb behavior in the Lyman continuum) in spite of the rapid increase of temperature with optical depth in that region of the atmosphere.

Noyes and Kalkofen (1970) have solved the equations of statistical equilibrium and radiative transfer for different models of the chromosphere in an attempt to match the above data. They include the first three levels of hydrogen in their calculations and treat all radiative and collisional transitions between these levels and the continuum (Ly-α and Ly-β are assumed to be in detailed balance because of their large optical depth at unit optical depth in the Lyman continuum). They also assume that the chromosphere is in hydrostatic equilibrium with a pressure at the top of the chromosphere sufficiently large to support the mass of the overlying corona – a requirement pointed out by Athay (1969).

The resulting model is shown in Figure 8. The top curve shows the run of temperature with height and the lower curve shows the run of particle densities and gas pressure with height. The height scale at the bottom of the graph is the height assuming no turbulent pressure, while that at the top is the height assuming a turbulent pressure equal to $\frac{1}{3}$ the gas pressure. Optical depth unity, at a temperature of about 8300 K, occurs close to the left-hand edge of the graph. We see that immediately above optical depth unity, the temperature undergoes a very sharp rise. This rise occurs at a height of about 2000 km, but the precise position depends on what value is assumed for the turbulent pressure and the pressure of the overlying corona. That a very sharp temperature rise must occur immediately above optical depth unity in the Lyman continuum stems directly from the fact that the electron temperature observed in the Lyman continuum is chromospheric, in other words, is below 10000 K. At these temperatures, if the gas pressure is great enough to support the overlying corona, the particle density must be greater than 10^{11} cm^{-3}. At such a high density, unit optical thickness in the Lyman continuum occurs in a geometrical distance of about 50 km (assuming the gas to be 50% ionized, as the model predicts for a temperature of 8300 K). This distance is much less than the pressure scale height, so gas at heights greater than 50 km above optical unity must have considerable mass yet be transparent to the Lyman continuum. In order for it to be transparent, the temperature must be raised high enough to cause essentially complete ionization of the hydrogen. The value to which it rises cannot be determined from the Lyman

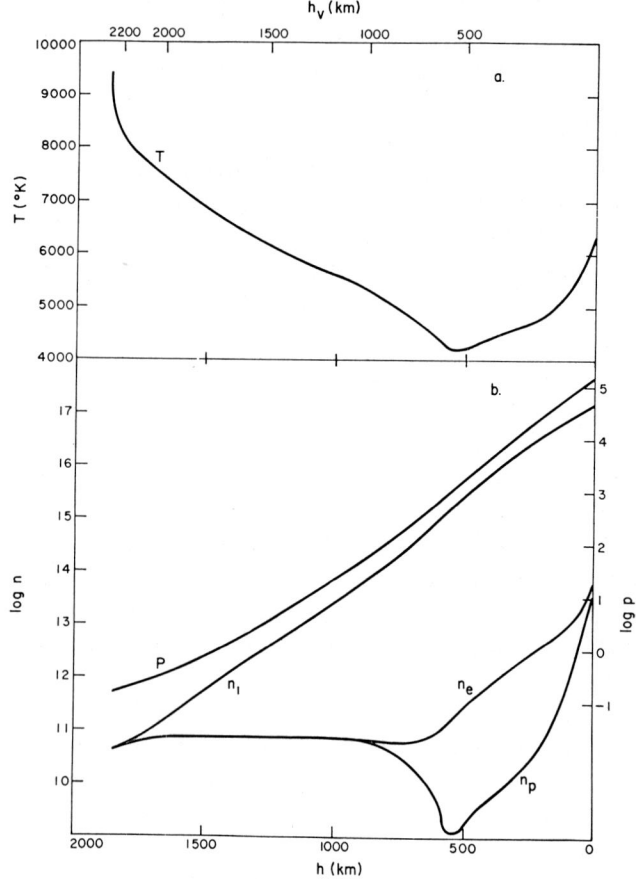

Fig. 8. Model of the mean chromosphere, from Noyes and Kalkofen (1970). (a) Kinetic temperature T versus height h above $\tau_{5000} = 1$. (The height h_v is for the turbulent pressure discussed in the text.) (b) Pressure P and the densities of ground-state hydrogen atoms n_1, of electrons n_e, and of protons n_p versus height.

continuum data – the temperature might rise directly to values characteristic of the transition zone or might level off slightly to a temperature of several tens of thousands of degrees. It is certain only that the rise from chromospheric temperatures occurs abruptly and within tens of kilometers of optical depth unity in the Lyman continuum.

We conclude our discussion of the quiet solar atmosphere by summarizing the results discussed above:

(1) The transition zone is an exceedingly sharp interface between the chromosphere and corona whose temperature profile appears to be governed by constancy of the conductive flux through the transition zone. The temperature gradient has the value

$$dT/dz \sim 10^{+12} T^{-5/2} \text{ K/cm}$$

which leads to a conductive flux $F_c = 6 \times 10^5$ erg/cm²/sec.

(2) The electron pressure in the equatorial transition zone is $n_e T = 1 \times 10^{14}$ K to within an accuracy of about a factor of 2.

(3) The temperature in the quiet corona above equatorial regions is about 2×10^6 K. It appears to vary only slightly for observations at different position angles around the limb, even at positions above active regions. The electron density on the other hand varies by a factor of 5 in the same regions.

(4) The strongest line in the low transition zone, namely C III 977, has an optical depth only of about unity, which rules out the possibility of an extended isothermal plateau at a temperature near 50 000 K in the low transition zone.

(5) From the known geometry of spicules and the observed spicular absorption of lines formed in the transition zone and immediately above it, the height of the steep transition zone at a temperature of about 10^5 K degrees is inferred to be at about 2000 km.

(6) The temperature in the high chromosphere at the level of optical depth unity in the Lyman continuum is about 8300 K. The height of this level above optical depth unity at 5000 Å is about 2000 km. The temperature rises with height in this region of the atmosphere, and in fact must exhibit a rather sharp rise immediately above the level of optical depth unity.

(7) The departure coefficient b_1 of hydrogen in the upper chromosphere is about 200 at the level of optical depth unity and increases rapidly with height.

3. Structure of the Active Sun

A. RESULTS FROM OSO-IV

Solar active regions can easily be recognized in the OSO-IV data as areas of enhanced emission relative to the quiet sun. Even with the 1 arc min resolution of the OSO-IV spectroheliometer, it is clear that the enhanced emission is extremely well correlated with the photospheric magnetic field. In Figure 9 we compare the emission observed in the Lyman continuum with a contour map of the photospheric magnetic field obtained by the Mt. Wilson magnetograph. The remarkably good correlation shows that the presence of a strong photospheric field is both necessary and sufficient for the existence of active region emission in the high chromosphere. Such a correlation is of course not surprising; the Lyman continuum is formed in nearly the same part of the atmosphere as the Ca II H and K lines observed from the ground, for which a similar one-to-one correlation has long been known to exist. The enhancement of active regions relative to the quiet sun in the Lyman continuum is about a factor of 5

– somewhat greater than the three-fold enhancement seen in the K line (Sheeley, 1967). The enhancement of lines formed higher in the chromosphere or in the transition zone and corona increases as the temperature of emission increases. It is possible to use the variation of enhancement with temperature to deduce the comparative structure of active and quiet regions. One important goal of such a study would be to determine how the conductive flux varies from quiet to active regions. This puts constraints on the amount of mechanical heating necessary to maintain the coronal temperature above active regions. Finally one might hope to relate the heating quantitatively with the magnetic field in the photosphere in order to determine the role of the photospheric field in the heating mechanism.

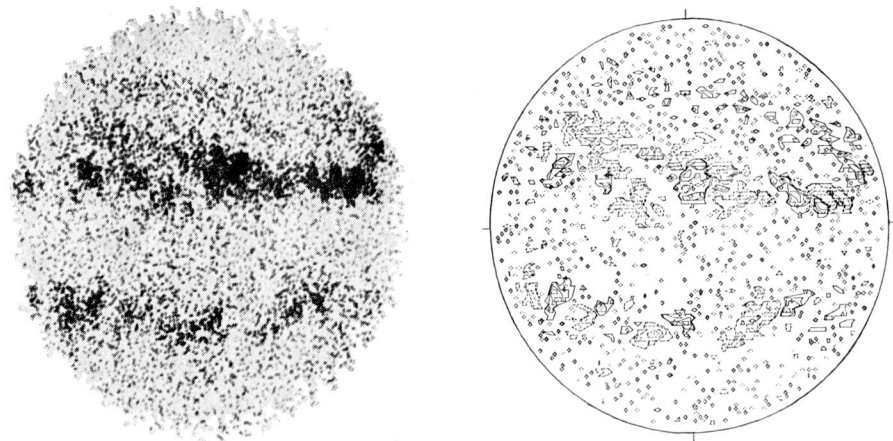

Fig. 9. (Left) OSO-IV spectroheliogram of the Lyman continuum at 897 Å. (Right) Mt. Wilson magnetogram. The two sets of data were obtained within a few hours of each other.

To date we have concentrated only on the first part of this sizable program, namely the determination of the physical structure of active regions without reference to the energy flux required to maintain them. Noyes *et al.* (1970) have taken the three parameter model of Withbroe described in Section 2 and extended it to active regions. Figure 10 presents the data which they have tried to match with an active region model: namely, the enhancement I_A/I_Q of active regions relative to the quiet sun for lines formed at different temperatures. The data were obtained by averaging enhancements measured for eight different active regions, and the results shown in Figure 10 therefore represent some fictitious mean active region. It is quite possible that there are large variations from point to point within the same active region. We shall return to this question later.

How could we alter the three parameters of Withbroe's quiet region model – namely, P_0, F_c, and T_{cor} – in order to produce the enhancement shown in Figure 10? Let us examine the influence of each parameter in turn.

A change in P_0 will cause every line formed at a temperature of 10^5 K or higher to be scaled by P_0^2, given the assumption that both the quiet and active regions are in hydrostatic equilibrium with the same temperature structure. Thus, the enhancement curve resulting from a simple increase of P_0 would be a horizontal line on Figure 10 with $I_A/I_Q = [P_0(A)/P_0(Q)]^2$. In order to produce the observed increase of enhancement with temperatures above $T = 10^5$ K, we must change either or both of the other two parameters.

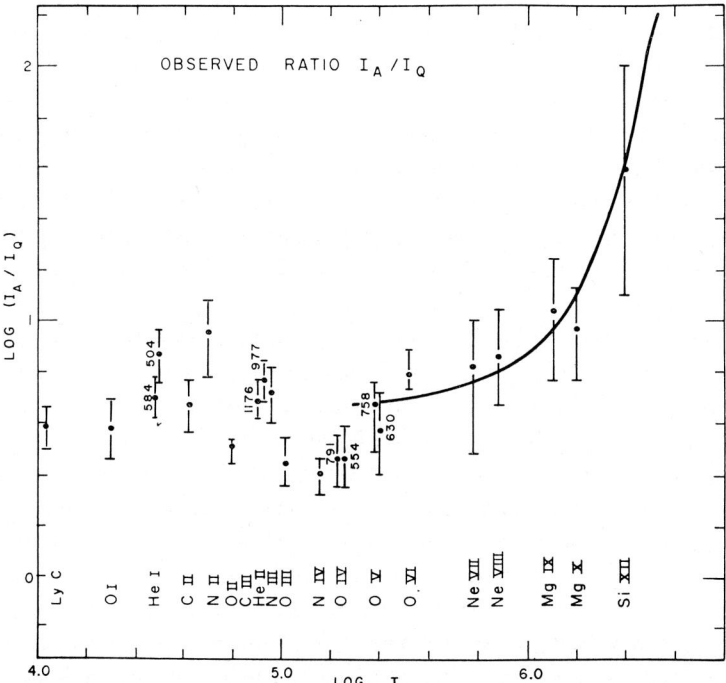

Fig. 10. Mean enhancement of eight individual active regions, relative to the quiet sun. The solid line is the enhancement calculated for the simple model described in the text. (From Noyes *et al.* 1970.)

A change in the conductive flux, or equivalently in C, where $1/C = T^{5/2} \, dT/dh$ (see Section 2), can produce a variation of enhancement with temperature. Let us suppose we increase the conductive flux by a factor of 5; in other words, that we decrease the thickness of the transition zone by a factor of 5. A line such as Ov which is formed entirely in the transition zone will decrease in intensity by a factor of 5. A line such as Neviii, which is formed partly in the transition zone and partly in the corona will then increase in intensity by a factor less than 5. Lines such as Sixii or Mgx, which are formed almost entirely in the corona, will be essentially unaffected. Therefore, the relative enhancement I_A/I_Q will increase steadily between temperatures of 10^5 K and 10^6 K.

An increase in T_{cor} in active regions will not affect lines formed purely in the transition zone. For coronal lines with most favorable temperature of emission below T_{cor} (such as Ne VII), an increase in T_{cor} will result in a decrease in intensity. Similarly coronal lines with most favorable temperature of emission above T_{cor} will be increased in intensity by an increase in T_{cor}. The increased enhancement of Si XII over Mg X implies a slight increase in T_{cor}, from about 2.0×10^6 K to 2.5×10^6 K.

The relative enhancements of the observed lines were calculated for various combinations of the above parameters and the best fit was found to be $P_0(A)/P_0(Q)=5$, $C(A)/C(Q)=0.2$; $T_{cor}(A)=2.5 \times 10^6$ K. The theoretical intensity ratio I_A/I_Q for these parameters is shown as a solid line in Figure 10. Again it should be emphasized that these results are for a fictional 'typical' active region and probably do not accurately describe any single region. We should also mention that there is no justification for assuming F_c=constant in active regions; although the data are consistent with that explanation of the enhancements, they are not sufficiently precise to differentiate the temperature distribution for constant conductive flux from other distributions with non-constant flux. The important result is that the transition zone is thinner above active regions: the constant conductive flux hypothesis is simply a convenient scheme for comparing active and quiet regions and has no physical significance. In fact, because of the strong magnetic fields in active regions and their probable inclination to the vertical, we must be cautious in concluding from the fivefold increase in temperature gradient of active regions that the conductive flux is also increased fivefold, much less that it is constant.

With these precautionary statements in mind, it would be interesting to note the implication if the conductive flux actually were increased fivefold. This factor is about the same as the enhancement seen in Figure 10 for chromospheric lines and continua. If indeed the main source of energy input to the chromosphere is the conductive flux downward from the corona, as Athay (1966a) and others have speculated for the quiet sun, the chromospheric enhancement may be due principally to the steepening of the temperature gradient above active regions. This in turn would presumably be a result of a fivefold increase in the rate of mechanical energy deposition in the corona; the gradient in the transition zone would steepen until the conductive flux were able to carry back most of this mechanical energy flux.

B. RESULTS FROM OSO-VI

The recent observations from OSO-VI have allowed us to make more detailed analyses of the structure of active regions. The OSO-VI spacecraft had the ability to point the Harvard telescope to any position desired on the solar disk, for instance, the quiet center of the sun, active regions near the center, active regions near the limb, or over the limb above an active region on the limb. The spacecraft could maintain the position of the 35 arc sec square aperture to within a few arc sec over the 60 min of satellite daytime, and during this time the Harvard spectrometer was able to make several scans of the spectral region between 300 and 1400 Å. Each spectral scan require 15 min. Thus, it is possible to compare data from different parts of the spectrum that were obtained from the same point on the sun and at the same time to within at most 15 min. This ability is to be contrasted with that of OSO-IV, in which a series of observations of the same active region at different wavelengths could only be obtained by successive spectroheliograms, which required several orbits to complete.

The ability to observe emissions from one point at several wavelengths and at nearly the same time opens up many new diagnostic possibilities for determining physical conditions in the active or quiet solar atmosphere. By way of illustration, we shall now give three examples of such techniques. In each case, the examples are of work presently underway and should not be interpreted as representing final results.

1. *The Lyman Continuum in Active Regions*

We recall from Section 1 that color temperature of the Lyman continuum as deduced from the variation of its intensity with wavelength, may be identified with the electron temperature at optical depth unity. The intensity in the Lyman continuum may be expressed in terms of its brightness temperature, and the relation between the brightness temperature and the electron temperature yields the departure coefficient of the ground state of hydrogen. The departure coefficient finally depends on the electron density in the region of formation of the Lyman continuum.

How do the electron temperature and the brightness temperature change in active regions? From the OSO-VI spectra one may deduce both with relative ease. Figure 11 shows some preliminary results derived from many spectra obtained at points on the disk varying from extremely quiet to extremely active locations. All points used had values of $\cos\theta$ greater than 0.8, so center-to-limb effects were minimal. For each spectrum the brightness temperature and electron temperature at optical depth unity were derived and the results plotted. In Figure 11 we see the rather surprising result that as the brightness temperature increases the electron temperature at optical depth unity actually decreases. Active regions have an electron temperature at optical depth unity about 500 K cooler than the quiet sun. This seems to be a somewhat paradoxical result, for we scarcely expect active regions to be emitting more energy yet be at a lower temperature than quiet regions. However, the situation is clarified somewhat by examining the variation in the departure coefficient b_1 with activity. Figure 11 shows that b_1 has a value of about 200 for quiet regions and 10 to 20 for active regions. This decrease of b_1 with increasing activity is a result of the increased density at optical depth unity in active regions. The departure coefficient may be written as: $b_1 = B_\nu(T_e)/S_\nu \sim B_\nu(T_e)/B_\nu(T_B)$. The fact that in active regions the electron temperature and the brightness temperature approach each other reflects an approach toward conditions of LTE, as expected in situations of high electron density.

The conclusion that the brightness and electron temperatures are converging does not of course specify whether the electron temperature increases or decreases in active regions. The fact that it actually decreases gives further information on the temperature and density structure, in addition to that given by the variation of b_1 alone.

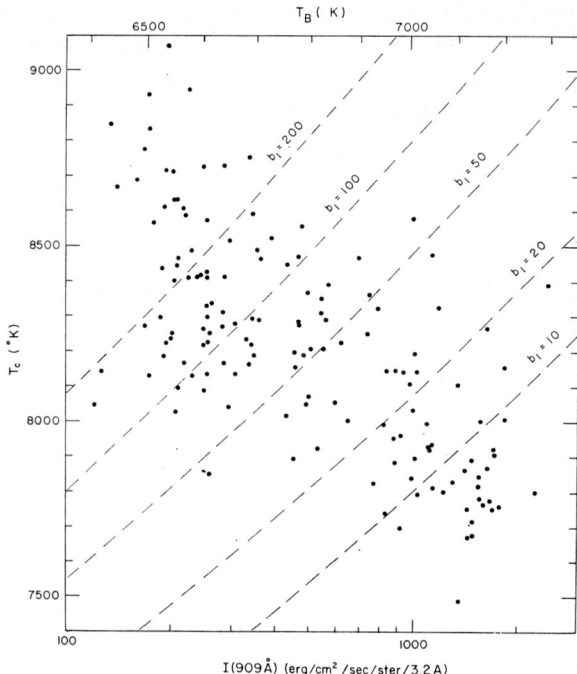

Fig. 11. Variations of T_B and T_C in the Lyman continuum with intensity at the head of the continuum. Dashed lines are curves of constant departure coefficient b_1.

It is important to note that the decrease in temperature at $\tau = 1$ need not imply a decrease in the energy content of the chromosphere in active regions. The effect can be due to a decrease in Lyman continuum opacity in active regions, such that $\tau = 1$ occurs in deeper, cooler layers. Such a decrease in opacity might come about through the decrease in b_1 mentioned above; if the gas is partially ionized, the number of ground-state atoms per unit mass decreases as b_1 decreases, and hence the opacity per unit mass decreases. Detailed model calculations of this phenomenon are presently underway at Harvard.

2. *Transition Zone Densities in Active and Quiet Regions*

A potentially very useful tool for determination of density in the solar chromosphere and corona is measurement of intensity ratios in ions whose level populations depend on the rate of electron collisions. The determination of densities in active regions and flares from intensity ratios in helium-like ions (Gabriel and Jordan, 1970) discussed elsewhere during this meeting is an outstanding example. Another example may be found in intensity ratios in beryllium-like ions (Munro et al., 1971). Figure 12 shows a diagram of the energy levels in beryllium-like O v. The ^1S ground state is connected to the ^1P level by a resonance transition at 630 Å. The metastable ^3P level is connected with a higher-lying $^3P^0$ level by a transition at 760 Å. In the solar transition zone, both of these spectral lines are excited by electron collisions from their lower levels. The relative intensity of the 760 Å and 630 Å lines thus depends on the relative population of the $^3P_{2,1,0}$ and $^1S_0$ levels. Radiative transitions between these two levels occur through the intersystem electric dipole transition $^3P_1 - {}^1S_0$ and the magnetic quadrupole transition $^3P_2 - {}^1S_0$, with the $^3P_0 - {}^1S_0$ transition strictly forbidden. For O v the first two transitions have transition probabilities of 3600 sec^{-1} and 3.1×10^{-3} sec^{-1} respectively. Those two transitions compete with electron collisions in establishing the population balance between the $^3P_{2,1,0}$ and $^1S_0$

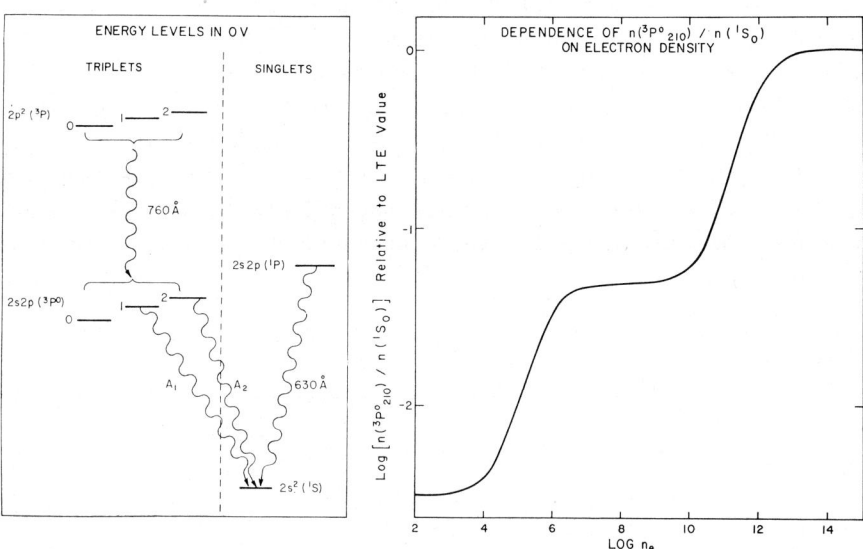

Fig. 12. (Left) Schematic diagram of O v energy level structure. (Right) Calculations (Munro et al., 1971) of the ratio of the population of the $^3P_{2,1,0}$ level to that of the $^1S_0$ level, relative to the LTE ratio, assuming a temperature of 2.5×10^5 K.

levels. At very low electron densities, electrons are ineffective in populating the $^3P_1$ and $^3P_2$ levels and the only sublevel with substantial population is the $^3P_0$ level. If we increase the electron density to a value greater than 10^4, electron excitation begins to balance the radiative decay in the $^3P_2 - {}^1S_0$ line, and the population of the $^3P_2$ level increases rapidly. The increase is reflected in the right hand curve of Figure 12, where we have plotted the total population of the $^3P_{2,1,0}$ level versus electron density. Above this critical density another plateau is reached which lasts until the density becomes high enough (greater than 10^{10}) that collisional excitation begins to balance radiative de-excitations through the $^3P_1 - {}^1S_0$ line. At this point a rapid increase in the population of the $^3P_1$ level occurs, which results in another increase of the total population of the $^3P_{2,1,0}$ level relative to the $^1S_0$ level. At even higher densities a plateau is reached where electron collisional rates dominate over radiative rates and the populations satisfy the Boltzmann equation.

If either of the two regions of steep dependence of line ratio upon electron density occurs at an electron density that is appropriate for the existence of the ion in question, then measurement of triplet-to-singlet line ratios in beryllium-like ions may be a powerful indicator of electron densities in the solar atmosphere. O v is formed at a temperature of about 250000 K. Since the product $n_e T$ in the transition zone is about 7×10^{14}, as discussed in Section 2 we expect densities in the region of O v emission to be 3×10^9 in the quiet sun and 1.5×10^{10} in the active sun. The curve on the right hand side of Figure 12 then suggests that the line ratio may be sensitive to electron density, especially in active regions. O v line ratios determined from many spectral scans of points on the sun varying in activity from extremely quiet to extremely active, are shown in Figure 13. Here the O v 760/O v 630 ratios are plotted against the intensity in O vi 1032 simply to indicate the extent of the activity. We see a definite trend in that the ratios increase from quiet to active regions, as we have predicted. Much of the scatter of the points is real; for instance the point representing the highest ratio was recorded shortly after the occurrence of a small subflare in the area of the entrance slit of the spectrometer. These results are exciting in that they provide a new type of measurement of the electron density in the transition zone in quiet and active regions.

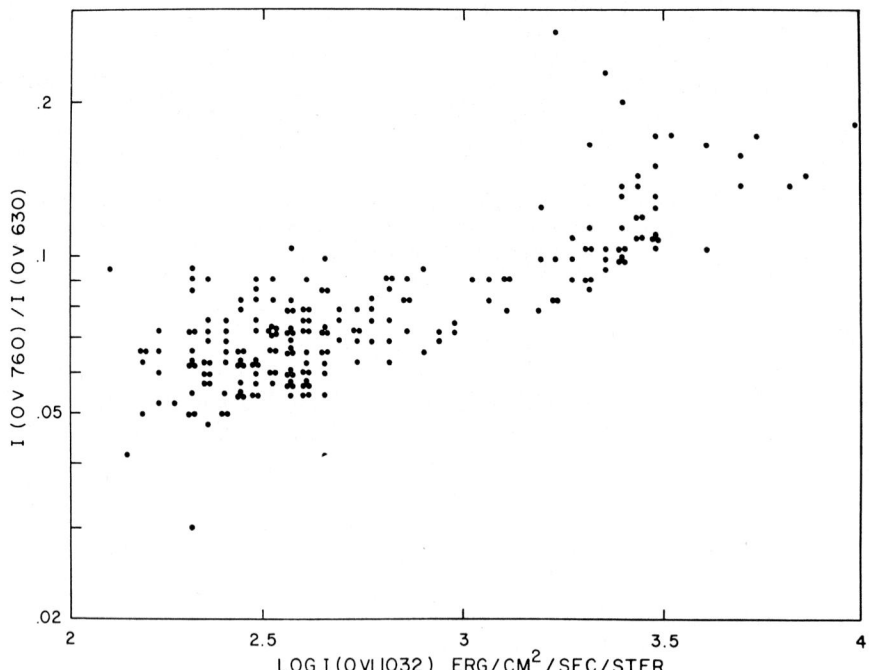

Fig. 13. OSO-VI observations of the line ratio $I(\text{O v } 760)/I(\text{O v } 630)$ versus intensity in O vi 1032. Only points with $\theta > 0.8$ are included.

3. Transition Zone Temperature Gradients in Quiet and Active Regions

As noted in Section 2, the intensity of a coronal line such as Mg x or Si xii is approximately proportional to P_0^2, where P_0 is the transition zone pressure, if the corona is approximately isothermal. On the other hand, the intensity of a line formed mainly in the transition zone, such as N v or O vi, is proportional to $P_0^2 \, (dT/dh)^{-1}$ where dT/dh is the mean temperature gradient at the temperature of emission of the line. Thus the intensity ratio $I(\text{Mg x})/I(\text{O vi})$, for instance, is roughly proportional to dT/dh at $T \sim 300000$ K.

In a preliminary analysis of OSO-VI spectra (Withbroe and Noyes, 1970) $I(\text{Mg x})/I(\text{O v})$ was plotted against Mg x (Figure 14). To the extent that the corona is isothermal, this may be considered a plot of dT/dh versus P_0^2. (It is possible to determine the coronal temperature for each individual spectrum from the Si xii/Mg x ratio and then find the pressure and temperature gradients that match each O vi and Mg x intensity; when the derived temperature gradient is plotted versus the square of the derived pressure, the results are actually quite similar to the data plotted in Figure 14.) We see that the temperature gradient increases with electron pressure from quiet regions to regions of moderate brightness, but that there is no significant trend with intensity (i.e., with pressure) between moderately bright and very bright active regions. In other words, dT/dz increases with P_0^2 until a maximum temperature gradient is reached. Beyond this level of activity dT/dz remains approximately constant even though P_0^2 increases by another order of magnitude.

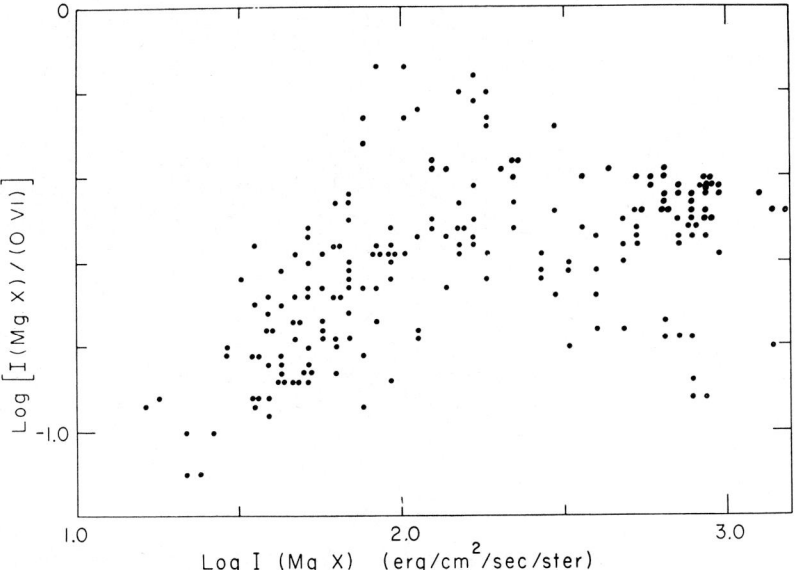

Fig. 14. The ratio $I(\text{Mg x } 625)/I(\text{O vi } 1032)$ versus intensity in Mg x 625, for points with $\cos \theta > 0.8$. As described in text, this is similar to a plot of dT/dh in the transition zone versus the sqare of the electron pressure P_0 in the transition zone.

These relations will be discussed in more detail in forthcoming papers by members of the Harvard Solar Satellite Project.

We conclude by applying the above treatments to a particular active region. The results, which are shown in Figure 15, are intended only to demonstrate that there may be significant variations in active region parameters within an individual region. The left-hand side of Figure 15 shows a crosssection of the intensity in an active region as observed in lines formed at different temperatures. The intensities were taken from a series of spectra recorded at points stepping sequentially from the edge of the active region (on the right) to the center (on the left). We immediately notice the interesting fact that the brightest point in the region, as seen in the transition zone lines of O VI and Ne VIII, is located further to the left than the maximum as seen in either the corona (Si XII and Fe XVI) or the chromosphere (He I and the Lyman

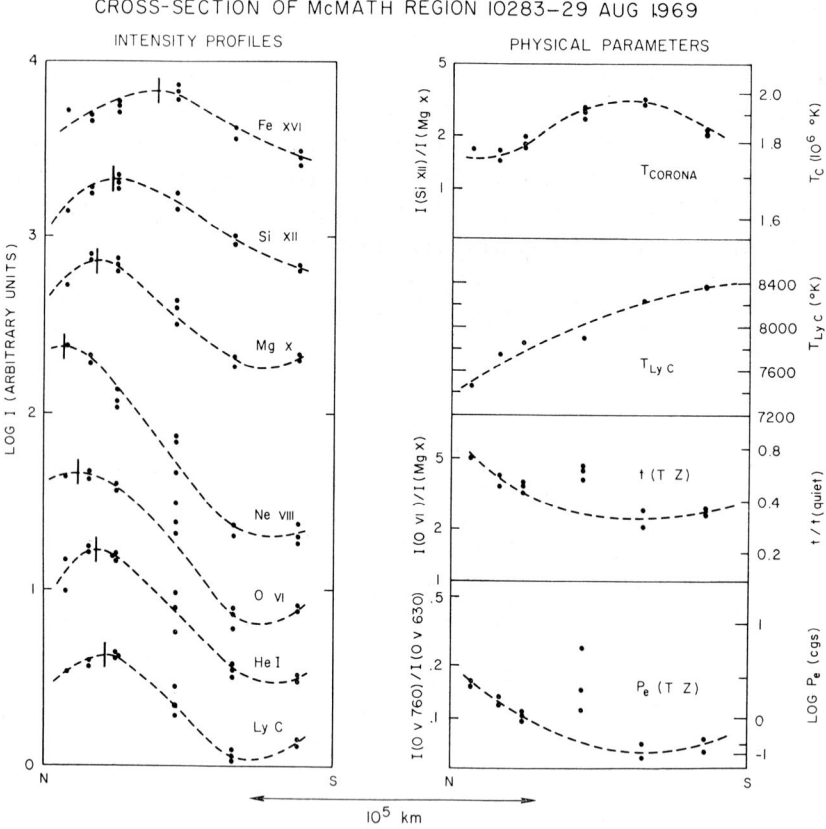

Fig. 15. (Left) Intensity profiles across an active region in lines of increasing temperature of emission from OSO-VI spectral. (Right) Deduced physical parameters of the active region. From top to bottom: (a) T_{cor} from the ratio $I(\text{Si XII})/I(\text{Mg X})$, (b) T_{Lyc} from slope of the Lyman continuum, (c) thickness of the transition zone from ratio $I(\text{O VI})/I(\text{Mg X})$, (d) electron pressure in the transition zone from ratio $I(\text{O V } 760)/I(\text{O V } 630)$.

continuum). A possible reason for this might be that the temperature gradient in the transition zone, while generally larger in the active than in the quiet sun, in addition shows a slight decrease from the right-hand to the left-hand edge of the cross-section shown. As we have just discussed, a rough measure of the temperature gradient in the transition zone is given by the ratio of the intensity of the two lithium-like lines O VI and Mg X. The third graph on the right-hand side of Figure 15 shows that the gradient deduced from this ratio is maximum near the center of the crosssection and decreases by perhaps a factor of 2 at the left-hand edge.

From the second curve on the right-hand side, we see that the temperature at optical depth unity in the Lyman continuum decreases from right to left. If, as we suggested earlier, this effect is correlated with an increase in pressure, we would also expect the pressure in the overlying transition zone to increase from right to left. The line ratio $I(O v 760)/I(O v 630)$ plotted on the lower graph on the right-hand side, clearly shows this increase.

Finally, one may use the line ratio $I(Si XII)/I(Mg X)$ to determine the temperature of the corona above the active region. The upper curve on the right-hand side of Figure 15 shows that this temperature is maximum near the center of the cross-section at a point somewhat distant from where the maximum emission occurs in the transition zone and in the corona.

It is clear from the above example that the structure of active regions varies in a complicated fashion from point to point. The local structure is no doubt influenced by factors such as the magnetic field configuration within the active region and the age or evolutionary state of the region. One of the principal goals of the analysis of the Harvard OSO-VI data is to investigate the detailed structure at many points within different active regions in order to understand the relation between observed structural details and the magnetic field and evolutionary state of the region.

This review of studies of active regions in the EUV at Harvard is by no means exhaustive. Analyses are underway of flares, filaments, and prominences. In addition there exist extended areas of very low coronal activity, which appear prominently in the Harvard data in the form of regions of extremely low emission of coronal lines. Study of these regions, plus regions of low activity at the poles, provide yet another way to study solar activity and its variations. We hope that such studies may eventually provide useful insights into the fundamental problem of mass and energy transport in the solar chromosphere and corona at all levels of solar activity from the most quiet to the most active.

The author gratefully acknowledges helpful discussions with A. K. Dupree, R. H. Munro, and G. L. Withbroe in the preparation of this review. None of the results reviewed here would have been obtained without the dedicated efforts of many members of the Harvard College Solar Satellite Project and the overall scientific and technical leadership of Drs. L. Goldberg, M. Huber, W. H. Parkinson, and E. M. Reeves. The OSO-IV and OSO-VI experiments are supported by the National Aeronautics and Space Administration through contracts NASw-184 and NAS 5-9274, respectively.

References

Athay, R. G.: 1966a, *Astrophys. J.* **145**, 784.
Athay, R. G.: 1966b, *Astrophys. J.* **146**, 223.
Athay, R. G.: 1969, *Solar Phys.* **9**, 51.
Beckers, J. M.: 1968, *Solar Phys.* **3**, 367.
Gabriel, A. H. and Jordan, C.: 1970, *Physics Letters* **32A**, 166.
Goldberg, L., Noyes, R. W., Parkinson, W. H., Reeves, E. M., and Withbroe, G. L.: 1968, *Science* **162**, 95.
Munro, R. H., Dupree, A. K., and Withbroe, G. L.: 1971, *Solar Phys.*, in press.
Noyes, R. W. and Kalkofen, W.: 1970, *Solar Phys.* **15**, 120.
Noyes, R. W., Withbroe, G. L., and Kirschner, R. P.: 1970, *Solar Phys.* **11**, 388.
Sheeley, N. R.: 1967, *Astrophys. J.* **147**, 1106.
Thomas, R. N. and Athay, R. G.: 1961, *Physics of the Solar Chromosphere*, Interscience Publishers, New York.
Van de Hulst, H. C.: 1950, *Bull. Astron. Inst. Neth.* **11**, 135.
Vernazza, J. E. and Avrett, E. H.: 1971, in preparation.
Vernazza, J. E. and Noyes, R. W.: 1971, submitted to *Solar Phys.*
Withbroe, G. L.: 1970a, *Solar Phys.* **11**, 42.
Withbroe, G. L.: 1970b, *Solar Phys.* **11**, 208.
Withbroe, G. L.: 1971, *Solar Phys.* **18**, 458.
Withbroe, G. L. and Noyes, R. W.: 1970, *Bull. Am. Astron. Soc.*, in press.

15. THE DETERMINATION OF CHROMOSPHERIC-CORONAL STRUCTURE FROM SOLAR XUV OBSERVATIONS

CAROLE JORDAN and R. WILSON
Astrophysics Research Unit, Culham, England

1. Introduction

The generally accepted model of coronal heating is that the hydrogen convection zone generates progressive waves in a hydrodynamic form (acoustic waves) which propagate outwards, increasing in amplitude because of the decreasing density, until they build up into shocks in the region of the upper chromosphere and corona where their energy is rapidly dissipated. The early theoretical investigations were carried out by Biermann (1946 and 1948) and Schwarzschild (1948) followed by Schatzman (1949) and Schirmev (1950). Since then many theoretical investigations have been carried out (e.g. Osterbrock, 1961; Kuperus, 1969) which show that the above mechanism is sufficient in broad energy terms to produce the observed heating but that the nature of the energy transport and dissipation processes is very complex due to the fine structure in the atmosphere and to the solar magnetic field whose pressure becomes equal to the gas pressure in the vicinity of the transition zone.

The structure of the upper solar atmosphere is determined entirely by the detailed energy balance in each layer, this balance being formed between the dissipation of mechanical energy by the wave motion and the loss or transport of energy by radiation, conduction and convection. The importance of these terms varies throughout the atmosphere as was shown by the considerations of Kopp (1968) and Kuperus (1969). This is demonstrated by Figure 1 which was compiled by Gabriel (1970) using the reference data given by Kuperus (1969). The model used is that of Jordan (1965) whose derivation is discussed below. In the region of the chromosphere up to the start of the transition zone the energy dissipation is balanced by radiation and other loss processes are negligible. In the region of the steep transition zone conduction from the corona becomes a very important term in establishing the local energy balance. In the inner corona the dominant local energy losses are due to radiation and the heat exchange with the chromosphere by conduction. For the far corona convection processes dominate the energy loss in the form of the solar wind. Since the rate of noise generated by the convection zone varies as the eighth power of the turbulent velocity, only order of magnitude estimates of the energy output can be made. A number of authors (De Jager and Kuperus 1961; Osterbrock 1961; Kuperus 1965) give a value $\sim 3 \times 10^7$ erg cm^{-2} sec^{-1}. This is adequate to explain the energy losses in Figure 1.

It is seen from Figure 1 that the energy dissipation rate required to maintain the corona is $\sim 4 \times 10^5$ erg cm^{-2} sec^{-1} and since the total energy contained by the corona is $\sim 2 \times 10^9$ erg cm^{-2} then the energy containment time of the corona is $\sim 5 \times 10^3$ sec.

Since this time is short it indicates that an energy supply must operate constantly even though severe perturbations occur due to localised and time-varying solar activity.

It is clear that advances in the understanding of the chromosphere and corona require a detailed knowledge of their structure. Such knowledge can be derived from observations of the solar XUV spectrum, many of which have been made and are continuing to be made from rockets and satellites. It is the purpose of this paper to describe and discuss the methods of interpreting such observations in terms of the structure of the upper solar atmosphere.

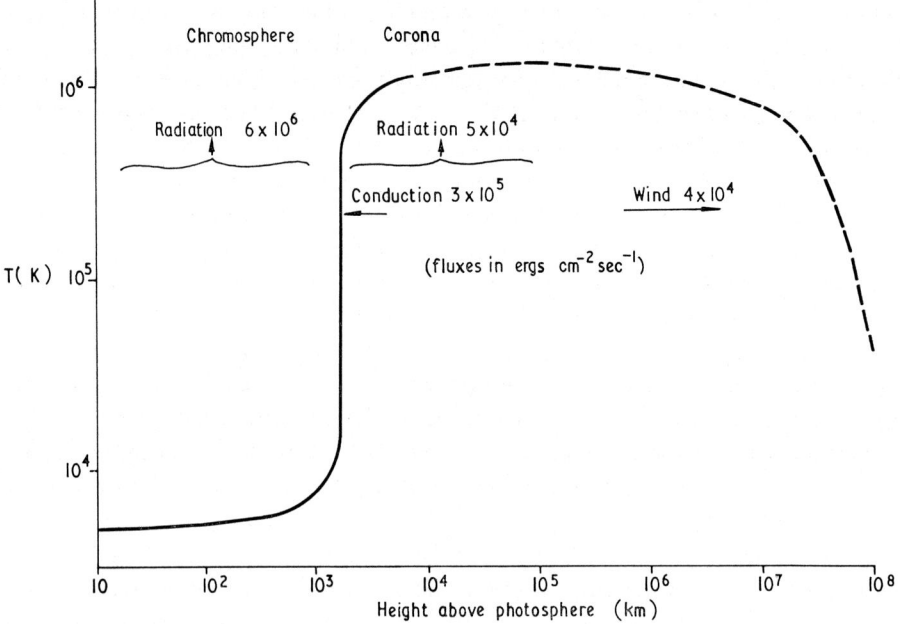

Fig. 1. Dominant energy loss processes in the chromosphere and corona.

2. Ionization Balance

The physical conditions in the chromosphere and corona are far removed from local thermodynamic equilibrium. The general expression for the ionization balance in a particular element is

$$\frac{N_{i+1}}{N_i} = \frac{S_i}{\alpha_{i+1}} \qquad (1)$$

where N_i and N_{i+1} are the ion densities in the particular stages of ionization, S_i is the total ionization rate coefficient for $i \rightarrow i+1$ and α_{i+1} is the total recombination rate function for $i+1 \rightarrow i$. In the early consideration of this problem (e.g. Biermann, 1947; Woolley and Allen, 1948) electron impact ionization from the ground level was used for S_i and radiative recombination was used for α_{i+1}. The ionization balance is then independent of density. Since then the importance of two other processes has been realised: these are dielectronic recombination (Burgess, 1964; Burgess and Seaton, 1964) and collisional excitation followed by autoionization (Goldberg *et al.*, 1965).

Jordan (1969) has computed the ionization distribution $N_i/N(E)$ as a function of temperature for abundant elements between carbon and nickel. In one set of tables only density independent rates are included. In the other set the contribution of density dependent stepwise processes to ionization and radiative recombination (Wilson, 1962, 1967) and to di-electronic recombination has been included, using a solar model for the electron densities. From these calculations it is possible to derive an 'ionization' temperature at which the particular ion has its maximum population. If the effect of excitation (see below) is taken into account, it is possible to determine the temperature at which a particular spectral line has its maximum intensity and such theoretically derived temperatures are the temperatures used to determine the structure of the upper atmosphere of the sun from interpretations of its XUV emission line spectrum.

3. Interpretation of the Total or Disk Spectrum

Early analyses of ultraviolet resonance line intensities were made by Allen (1961) and Ivanov-Kholodnyi and Nikol'skii (1962). Allen computed the ultraviolet emission expected from the radio emission model of Oster (1956), and compared his calculations with available observations. His conclusion that the results supported a low temperature gradient has been superceded by later improved atomic and intensity data. Ivanov-Kholodnyi and Nikol'skii pointed out that each emission line can be treated as if it originates in the temperature region where its production is most efficient. This optimum temperature, T_m, is not necessarily equal to that given by the peak of the $N_i/N(E)$ distribution, but should include both the temperature dependence of the line excitation rate and the electron density variation. (Their method of deriving the temperature structure is discussed below.) An extensive analysis of the extreme ultraviolet line emission has been carried out by Pottasch (1963). We will adopt his formulation since it has been generally used by other authors.

a. DERIVATION OF RELATIVE ABUNDANCES

The flux E (in erg cm^{-2} sec^{-1}) observed at the distance of the earth, in an emission line is given by

$$E = \tfrac{1}{2}h\nu \frac{1}{4\pi L^2} \int N_2 A_{21} \, dV \qquad (2)$$

$$= \tfrac{1}{2}h\nu \frac{R^2}{L^2} \int N_2 A_{21} \, dh \qquad (3)$$

where $h\nu$ is the quantum energy, N_2 is the population density of the excited level, A_{21} is the spontaneous transition probability, R is the solar radius, L is the earth-sun distance, and the integral is over the whole atmosphere. The factor of one half is the approximate fraction of the atmosphere observed.

Under the physical conditions existing in the upper chromosphere and corona, for resonance lines the depopulation rate $N_2 A_{21}$ is balanced by collisional excitations from the ground state, given by

$$C_{12} N_g N_e$$

where C_{12} is the collisional rate coefficient, N_g is the ground state population of the ion, N_e is the electron density.

N_g may be expressed in terms of the ionization distribution $N_i/N(E)$, the abundance of the element relative to that of hydrogen, $N(E)/N(H)$, and N_e, since $N(H) \approx 0.80 N_e$ for $T_e > 20000$ K. If it is assumed that all the ion is in the ground state, i.e. $N_g = N_i$, then substituting in equation (3) gives

$$E = \tfrac{1}{2} \times 0.80 \, hv \, \frac{N(E)}{N(H)} \int \frac{N_i}{N(E)} C_{12} N_e^2 \, dh \qquad (4)$$

The collisional excitation rate coefficient derived from Van Regemorter (1962) is

$$C_{12} = 1.7 \times 10^{-3} \, T^{-1/2} W^{-1} f_{12} 10^{-5040 W/T_e} P\left(\frac{W}{kT_e}\right) \text{cm}^3 \text{ sec}^{-1} \qquad (5)$$

where T_e is the electron temperature, W is the excitation energy of the transition in electron volts, f_{12} is the absorption oscillator strength, $P(W/kT_e)$ is the integrated gaunt factor.

The function

$$g(T) = T_e^{-1/2} 10^{-5040 W/T_e} \frac{N_i}{N(E)} \qquad (6)$$

as defined by Pottasch, peaks sharply at a certain temperature. Pottasch assumed that $g(T)$ has a constant value, 0.7 times its maximum value over a temperature range ΔT such that

$$\Delta T \langle g(T) \rangle = \int g(T) \, dT$$

where

$$\langle g(T) \rangle = 0.70 \, g(T_m)$$

$g(T_m)$ being the maximum value of $g(T)$. The function $g(T)$ can then be removed from inside the integral in Equation (4), and the integral limited to the region R where the line is formed.

This approximation leads to a different ΔT for each line and could introduce errors when deriving relative abundances from lines of different elements formed at the same temperature (see below). An alternative procedure is to choose a constant logarithmic temperature width $\Delta \log T = \pm 0.15$ dex and compute the correct normalization factor G such that

$$\int_{\log T_1}^{\log T_2} g(T) \, d \log T = G g(T_m)$$

where $\log T_1$ and $\log T_2$ are chosen to be well outside the limits $\log T_m \pm 0.15$. Substituting for constants in Equations (4) and (5) the flux becomes

$$E = 2.4 \times 10^{-20} f_{12} P \left(\frac{W}{kT}\right) Gg(T_m) \frac{N(E)}{N(H)} \int_R N_e^2 \, dh \tag{7}$$

Further refinements can be applied in practise. These are (i), $N_g = N_i$ can be replaced by N_g/N_i the true fraction of the ion in the ground state; (ii) N_1/N_g, the relative population of the level (or levels) from which excitation takes place can be estimated. These factors are important in low ions where metastable populations can be comparable to that of the ground state, and in high ions where the levels of the ground term do not have a statistical population. (iii) at temperatures below 2.0×10^4 K the actual population N_H/N_e can be used.

Thus for each line observed, the quantity $N(E)/N(H) \int_R N_e^2 \, dh$ can be calculated, provided the atomic data are known. For a given element, the series of points obtained from the various stages of ionization define the variation of $\int_R N_e^2 \, dh$ with temperature, since it is assumed that the abundance of an element is constant throughout the atmosphere. The curves for different elements can be fitted together to give the relative abundances of the elements, and if one abundance is known absolutely, the mean distribution of $\int N_e^2 \, dh$ as a function of temperature can be found. Figure 2, obtained by Jordan (1965) from the intensity data of Hall et al. (1963), illustrates such a distribution. (The refinements discussed above are not included in this plot.) The position of the high temperature part of the curve relative to the low temperature part of the curve is provided by the silicon ions since both low and high ions of this element are observed.

The main result from such abundance determinations has been that they agree with photospheric values within a factor of two, except for iron and nickel which are an order of magnitude more abundant, when compared with silicon, than they are in the photosphere (Jordan, 1966; Pottasch, 1967). The low iron to silicon ratio obtained by Dupree and Goldberg (1967) arose from their use of only Fexv and Fexvi lines which originate predominantly in active regions. The discrepancy between the iron to silicon ratio in the photosphere and corona has now been resolved. The new oscillator strengths measured by Garz and Kock (1969) lead to a photospheric abundance of iron which is an order of magnitude greater (Garz et al., 1969) than the previously accepted value.

It is difficult to put the relative abundances on to an absolute scale. Pottasch (1964) computed the radio brightness temperature as a function of wavelength expected from the mean distribution of $\int_{h_0}^{\infty} N_e^2 T_e^{-3/2} \, dh$ as derived from ultraviolet observa-

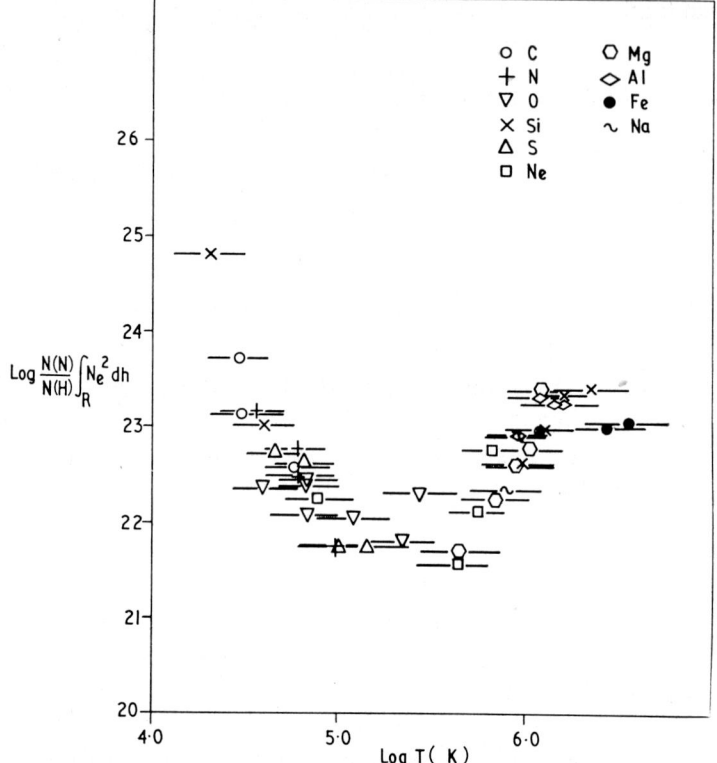

Fig. 2. Values of $\log N(N)/N(H) \int_R N_e^2 \, dh$, from intensities of Hall et al. (1963).

tions and compared it with that observed. He found a satisfactory agreement by adjusting the absolute abundance of oxygen used in the calculation, but assumed the optical depth to be constant as a function of radio wavelengths. Jordan (1965) and Dupree and Goldberg (1967) have pointed out that when di-electronic recombination is included, a satisfactory fit can only be made by decreasing the optical depth between the wavelengths 100 cm and 10 cm. Below 10 cm there is considerable disagreement between the two distributions. However, the radio emission does not necessarily originate in the same region of the atmosphere as the extreme ultraviolet emission and the value of such comparisons is limited.

b. STRUCTURE DETERMINATIONS

The mean distribution of $\int_R N_e^2 \, dh$ as a function of temperature is the basis of determinations of the temperature and density gradients in the transition region and inner corona. But in order to derive N_e and T_e as a function of height some additional information is needed, which implicitly or explicitly introduces assumptions.

Ivanov-Kholodnyi and Nikol'skii (1962) calculated $\int_h^\infty N_e^2 T_e^{-3/2} \, dh$ as a function of temperature from the extreme ultraviolet intensities. They then used the values of

$\int_h^\infty N_e^2 T_e^{-3/2} \, dh$ as a function of *height*, derived from eclipse observations of the hydrogen Balmer emission and the continuum emission beyond the Balmer limit, and by combining the two distributions calculated the temperature and density as a function of height. Their model has a slow temperature gradient and a greater electron pressure in the corona than in the upper chromosphere which if real would be difficult to explain. However, the eclipse results are effected by inhomogeneities, such as spicules, above the limb and the distribution of $\int_h^\infty N_e^2 T_e^{-3/2} \, dh$ as a function of height should not be combined with whole disk extreme ultraviolet intensity data.

Pottasch (1964) also used eclipse measurements of electron density as a function of height as the additional information, and the above criticism applies also to his method. His model also has a higher electron pressure in the corona than in the upper chromosphere but has a fairly steep temperature gradient (see Figure 4). Pottasch points out that the positive pressure gradient with temperature is unlikely to be real and that a negative pressure gradient would produce an even steeper temperature gradient.

Jordan (1965) has used the equation of hydrostatic equilibrium as additional information. The justification for doing so is that the low chromosphere and inner corona are in hydrostatic equilibrium and because the distance between the two regions is small on any model, it is likely that the transition region is also in hydrostatic equilibrium.

The integral in Equation (7) can be re-written as

$$\int_R^\infty N_e^2 \, dh = \int_{\log T_1}^{\log T_2} N_e^2 \frac{dh}{d \log T_e} d \log T_e \qquad (8)$$

Initially it is assumed that

(i) $\quad \dfrac{dh}{d \log T_e}$ is constant over $\Delta T = T_2 - T_1$

(i) $\quad P_e = N_e T_e$ is constant over ΔT.

Then

$$\int_R^\infty N_e^2 \, dh = \frac{|P_e^2|_{\Delta T}}{T_m^2} \left| \frac{dh}{d \log T_e} \right|_{\Delta T} 0.30 \qquad (9)$$

where T_m is the temperature at which $g(T)$ has its maximum value and $\Delta \log T = \text{const} = 0.30$.

The equation of hydrostatic equilibrium can be written as

$$\frac{d \log P_e}{d \log T_e} = -1.7 \times 10^{-4} \frac{1}{T_e} \frac{dh}{d \log T_e} \qquad (10)$$

The temperature gradient and pressure gradient can then be found as a function of height from Equations (9) and (10), by successive iterations, if a starting value for

P_e at some T_e is known. The range of possible electron pressures is about

$$10^{14} \text{ cm}^{-3} \text{ K} < N_e T_e < 10^{15} \text{ cm}^{-3} \text{ K},$$

the lower limit being a typical coronal value and the upper limit being a typical low chromospheric value – from other types of observations. Figure 3 shows models computed for a range of initial electron pressures. There are boundary conditions

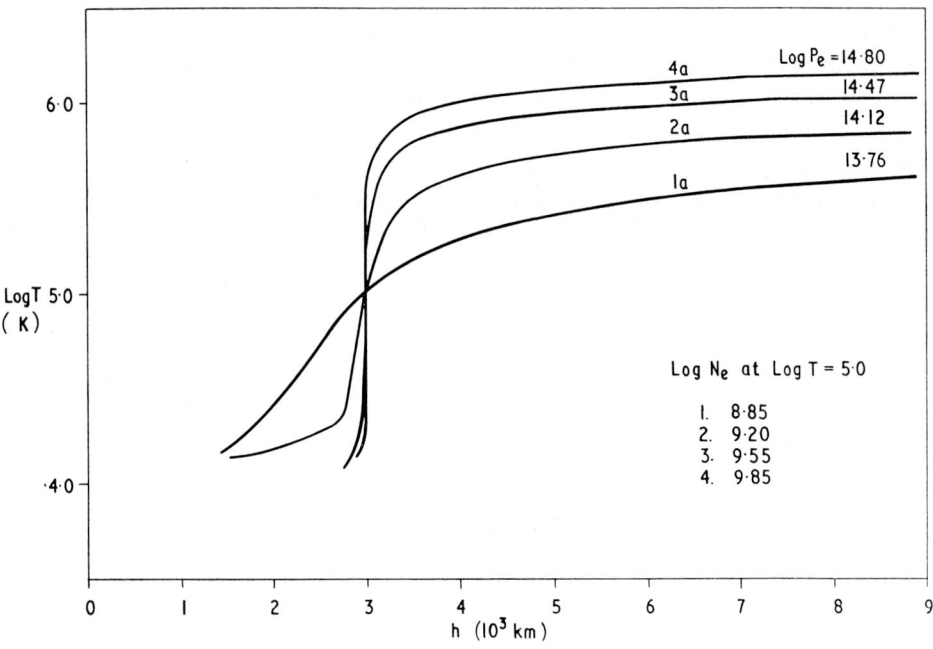

Fig. 3. The variation of temperature with height from models with different values of $P_e = N_e T_e$

imposed by other types of observations. These are:

(i) The maximum temperature (excluding active regions) in the corona occurs between about 20000 km and 60000 km, which excludes very high pressure models. The maximum temperature in the quiet corona, or rather the temperature at which most coronal material exists in the absence of active regions, is close to the temperature at which $\int_V N_e^2 \, dV$, has its maximum coronal value. From Figure 2 this is $T_e \simeq 1.8 \times 10^6$ K. More precisely, if Equation (9) is applied to the decreasing part of the distribution above the maximum, the gradient derived begins to increase again as T_e increases; the temperature at which this occurs can be taken as the quiet coronal temperature.

(ii) The forbidden line of Fe XI is observed undiminished in intensity to heights below 10000 km. Fe XI is formed at $T_e \sim 1.4 \times 10^6$ K, hence the temperature must reach $\sim 10^6$ K by 10000 km.

These boundary conditions limit the pressure to within a factor of two of $\log P_e = 14.75$ at $\log T_e = 5.0$.

The model derived by Jordan (1965, 1966) is shown in Figure 4 with several other models as a comparison. These are, the model derived by Pottasch (1964), the model derived from thermal conduction considerations by Woolley and Allen (1950), and the radio-emission model tabulated by Allen, (1963).

The absolute height of the transition region cannot be found from the extreme ultraviolet emission alone. For the model by Jordan shown in Figure 4, the height

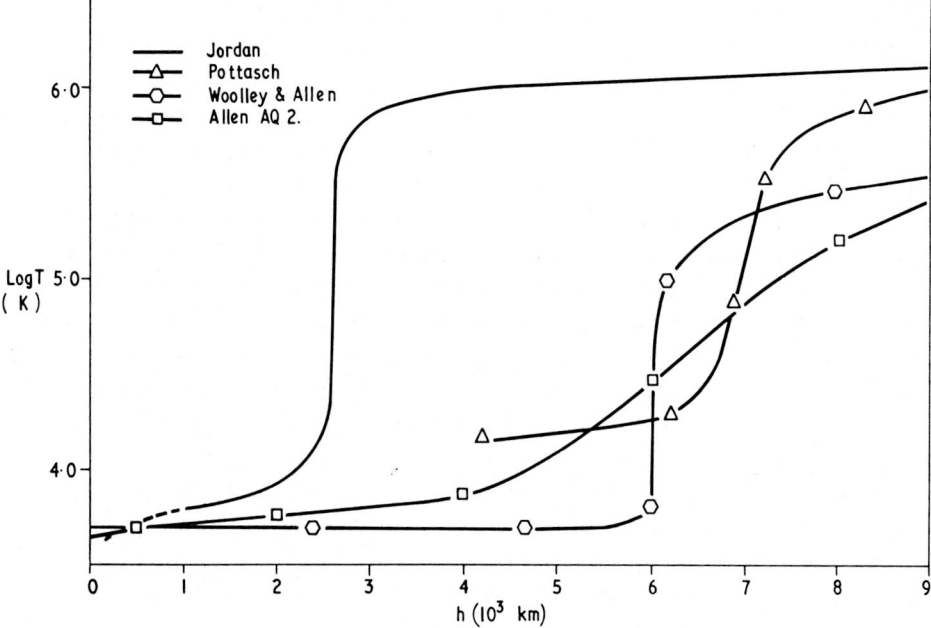

Fig. 4. The variation of temperature with height in several models of the transition region.

above $T = 5000$ K was found by assuming that the total gas pressure (found from various models in the literature), follows hydrostatic equilibrium.

Once a model is established, the initial assumptions concerning the $g(T)$ functions can be improved. At low temperatures, the points for singly charged ions can be replotted at the temperature at which $\int N_e^2 g(T) dh$ has its maximum value, this being lower than T_m. A similar correction should be applied to ions found just below and above the peak of the $\int N_e^2 dh$ distribution at $\sim 10^6$ K, but this was not done for the model in Figure 4.

Athay (1966) used a different approach by re-writing the integral in Equation (7) as

$$\int_R g(T) N_e^2 \, dh = P_e^2 \left(\frac{dh}{dT}\right) \int_R g(T) T_e^{-2} \, dT \qquad (11)$$

where P_e is assumed to be constant over the whole atmosphere, and dh/dT constant over the region where the line is formed. By plotting dT/dh against temperature

Athay noticed that over the temperature range 10^5 K–10^6 K, the gradient approximately satisfies the equation

$$T_e^{5/2} \frac{dT}{dh} = \text{const}. \tag{12}$$

which in the absence of a magnetic field corresponds to constant conductive energy flux. With $P_e = 6 \times 10^{14}$ cm^{-3} K the constant gives $F = 5 \times 10^5$ erg cm^{-2} sec^{-1} in reasonable agreement with the values discussed in Section 1.

Athay's model also shows a very steep temperature gradient between 5×10^4 K and 3×10^5 K. He did not include di-electronic recombination in his calculations but its inclusion does not alter this main result. Athay's method has been applied by Dupree and Goldberg (1967), with di-electronic recombination included. Their model is similar to that of Athay and they also find that the gradient fits a constant conductive flux. However, if the atomic data were reliable the gradient should be found from the distribution of ions from one element only, and not from ions of different elements, unless these overlap continuously in temperature range. For example, a relative error of a factor of 2 in the relative positions of ions from different elements will cause a factor of 2 error in the gradient derived. The gradient from ions of any one element appears to be larger than that from a constant conductive flux, which implies deposition of energy in the upper transition region and inner corona as well as close to the region of maximum temperature. In view of the overall factor of two uncertainty in intensity data, relative abundances, and atomic data, it is perhaps premature to conclude that the temperature gradients agrees precisely with those expected from constant conductive flux.

In conclusion, the extreme ultraviolet absolute intensity results show that the temperature gradients between the low chromosphere and corona are extremely steep, with over an order of magnitude rise in temperature over 100 km or less. The steep gradients will also occur around inhomogeneities such as spicules. To investigate the spatial distribution of the transition region other types of observations such as limb to disk intensity ratios must be used.

4. Interpretation of Limb/Disk Ratios

The method of deriving a structure from an interpretation of the solar disk spectrum, described above, suffers from a number of disadvantages. It depends on absolute photometry, a knowledge of atomic rate functions, the assumption of constant chemical abundance through the atmosphere, and a knowledge of the variation in electron density obtained either from internal assumptions or other observations. These disadvantages are removed if the interpretation is based on the relative intensities of spectral lines observed on the disk and at the limb. Other disadvantages accrue in that it is necessary to select quiet regions of the limb and disk where the atmosphere can be assumed to be the same; further, many strong lines become optically thick at the limb with consequent complication to the theory. In principle, the former can be overcome with a suitable pointing control and the latter by probing the spectrum deep enough to reveal optically thin lines, such as the intercombination transitions. With the appropriate instrumentation, observations of limb spectra have been obtained (Black et al., 1965; Burton and Wilson, 1965; Burton et al., 1967) leading to observations of limb and disk spectra and their interpretation (Burton et al., 1970).

In the analysis of the limb/disk ratios it is assumed that the atmosphere is spherically symmetric, but departures from this do not affect the results significantly in the case of optically thin lines if the size of the inhomogeneities is less than the width of the slit function. Analysis of optically thick lines can reveal any inhomogeneities.

a. LINES WHICH ARE OPTICALLY THIN ON THE DISK AND AT THE LIMB

Consider a fine slit parallel to the y-axis of width dx at projected height x above the limb, and consider the emission from a layer at radial distance r from the limb. Then the slit isolates a ring of thickness dy and radius y. The emission from this ring is given by

$$2\pi\varepsilon(r)\, y\, dy\, dx \quad (r > x) \quad \text{or zero} \quad (r < x)$$

where $\varepsilon(r)$ is the volume emission coefficient for the spectral line under consideration.

The total line emission from the plane parallel slab of thickness dx is then

$$dE = 2\pi\, dx \int_x^\infty \varepsilon(r)\, y\, dy$$

With a solar radius R, geometry gives

$$y\, dy = (R + r)\, dr$$

and hence

$$dE = 2\pi\, dx \int_x^\infty \varepsilon(r)\, (R + r)\, dr$$

$$\simeq 2\pi R\, dx \int_x^\infty \varepsilon(r)\, dr \quad \text{since} \quad r \ll R$$

In practise, due to aberrations and pointing errors the slit function projected at the source is not rectangular but is smeared out into some function $f(x - x_0)$, where x_0 is the centre of the distribution but is essentially an arbitrary constant. The function is normalized in terms of the slit width Δx so that

$$\int_{-\infty}^{+\infty} f(x - x_0) = \Delta x$$

The observed spectral line intensity at the earth (distance L) is then given by

$$I(\text{limb}) = \frac{R}{2L^2} \int_{-\infty}^{+\infty} \int_{x}^{\infty} f(x - x_0) \, \varepsilon(r) \, dr \, dx \tag{13}$$

As each emission line is formed close to the temperature at which the ion has its maximum abundance, the integral over dr can be replaced by a step function to give

$$\int_{x}^{\infty} \varepsilon(r) \, dr = \varepsilon_h \, \Delta h \quad \text{for} \quad h > x$$

$$= 0 \quad \text{for} \quad h < x$$

where Δh is the effective width of the layer located at height h. Hence,

$$I(\text{limb}) = \frac{R}{2L^2} \varepsilon_h \, \Delta h \int_{-\infty}^{h} f(x - x_0) \, dx \tag{14}$$

The disk flux, from an area covered by a slit of width Δx and length R is

$$I(\text{disk}) = \frac{R}{4\pi L^2} \Delta x \int_{0}^{\infty} \varepsilon(r) \, dr \tag{15}$$

$$\simeq \frac{R}{4\pi L^2} \Delta x \, \varepsilon_h \, \Delta h \tag{16}$$

From Equations (14) and (15), the limb/disk ratio is

$$\frac{I(\text{limb})}{I(\text{disk})} = \frac{2\pi}{\Delta x} \int_{-\infty}^{h} f(x - x_0) \, dx \tag{17}$$

b. LINES WHICH ARE OPTICALLY THICK ON THE DISK AND THE LIMB

If the optical depth is large then for a layer with constant source function the relative intensity at the centre of the disk and at the limb is given by the relative areas observed.

The area at the limb is the area of the segment of a circle of radius $(R+h)$ cut by a chord at distance $(R+x)$ from the centre. If θ is the angle between the equator and the intersection of the chord and the circle,

$$I(\text{limb}) \propto 2 \times \tfrac{1}{2}(R + h)^2 \, (\theta^c - \sin\theta \cos\theta)$$

In practise θ is small. Hence

$$I(\text{limb}) \propto \tfrac{4}{3} (2R)^{1/2} (h - x)^{3/2}$$

Introducing the slit function, the area viewed at the limb

$$= 2(2R)^{1/2} \int_{-\infty}^{h} (h-x)^{1/2} f(x-x_0) \, dx$$

The area of the slit on the disk is $R \, \Delta x$. Hence

$$\frac{I(\text{limb})}{I(\text{disk})} = 2\left(\frac{2}{R}\right)^{1/2} \frac{1}{\Delta x} \int_{-\infty}^{h} (h-x)^{1/2} f(x-x_0) \, dx \tag{18}$$

C. LINES WHICH ARE OPTICALLY THIN ON THE DISK BUT THICK ON THE LIMB

Since the emission included in the limb observations does not come from a simple plane parallel slab, and features such as spicules may be important, the approach adopted is that of considering the emission at the limb as coming from the depth at which $\tau_L = 1$. If this distance is $L(\tau=1)$ then the limb emission, restricting the region of emission in the radial direction to a layer of width Δh as before, is

$$I(\text{limb}) = \frac{(2R)^{1/2}}{2\pi L^2} \varepsilon_h \, L(\tau=1) \int_{-\infty}^{h} (h-x)^{1/2} f(x-x_0) \, dx \tag{19}$$

The disk emission is given by Equation (16). Hence

$$\frac{I(\text{limb})}{I(\text{disk})} = 2\left(\frac{2}{R}\right)^{1/2} \frac{1}{\Delta x} \frac{1}{\tau_h} \int_{-\infty}^{h} (h-x)^{1/2} f(x-x_0) \, dx \tag{20}$$

Hence, in this intermediate case, the height of the emitting layer can only be determined if the optical depth of the line at the centre of the disk is known.

The above methods of analysis have been applied by Burton et al. (1970) to data obtained during the flight of a sun-pointing Skylark rocket in April 1969. Normal-incidence optics were used and the collector mirror was finely controlled by a servo system which also had a programming capability for limb and disk setting during flight. Limb/disk ratios were measured for about 100 lines between 900 Å and 2000 Å. Absolute fluxes were also derived, the system being calibrated in the manner described by Burton et al. (1968). The photometric quality of the data was impaired by deterioration of the exposed film (Kodak 101-02) during a week's delay in locating the payload after flight. However, it is considered that ratios and absolute intensities are accurate within a factor of two. The slit function was determined from pre-flight laboratory measurements of a stationary image which were then convoluted with the flight pointing noise derived from the telemetry records, and the slit width of 8 arc sec. The position of the slit was set before launch but was not measured in flight. The heights derived are therefore relative positions of regions emitting the different lines.

Ideally the position of the slit would be derived from limb/disk ratios in the continuum, making use of known limb darkening coefficients.

The data were divided into groups corresponding to optically thin, intermediate and optically thick lines and analysed by the expressions outlined above. The absolute disk measurements were also analysed by the method described in the previous section. Figure 5 shows the results obtained from the optically thin and optically thick lines. These are presented without any smoothing. The steep form of the transition zone between 10^4 and 3×10^5 K from the disk intensities alone is illustrated by a vertical line.

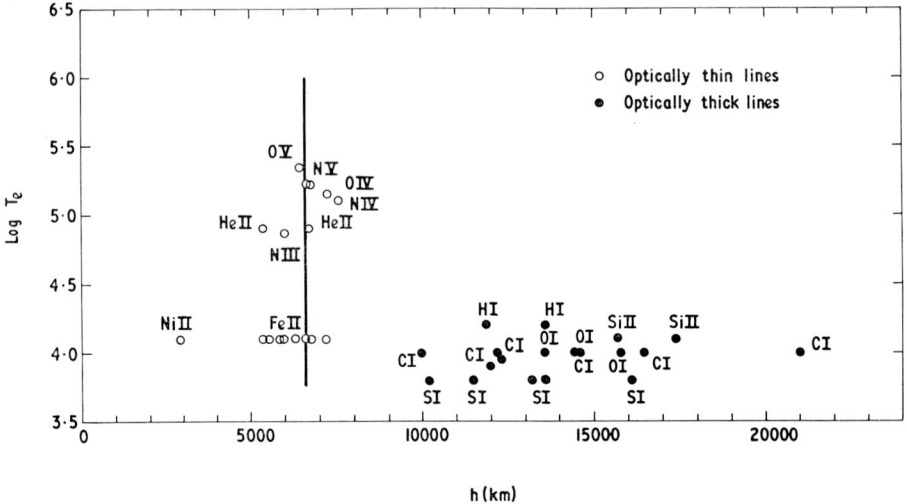

Fig. 5. The temperature structure derived from optically thin and optically thick lines.

In the case of the optically thin analysis, the height from each line is the average height of emission for that line. Inhomogeneities will not affect the derived temperature structure if the height of the inhomogeneities is less than the width of the slit function, in this case ~ 30 arc sec or 21700 km. It is apparent that the results are consistent with a steep rise in temperature between 1.2×10^4 K and 2.2×10^5 K. The uncertainty in the height for each line, at $h \sim 6500$ km (on the relative scale), assuming that the limb/disk ratios are correct to within a factor of two, is ± 700 km (i.e. ± 1 arc sec). Within this uncertainty the limb/disk results support the steep temperature gradients found from absolute intensity analysis.

In the case of the optically thick analysis, it is clear that the results are quite different. These lines are generally representative of much lower temperatures than the optically thin lines and indicate low temperature material extending well above the transition zone. Consider the effect of spicules at the limb. If $\tau_h \sim 50$, the distance to $L(\tau = 1)$, is very small, about 1 km for the hydrogen lines. This is less than the diameter of one spicule so that spicules superimposed in the line of sight would raise the average height of the top of the emitting layer. Lines of neutral ions would then apparently be formed above lines of the transition region ions in which the spicules are optically thin.

Although the optical depth in a line of a neutral atom is certainly very large if the integration is continued down to the photosphere, it is difficult to find the optical depth in the region from which the observed emission originates without complex calculations of the source function. However, the H I, O I and strong C I lines certainly have $\tau_h > 1$. The heights of the weaker lines may be overestimated if $\tau_h < 1$, and then the intermediate analysis would be appropriate. But these corrections would not be sufficient to give neutral emission from heights below the optically thin Fe II lines. Another source of error for lines in which spicules are optically thin, is that the limb emission could originate in a region where the optical depth is smaller and the temperature is greater than in the region where the disk emission originates. This will not affect lines for which $L(\tau=1) \lesssim 700$ km.

Thus the model derived is one in which cool material extends into the corona to a height of $\sim 10^4$ km, presumably in the form of spicules, with the steep transition region existing between and around the spicules.

Although the limb/disk results from optically thin lines support a steep rather than slow temperature rise between the low chromosphere and inner corona, from Figure 3 it can be seen that with a height resolution of ± 700 km it is not possible to distinguish between models with P_e greater than 2×10^{14} cm^{-3}K.

The lines of 'intermediate' optical depth provide a method for finding the correct electron pressure.

The optical depth can be expressed in terms of the absorption coefficient at the line centre, in the form

$$\tau_h = 1.2 \times 10^{-15} \lambda_{12} f_{12} M^{1/2} \frac{N(E)}{N(H)} \int_{\Delta h} \frac{N_i}{N(E)} N_H T_e^{-1/2} \, dh \qquad (21)$$

where M is the atomic weight of the atom. For $T_e > 2 \times 10^4$ K $N_H \simeq 0.80 N_e$. The quantity inside the integral is similar to that in the equation for the absolute intensities. If τ_h is computed from the model derived from the absolute intensities then the value depends mainly on the absolute intensity of the line and the value of $N_e T_e$ chosen for the model.

For lines which are optically thin at the centre of the disk but optically thick at the limb the optical depth is used in calculating the height of the emitting layers (see Equation 20). If the incorrect value of $N_e T_e$ has been used the heights derived will not agree with those found from the optically thin or thick lines. Also variations in $N_e T_e$ as a function of T_e will be apparent. The heights derived for C IV, Si IV and O VI in Figure 6 are systematically lower than those for He II, N IV and O V in Figure 5, the difference corresponding to a factor three in the optical depths, in the sense that the computed values are too small. Comparison of SL 606 absolute intensities with those of Hall et al. (1963) suggests that the former are underestimated by about a factor of 3. This suggests that the chosen value of $N_e T_e = 5.6 \times 10^{14}$ cm^{-3} K is correct to well within 50%.

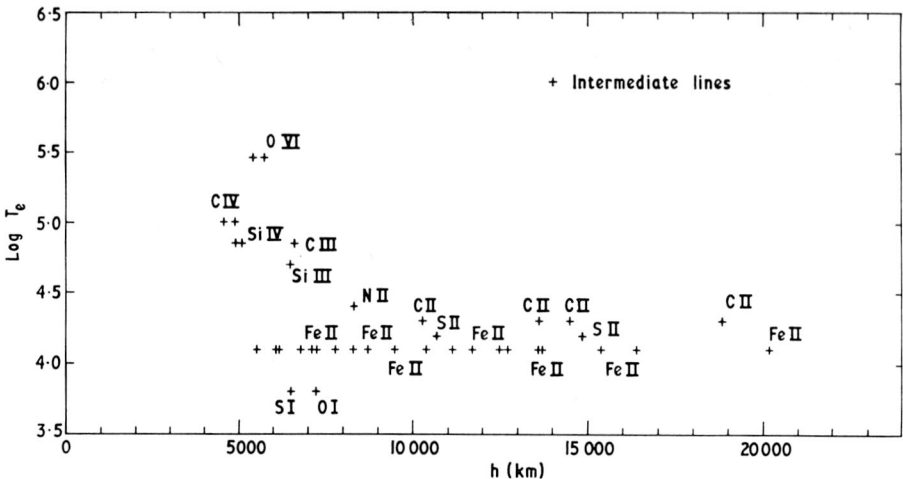

Fig. 6. The temperature structure derived from lines of 'intermediate' optical depth.

Future work on the limb/disk method is in progress with the aim of reaching fainter lines and improving the photometry. In principle it is possible to determine the relative intensities to $\pm 25\%$, and it is hoped that this will be achieved in the next flight. The uncertainty in the height would then be only ± 350 km and the electron pressure could be determined with greater precision. It is also planned to extend the method into the grazing incidence region to include lines formed in the upper part of the transition zone and the inner corona.

5. Interpretation of XUV Flash Spectra

In addition to the methods described above, another category of observations which are important to the study of chromospheric and coronal structure are those which have an intrinsically high spatial resolution. The accomplishment of this in the extreme ultraviolet to the kind of accuracy desirable (<1 arc sec) presents formidable problems of optical, mechanical, thermal and pointing stabilities in space vehicles. However, high spatial resolution can be extracted from observations during a solar eclipse due to the relative motions of the moon and sun. The application of such techniques to the extreme ultraviolet also presents considerable technical difficulties but recently groups from Imperial College, Harvard, Toronto and Culham carried out a successful experiment during the total solar eclipse of 7 March 1970 when flash spectra throughout all phases of the eclipse were obtained in the extreme ultraviolet for the first time (Speer *et al.* 1970). The experiment was carried out on a stabilized Aerobee rocket launched into the eclipse path from Wallops Island and fifty spectra were obtained in the range 900 Å–2200 Å covering a period from before second contact to mid-totality. The spatial separation of successive frames due to the apparent motion of the moon was estimated as 1.3 arc sec. Speer *et al.* (1970) give a brief description of the data which indicates the wealth of new information they contain. Just after second contact the chromospheric flash spectrum dominates but has disappeared by mid-totality when coronal lines, appearing as complete rings are the dominate feature of the spectrum. Forbidden lines are seen and an extensive corona in Ly-α produced by resonance fluorescence. The data also isolates the spectra of several limb prominences and reveals regions of activity. It is clear that these observations will contribute significantly to the knowledge of the chromosphere and corona. Here we will consider the aspect with which this paper is concerned i.e. the means by which a knowledge of the structure of the quiet atmosphere can be derived from such observations.

The method of analysis bears some similarity to that of the limb spectra discussed above. Using the same solar coordinate system, i.e. the z-axis towards the earth and the x-axis through the point at the lunar limb to be investigated (say at some positional angle θ), then the total observed intensity of a spectral line S_λ above the moon's limb (located at position x_n) over an element of length dy, is, for optically thin transitions:

$$I(S_\lambda, \theta, x_n) = \frac{dy}{4\pi L^2} \int_{x_n}^{\infty} \int_{-\infty}^{+\infty} \varepsilon(x, y, z) \, dx \, dz \tag{22}$$

where $\varepsilon(x, y, z)$ is the volume emission coefficient at position (x, y, z). The observations will give values of I for all optically thin lines S_λ, at various position angles θ in the spectral arcs and for the various positions x_n of the moon's limit. Any solar model can then be used to calculate $\varepsilon(x, y, z)$ and hence the double integral in Equation (22) for comparison with observation. In principle, a model consistent with the observations can then be derived.

In practice, most models will be in a spherically symmetric form and in this case the expression for the total intensity above the moon's limb becomes

$$I(S_\lambda, \theta, x_n) = \frac{dy}{2L^2} \int_{x_n}^{\infty} \int_{x}^{\infty} \frac{(R+r)\varepsilon_r}{\sqrt{(R+r)^2 - (R+x)^2}} \, dr \, dx \tag{23}$$

where ε_r is the emission coefficient of the line at radial height r and R and L are the solar radius and earth distance, respectively, as before.

Making the same approximations as in the limb analysis, i.e. that each line emission can be represented as coming from a narrow layer of height h, width Δh and volume emission coefficient ε_h, together with $h \ll R$, allows equation (23) to be solved approximately to give

$$I(S_\lambda, \theta, x_n) = \frac{dy}{2\pi L^2} \varepsilon_h \Delta h \sqrt{2R(h - x_n)} \quad (h > x_n) \tag{24}$$

This expression can be used to easily derive approximate expressions for heights against temperature and hence form the start of a possible iteration to find a final solar model.

This method of determining structure is directly susceptible to inhomogeneities such as spicules and will reveal their extent into the corona. It is therefore complimentary to the methods discussed above and, taken altogether, they present a considerable hope for advancing the knowledge of the structure of the upper solar atmosphere.

References

Allen, C. W.: 1961, *Mem. Soc. Roy. Liège* **4**, 241, 1961.
Allen, C. W.: 1963, *Astrophysical Quantities*, Second Edition, University of London, The Athlone Press, p. 174.
Athay, R. G.: 1966, *Astrophys. J.* **145**, 784.
Biermann, L.: 1946, *Naturwiss.* **33**, 118.
Biermann, L.: 1947, *Naturwiss.* **34**, 87.
Biermann, L.: 1948, *Z. Astrophys.* **25**, 161.
Black, W. S., Booker, D., Burton, W. M., Jones, B. B., Shenton, D. B., and Wilson, R.: 1965, *Nature* **206**, 254.
Burgess, A.: 1964, *Astrophys. J.* **139**, 776.
Burgess, A. and Seaton, M. J.: 1964, *Monthly Notices Roy. Astron. Soc.* **127**, 355.
Burton, W. M. and Wilson, R.: 1965, *Nature* **207**, 61.
Burton, W. M., Ridgeley, A., and Wilson, R.: 1967, *Monthly Notices Roy. Astron. Soc.* **135**, 207.
Burton, W. M., Ridgeley, A., and Hatter, A. T.: 1968, ESRO SP-33, 145.
Burton, W. M., Jordan, C., Ridgeley, A., and Wilson, R.: 1970, *Royal Society Discussion on Solar Physics*, London (to be published) 1970.
De Jager, C. and Kuperus, M.: 1961, *Bull. Astron. Inst. Neth.* **16**, 71.
Dupree, A. K. and Goldberg, L.: 1967, *Solar Physics* **1**, 229.
Gabriel, A. H.: 1970, private communication.
Garz, T. and Kock, M.: 1969, *Astron. Astrophys.* **2**, 274.
Garz, T., Holweger, H., Kock, M., and Richter, J.: 1969, *Astron. Astrophys.* **2**, 446.
Goldberg, L., Dupree, A. K., and Allen, J. W.: 1965, *Ann. Astrophys.* **28**, 589.
Hall, L. A., Damon, K. R., and Hinteregger, H. E.: 1963, *Space Research*, 3, North-Holland Publ. Co., Amsterdam, 745.
Ivanov-Kholodnyi, G. S. and Nikol'skii, G. M.: 1962, *Astron. Zh.* **39**, 777.
Jordan, C.: 1965, Ph.D. Thesis, University of London.
Jordan, C.: 1966, *Astron. J.* **71**, 860.
Jordan, C.: 1966, *Monthly Notices Roy. Astron. Soc.* 463.
Jordan, C.: 1969, *Monthly Notices Roy. Astron. Soc.* **142**, 501.
Kopp, R. A.: 1968, Ph.D. Thesis, Harvard University.
Kuperus, M.: 1965, *Rech. Astron. Obs. Utrecht* **17**, 1.
Kuperus, M.: 1969, *Space Sci. Rev.* **9**, 713.
Oster, L.: 1956, *Z. Astrophys.* **40**, 28.
Osterbrock, D. E.: 1961, *Astrophys. J.* **134**, 347.
Pottasch, S. R.: 1963, *Astrophys. J.* **137**, 945.
Pottasch, S. R.: 1964, *Space Sci. Rev.* **3**, 816.
Pottasch, S. R.: 1967, *Bull. Astron. Inst. Neth.* **19**, 113.
Schatzman, E.: 1949, *Ann. Astrophys.* **12**, 203.
Schirmev, H.: 1950, *Z. Astrophys.* **27**, 132.
Schwarzschild, M.: 1948, *Astrophys. J.* **107**, 1.
Speer, R. J., Garton, W. R. S., Goldberg, L., Parkinson, W. H., Reeves, E. M., Morgan, J. F., Nicholls, R. W., Jones, T. J. L., Paxton, H. J. B., Shenton, D. B., and Wilson, R.: 1970, *Nature* **226**, 249.
Wilson, R.: 1962, *J. Quant. Spectr. Radiative Transfer* **2**, 477.
Wilson, R.: 1967, Plasmas in Space and in the Laboratory, ESRO SP-20, 373.
Woolley, R. v. d. R. and Allen, C. W.: 1948, *Monthly Notices Roy. Astron. Soc.* **108**, 292.
Woolley, R. v. d. R. and Allen, C. W.: 1950, *Monthly Notices Roy. Astron. Soc.* **110**, 25.

16. X-RAY SPECTROSCOPY OF SOLAR ACTIVE REGIONS AND FLARES

WERNER M. NEUPERT

Solar Plasmas Branch, Laboratory for Solar Physics, NASA-Goddard Space Flight Center, Greenbelt, Md 20771, U.S.A.

1. XUV Spectroscopy of Active Regions

The study of emission lines from highly ionized heavy ions has, since the pioneering work of Grotrian and Edlén, been a basic method for probing the physical characteristics of the lower solar corona where these lines originate (Billings, 1966; Zirin, 1966). The method of observing forbidden lines in the visible spectrum has been difficult because of the photospheric continuum which limits observations to the solar limb either with sophisticated coronographs or during times of total solar eclipse. In addition, not all elements or stages of ionization of an element have forbidden transitions conveniently located in the visible range of wavelengths where spectral analysis can be made on the ground. With the advent of rockets and satellites, the possibility of recording resonance lines of highly stripped ions in the X-ray and EUV spectrum gives us an even more powerful tool for studying the solar corona. Whereas, for example, there are no emission lines from highly ionized silicon in the visible spectrum, the XUV spectrum has emission lines from Si IX through Si XIV. Instead of observing five stages of ionization of iron, Fe X, Fe XI, Fe XIII, Fe XIV, and Fe XV, one can record resonance lines from 18 consecutive stages of ionization, from Fe IX through Fe XXVI. The implications, so far as a study of active regions and flares is concerned, can be appreciated by studying Figures 1 and 2, which give ionization equilibria for silicon and iron, respectively, as calculated by Jordan (1969). In the electron temperature range from 1 to 2 million deg, which includes the temperature of the 'undisturbed' corona, we expect silicon to exist as Si X, Si XI, and Si XII, i.e., atoms from which 9, 10, and 11 electrons have been removed. Note that Si XII exists at more than 4 million deg and its appearance gives evidence for the existence of hotter regions in the corona. In the range from 2 million to 10 million deg helium-like Si XIII is prominent and at higher temperatures Si XIV appears. Thus, the observations of Si XIV gives evidence for regions having electron temperatures of between 5 million and 20 million deg.

Ions of a specified iso-electronic sequence become increasingly more difficult to ionize with increasing atomic number. Thus, C V (a carbon atom with 4 electrons removed) will be the predominant ion of that element in a plasma having an electron temperature between 160000 K and 600000 K. O VII will exist principally between 400000 K and 1200000 K, and is therefore expected in the quiet corona, for which the electron temperature lies in the range from 1.2 to 1.8×10^6 K. As we have said already, Si XIII exists between 2 and 10×10^6 K and will therefore be associated with regions of enhanced electron temperature.

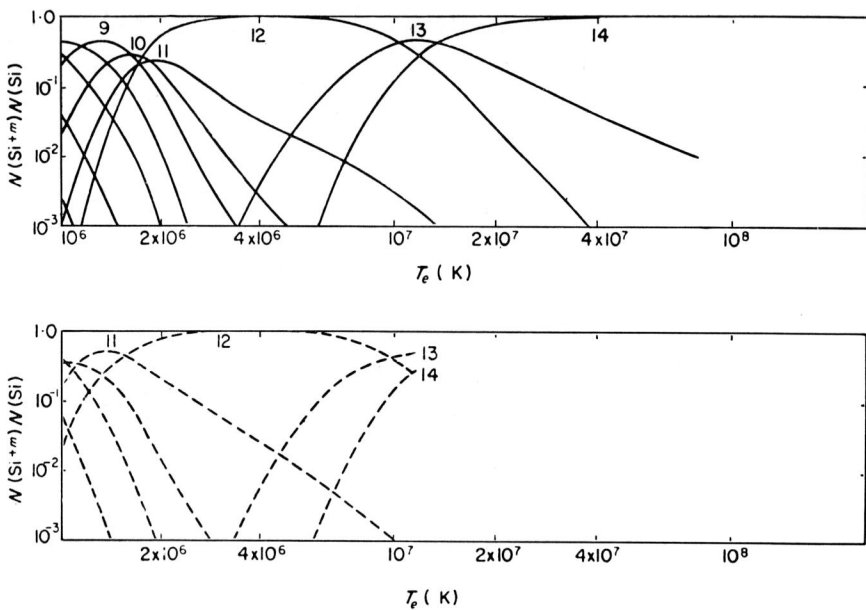

Fig. 1. The ionization equilibrium of Si ions applicable to the solar corona and chromosphere (Jordan, 1966). The dashed lines show the results of House which did not include di-electronic recombination or ionization via auto-ionizing levels (House, 1964). The charge on each ion is indicated.

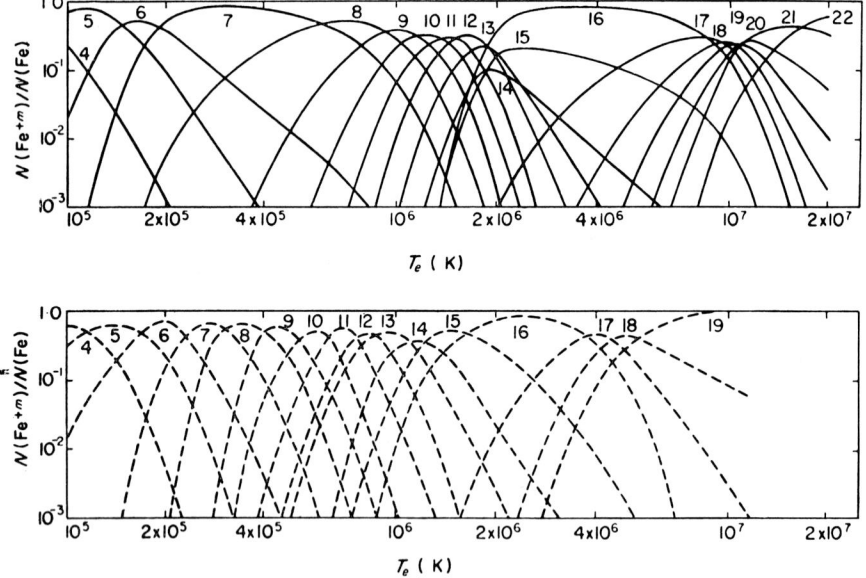

Fig. 2. The ionization equilibrium of Fe ions. Jordan's (1966) results are again compared with those of House (1964).

Figure 2 shows the distribution of ions for each stage of ionization of iron as a function of the electron temperature of the plasma. At electron temperatures of 1×10^6 K we expect to find predominantly Fe IX, Fe X, and Fe XI. The ionization balance passes rapidly through these stages of ionization with increasing electron temperature so that observation of emission lines from one of these ions is a sensitive indication that plasma is present with an electron temperature of approximately 1 million deg. With increasing electron temperature, the ionization balance passes to yet higher stages of ionization. Because we can observe only up to Fe XV in the visible spectrum, it is not possible to measure the temperature in very active regions. This is no problem when observing in the XUV spectrum where we can record radiation from more highly stripped iron atoms up to Fe XXVI, in the hydrogen-like isoelectronic sequence, which will exist in plasmas at electron temperatures of $50-100 \times 10^6$ K. Note the broad distribution with temperature of Fe XVII (+16) of the neon-like isoelectronic sequence. This ion exists over a wide range of electron temperature because it is very stable against further ionization, 1261 eV being required to ionize it. It therefore exists over a wide range of coronal conditions but especially in association with active regions. Even near solar minimum, in 1964, Blake *et al.* (1965) found the lines of Fe XVII to be prominent in the solar X-ray spectrum at 15 Å and 17 Å. Ions of the helium-like isoelectronic sequence are also difficult to ionize further and therefore also exist over a wide range of coronal conditions. Prominent emission lines from these ions are found in the X-ray spectrum at 22 Å (O VII), 13.5 Å (Ne IX), 9.2 Å (Mg XI) and 6.7 Å (Si XIII). A recent spectrum of this region, taken from OSO-5 when no flares were in progress, is given in Figure 3. Of particular interest are strong features found on the long wavelength side of the resonance $(1\,^1S_0 - 2\,^1P_1)$ and intersystem $(1\,^1S_0 - 2\,^3P_1)$ transitions of O VII, Ne IX and Mg XI. The identification of these features as the forbidden transition $1\,^1S_0 - 2\,^3S_1$ in the helium-like ion was first made by Gabriel and Jordan (1969a) for O VII by establishing that the unidentified

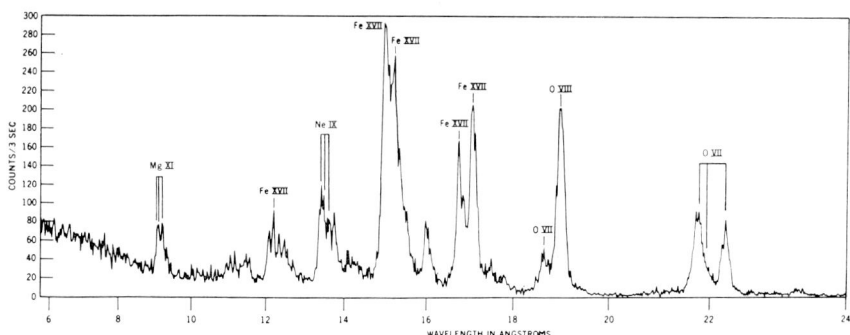

Fig. 3. The solar spectrum between 6 Å and 25 Å recorded in the absence of flare activity by the Orbiting Solar Observatory, OSO-5, January 27, 1969.

solar feature at 22.09 Å did not coincide in wavelength with any known satellite lines of oxygen measured in the laboratory, but did correspond to the predicted position of the forbidden transition. From their analysis of existing solar data they also found this transition in Cv, Neix, Mgxi. The transition was subsequently identified in Sixiii and Sxv by Walker and Rugge (1970) and in Caxix and (probably) in Fexxv by Neupert and Swartz (1970). The Grotrian diagram for the helium-like ion is shown in Figure 4 in which transitions can be identified by arrows between energy levels (wavelengths given are for silicon and iron).

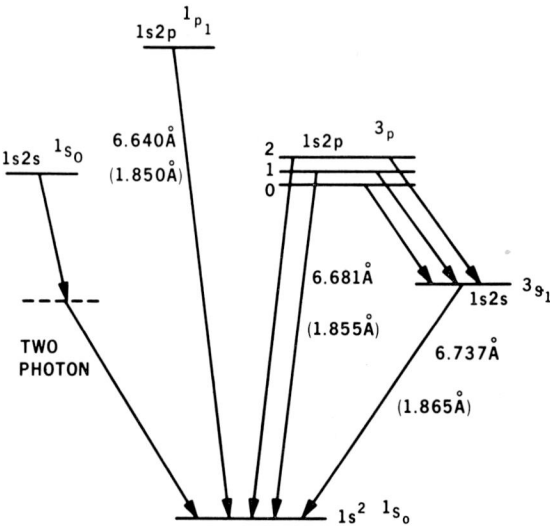

Fig. 4. Grotrian diagram for the helium-like ion. Wavelengths indicated are those observed for Sixiii and Fexxv (Neupert and Swartz, 1970).

Before proceeding further into the X-ray spectrum, let us mention the resonance transitions of Feix through Fexvi which produce spectral lines at wavelengths between 170 Å and 400 Å. As we see from Figure 1, the populations in these stages of ionization are sensitive to small changes in electron temperature in the range of typical coronal temperature. Boardman and Billings (1969) have used this fact to analyze the physical characteristics of a coronal enhancement observed in the extreme ultraviolet with a slitless spectrograph by the US Naval Research Laboratory during a rocket flight (Tousey, 1967). They established the relative amounts of material within the enhancement at different temperatures, finding a hot 'core' of 20000 km height and 30000 km radial distance (the instrumental resolution corresponded to a solar scale of 7000 km) within which essentially all the emission from temperatures above 3×10^6 K took place. This core was surrounded by a shell of material at lower temperatures.

In order to establish that changes in such physical conditions take place during the evolution of active regions, one requires data over an extended period of time, which can only be acquired by an orbiting spacecraft. An example of such data, taken by the first grating spectrometer pointed at the Sun from a satellite, OSO-1 (Behring *et al.*, 1963), is shown in Figure 5. The fact that successive stages of ionization have slightly

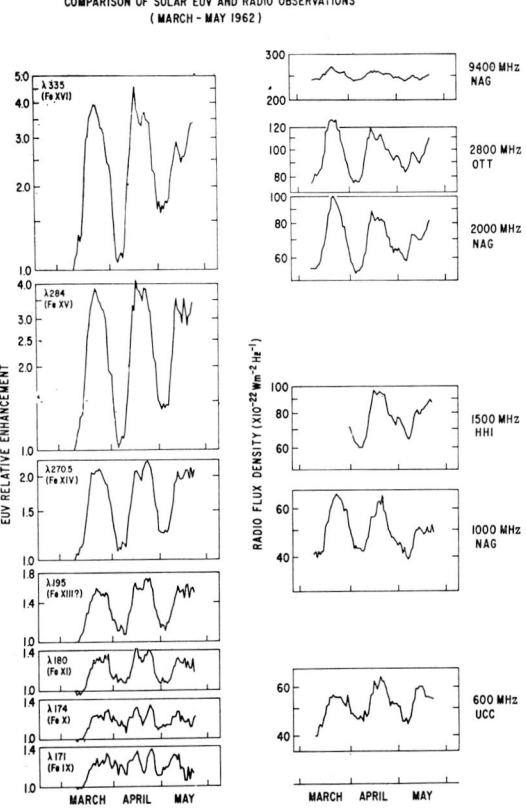

Fig. 5. Variation of the EUV line emission and radio emission integrated over the solar disk during period of nearly three solar rotations. These variations are associated with the appearance and disappearance of solar active regions as a result of the Sun's rotation.

different associations with solar active regions can be established by comparing the variability of their emission lines over several solar rotations. Although the OSO-1 instrument had no capability for spatially resolving active regions on the Sun's disk, the repeated observation of only one or two active regions over several solar rotations, separated by periods of very low solar activity (Zurich Provisional Relative Sunspot Number equal to zero), made it possible to establish changes in the ionization balance of the corona above active regions as these regions evolved on the solar disk (Neupert, 1967). The observations obtained from the OSO-1 data are summarized in Figure 6.

Notice that the highest stages of ionization are most prominent early in the lifetime of an active region when solar flares are most frequent. With the passage of time, as activity subsides, Fe XIV and lower stages of ionization become predominant. At this stage of evolution of an active region the sunspot area has already substantially decreased but the associated Ca II plage area is still high. This is probably the underlying physical basis for the argument made by Xanthakis (this meeting) that an 'Areas Index' including both sunspot and plage areas more closely matches the integrated intensity of the Fe XIV green line emission than does sunspot area alone. The OSO-1 data was used (Neupert, 1967) to derive an electron temperature of 4.2×10^6 K for

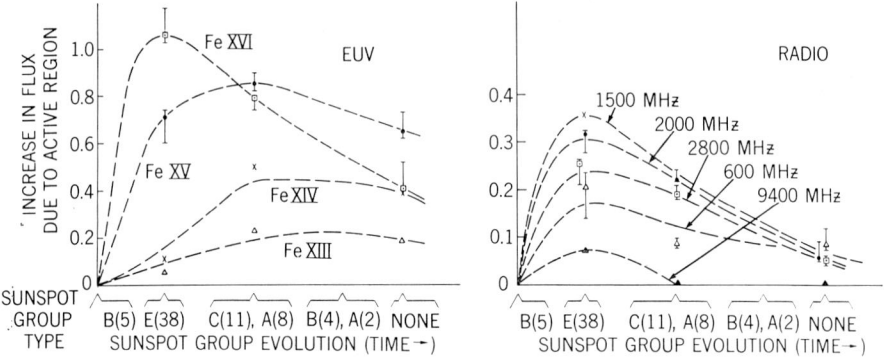

Fig. 6. Growth and decline of EUV line emission and radio emission during the evolution of a typical active region as observed by OSO-1. The increases in emission are normalized to the total solar emission in each line recorded in the absence of appreciable solar activity. Note that maximum emission in the highest observed stages of ionization of iron (Fe XVI) coincides with maximum complexity and size of the sunspot group.

the bulk of the material above an E-spot group and a ten-fold increase in coronal electron densities (9×10^9 cm^{-3} at a height of 20000 km). With the cessation of activity the average temperature decreased to $2.5-3.0 \times 10^6$ K and the average density to approximately 2.5×10^9 cm^{-3} at 20000 km. For an intermediate level of activity, Evans and Pounds (1968) derived a two component model having $N_e^2 V = 1.7 \times 10^{48}$ cm^{-3} at 3×10^6 K and $N_e^2 V = 1.0 \times 10^{47}$ at 4×10^6 K. For a region of 1 arc min extent in diameter and height they calculated a density of 5×10^9 cm^{-3}. We know, of course (Dollfus, this meeting), that coronal features are often complex and highly structured, so that even the most recent XUV spectroheliograms and X-ray photographs (which we will mention later), with a resolution of 3–5 arc sec, permit us to place only lower limits on the electron density, and the figures just given must be understood in that way. Recently, however, Gabriel and Jordan (1969b) in connection with their work on the identification of the $1\,^1S_0-2\,^3S_1$ transition in ions of the helium-like isoelectronic sequence, have found that the intensity of this line relative to the intersystem transition ($1\,^1S_0-2\,^3P_1$) can be used to measure actual electron density in the region in which the line is generated without the need for spatially resolving the region.

They point out that the populations of the $2\,^3P$ and $2\,^3S$ terms are dependent on processes between these excited states. These processes are collisional excitation of $2\,^3P$ from $2\,^3S$, radiative decay from $2\,^3P$ to $2\,^3S$ and photoexcitation of $2\,^3P$ from $2\,^3S$ (which is only important for C v). From a solution of the coupled rate equations for populating and depopulating the $2\,^3P$ and $2\,^3S$ terms they derive the ratio of forbidden to intersystem line intensities which has a density dependent term – the collisional excitation rate from $2\,^3S$ to $2\,^3P$. This ratio is sensitive to changes in density when the density is above a limit given by

$$N_e \sim \frac{A(2\,^3S \to 1\,^1S)}{C(2\,^3S \to 2\,^3P)}$$

where A is the transition probability and C the collisional excitation rate between the terms indicated. For C v the lower limit is 1×10^9 cm^{-3}. This limit is already higher than the densities at which the C v ion exists, so that no real changes in the ratio reflecting changes in electron density have been confirmed. For O vii and heavier ions the observed ratios imply that electron densities in active regions range from 3×10^{10} cm^{-3} for O vii up to 10^{12} cm^{-3} or even more for Si xiii (Gabriel and Jordan, 1970). We shall return to this valuable technique in our discussion of X-ray flare events.

We must realize that these data are still far from an accurate description of the complex features which compose the corona above active regions. This is best illustrated by examining an actual image of the corona, recorded by Van Speybroeck *et al.* (1970) on March 7, 1970, the day of the total solar eclipse. The image in Figure 7 was taken with a nominal passband of 3–30 Å and 44–55 Å. It therefore represents emission not only from the hotter portions of the corona which at temperatures of 4–6×10^6 K are strongly emitting Fe xvii, Mg ix and Si xiii in the 3–30 Å range, but also cooler portions of the corona, which are represented by emission lines from Si ix, Si x, Si xi, in the band from 44 Å to 55 Å. The photograph shows that the coronal structures above action regions often have a loop-like appearance, these loops sometimes arching from one active region to another. In a shorter exposure, the X-ray plages associated with active regions often appear to have one or more bright narrow cores, typically 10 to 20 arc sec in length and having a striated appearance with an apparent width comparable to the resolution of the X-ray telescope (3 to 5 arc sec). In some cases these cores appear to take the form of tubes which link the preceding and following magnetic polarity portions of the active regions. These may be the regions of high electron density inferred from the ratio method and therefore may have densities of 10^{12} cm^{-3} or even greater.

The ability to record the corona against the disk of the Sun, and consequently the ability to associate coronal structures with photospheric magnetic fields has also enabled Van Speybroeck *et al.* to identify many bright diffuse X-ray emitting regions with remnants of old active regions of the type giving rise to unipolar magnetic regions. Such regions probably have electron temperatures between 2.0 and 3.0×10^6 K and densities of the order of 3×10^9 cm^{-3}.

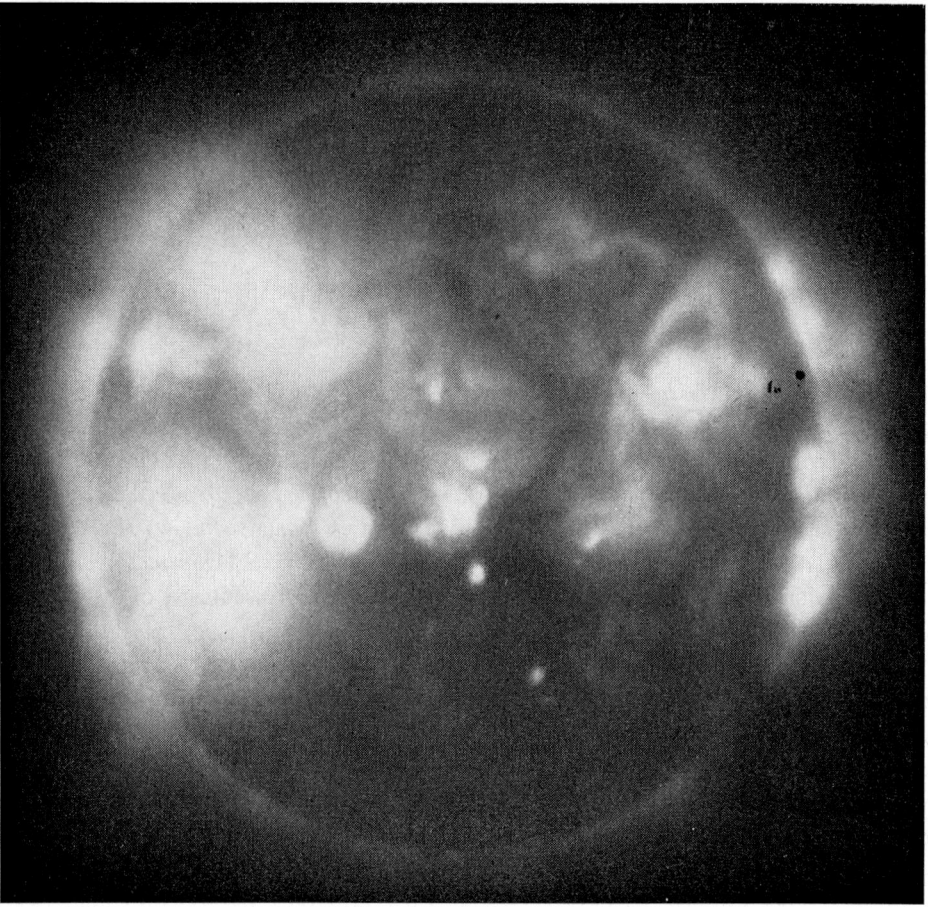

Fig. 7. The appearance of the corona in X-rays on March 7, 1970. This image, recorded by Van Speybroeck *et al.* (1970) represents solar emission in the wavelength bands of 3–30 Å and 44–55 Å. Courtesy of L. Van Speybroeck, *American Science and Engineering*.

2. X-Ray Spectroscopy of Flares

It has long been established, on the basis of broad-band X-ray observations, that the solar flux between 1.0 Å and 20 Å is enhanced, sometimes by many orders of magnitude, during solar flare events. At wavelengths below 1 Å where the radiation often appears in short bursts in good temporal correspondence to impulsive microwave bursts, we are apparently observing a continuum produced as a result of the interaction of high-energy non-thermal electrons with the ambient matter and magnetic field in the lower corona (Frost, 1969). At longer wavelengths, above 1 Å, we find that the emission is composed not only of a continuum, but of line emission as well (Neupert *et al.*, 1967; Meekins *et al.*, 1968). Examples of marked changes which appear in the spectrum are depicted in Figure 8 and Figure 9, both being records of

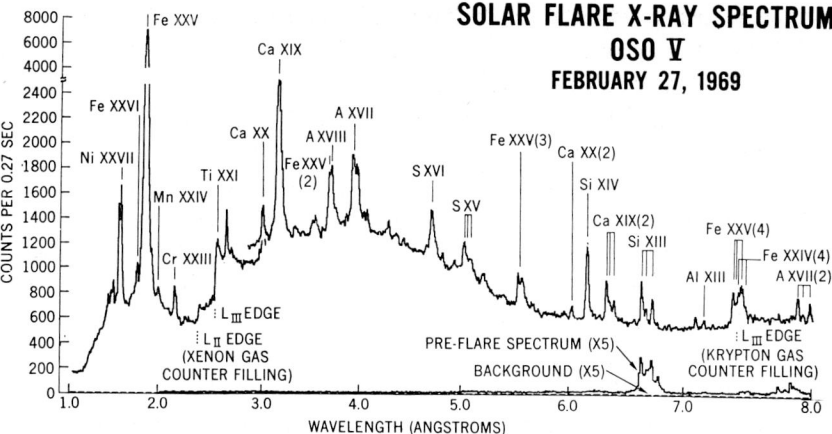

Fig. 8. X-ray spectrum associated with a large solar flare on February 27, 1969. The spectrum is a composite of four spectral scans taken with each of two crystal spectrometers (covering the ranges 1–3 Å and 3–8 Å) during the period of maximum X-ray emission. Higher orders of diffraction are indicated in parentheses.

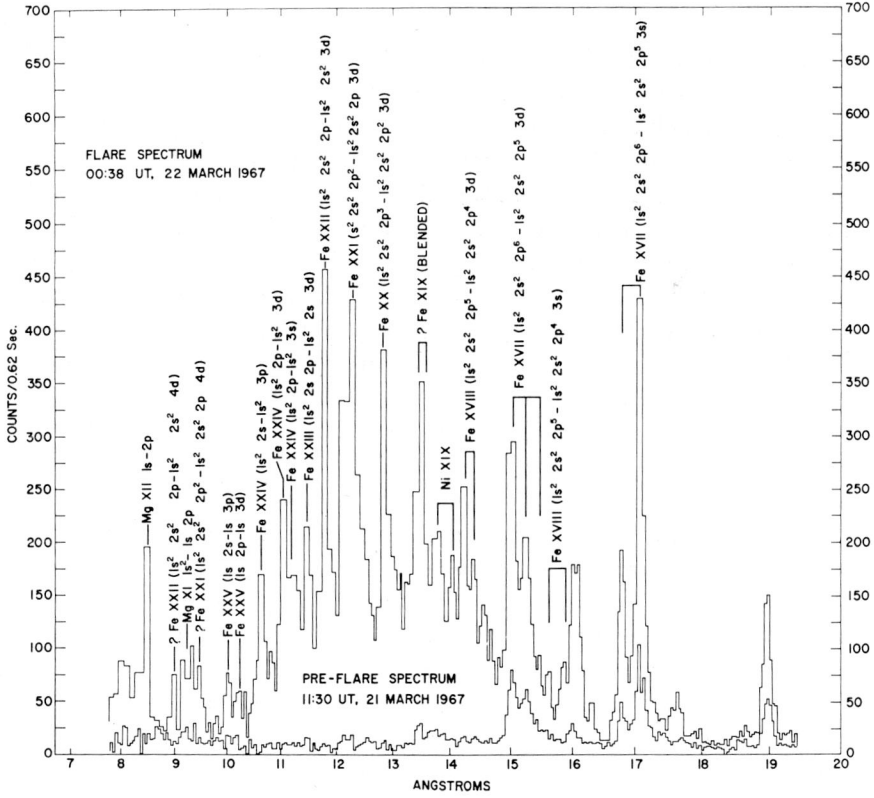

Fig. 9. Comparison of a solar spectrum between 6.6 Å and 20 Å obtained during a flare on March 22, 1967, with a spectrum obtained on the previous day when no flares were in progress. Identification of transition arrays of Fe xxv–Fe xx observed only during the flare are indicated.

spectra taken during Importance 2B flares. The intense feature between 1.85 and 1.88 Å (Figure 8) is apparently a blend of emission lines from highly ionized iron, being primarily Fe XXV and Fe XXIV. The component lines can be identified by examining the feature in higher spectral orders (Neupert and Swartz, 1970) which appear at multiples of the actual wavelength. Such higher orders are noted in parentheses in Figure 8. At 1.78 Å there appears a weak feature (observed with certainty only during intense soft X-ray events) which can be attributed to the Ly-α line of Fe XXVI. Additional radiation also appears between 1.88 Å and 1.94 Å. This probably arises as the result of transitions of the type $1s^2\,2s^2\,2p^n-1s2s^22p^{n-1}$ ($1 \leq n \leq 6$), often referred to as inner-shell or Kα type transitions. In contrast to emission lines from lighter elements such as calcium and silicon, where satellite lines of the lithium-like ion represent less than 5% of the intensity in the $1\,^1S_0-2\,^1P_1$, $1\,^1S_0-2\,^3P_1$ and $1\,^1S_0-2\,^3S_0$ transitions of the helium-like ion, lithium-like Fe XXIV lines have about 50% of the intensity of the lines of Fe XXV. Satellite lines of stages of ionization below Fe XXV contribute 45% of the total line emission between 1.85 and 1.94 Å.

Toward longer wavelengths from 2 Å, one finds the spectra of lighter elements stripped of all but one or two of their electrons. Note that with decreasing atomic number the hydrogen lines become steadily more intense relative to the helium-like emission from the same element. This is due to the fact that less energy is required for ionization of the helium-like ion as the atomic number decreases. Since each stage of ionization of an element exists predominantly over a restricted range of electron temperature, the simultaneous appearance of emission lines originating from the hydrogen-like and helium-like ions from neon through iron requires a range of temperatures from 6 million to about 40 million degrees. Each of the emission lines provides a probe with which to study the physical conditions in a limited region of the flare plasma. The three emission lines corresponding to the resonance ($1\,^1S_0-2\,^1P_1$), intersystem ($1\,^1S_0-2\,^3P_1$) and fobidden ($1\,^1S_0-2\,^3S_1$) transitions in the helium-like ion can be clearly identified for silicon, sulfur, argon, and calcium.

Spectra for the region 6.6 Å–20.0 Å observed from OSO-III are given in Figure 9 (Neupert *et al.*, 1967). Spectra similar to the 'pre-flare' spectrum depicted in this figure have been obtained in the past from rockets and a satellite above the Earth's atmosphere (Blake *et al.*, 1965; Evans and Pounds, 1968; Rugge and Walker, 1968); the data presented here are the first observations of the spectrum during a solar flare. In addition to increased intensity of all lines observed in the pre-flare spectrum, we observe a new group of lines between 9.0 Å and 14.0 Å which appear not to have been present prior to the flare. Optical transitions between quantum levels $n=3$ and $n=2$ in highly ionized iron are expected to produce lines between 10 Å and 20 Å. The most intense lines are apparently transitions from $3d$ levels to the ground state, being similar in this respect to strong emission lines of silicon from the same isoelectronic sequences which appear in the solar spectrum at longer wavelengths (40 Å–80 Å). Similar spectra taken from OSO-V with higher resolution and lower background reveal a total of sixty spectral lines between 8.0 Å and 15 Å.

An example of the temporal relationships which have been observed among several of the flare-associated emission lines is shown in Figure 10. Three lines, at 1.87 Å, 14.25 Å and 303.78 Å were observed simultaneously, with data readouts every 0.64 sec, during a flare on August 27, 1967. The He II (303.78 Å) enhancement appears to begin slowly, one or two minutes before the impulsive microwave burst. At the time of the burst it rises much more rapidly and reaches maximum emission not more than 25 sec after maximum microwave emission. An impulsive hard X-ray burst, if it occurred, would be coincident in time with the microwave burst (Frost, 1969). At the time of the impulsive burst, emission lines at 1.87 Å and 14.25 Å, originating in Fe XXV and Fe XVIII, respectively begin to increase rapidly in intensity, the shorter wavelength

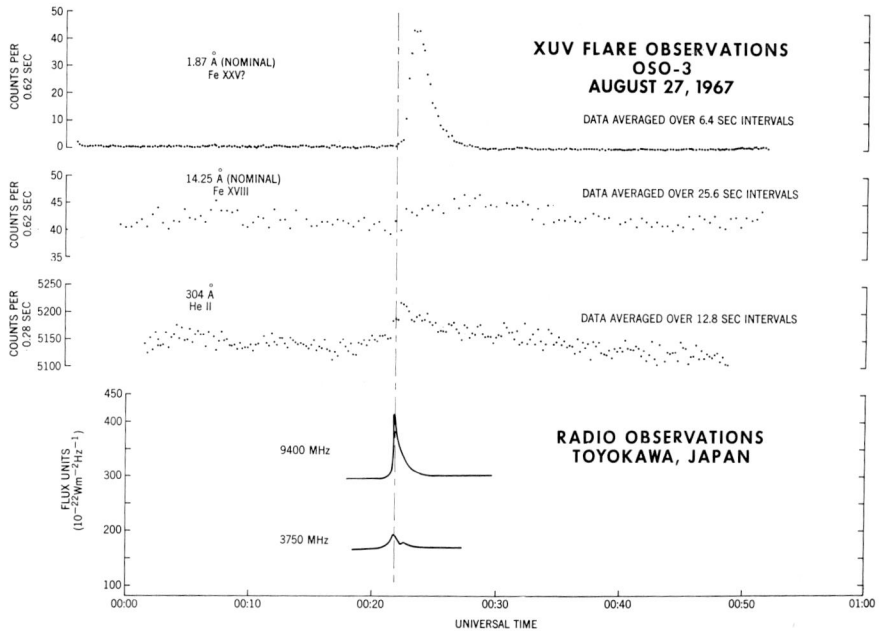

Fig. 10. Simultaneous observations of flare-associated emission at 1.87 Å (Fe XXV), 14.25 Å (Fe XVIII) and 304 Å (He II) and a comparison with concurrent radio microwave emission. Radio data courtesy of Professor H. Tanaka, Nagoya University, Toyokawa, Japan.

reaching maximum emission first. Fe XVIII emission continues to increase even after the 1.87 Å emission has reached maximum and begun its decline. This may be the result of recombinations from higher stages of ionization in a slowly cooling plasma. It therefore appears that a hot plasma of electrons and highly ionized atoms is produced in the duration of the non-thermal microwave emission. Evidence for this is found not only in the line emission but also in the soft X-ray continuum (Figure 11). The spectral distribution of this continuum changes only slowly during most events and becomes 'softer' after the flare maximum, indicating a plasma which is slowly cooling (Neupert et al., 1969; Culhane et al., 1970). Typical electron temperatures derived from the continuum, assuming it to be free-free emission, range from 10 to

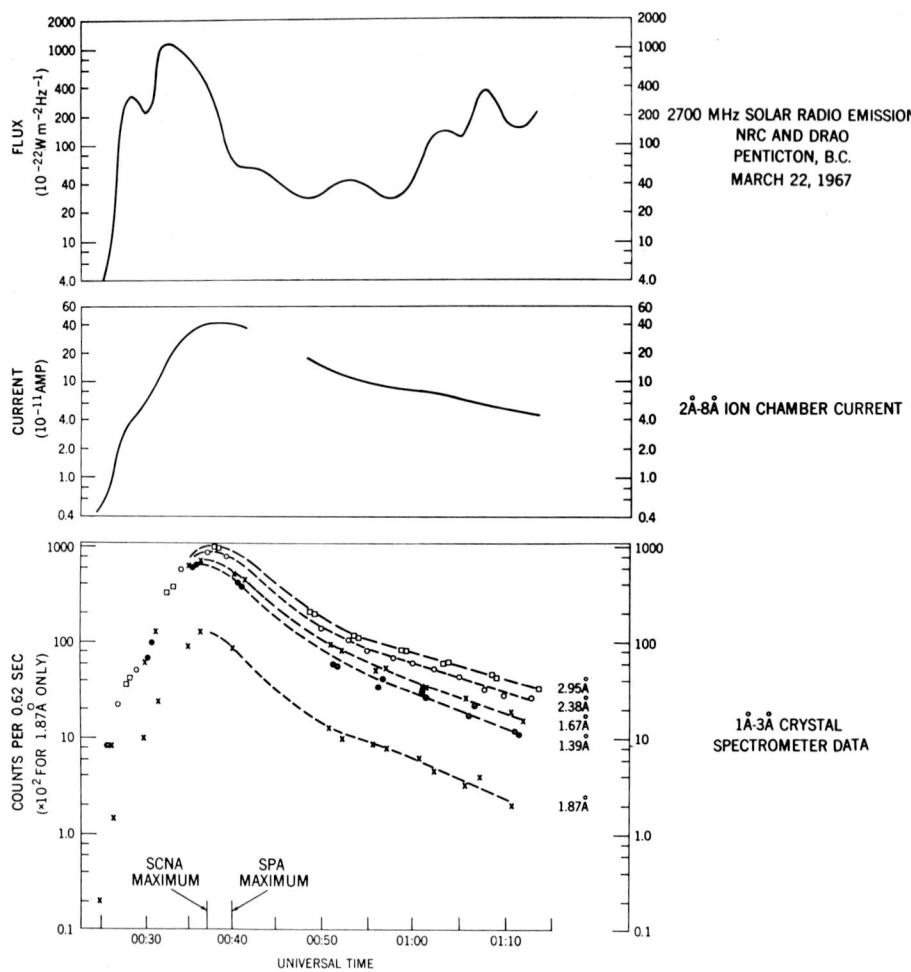

Fig. 11. Comparison of line emission at 1.87 Å, continuum emission at selected wavelengths between 1 Å and 3 Å and the solar soft X-ray emission integrated from 2 Å to 8 Å during a large flare on March 22, 1967.

40 million deg (Culhane and Phillips, 1970; Neupert et al., 1969). From the observation that the Fe xxv emission tracks the continuum perfectly, within the limitations of the data in Figure 11, we deduce that the ionization to Fe xxv takes place very rapidly, that is, there is no discernable difference between the ion temperature and the electron temperature. If the time, τ, required to ionize Fe xxiv is 1 sec, for example, then the electron density can be estimated from

$$N_e = \frac{1}{\tau q} = \frac{1}{1 \times 6.5 \times 10^{-12}} = 1.6 \times 10^{11} \text{ cm}^{-3}$$

In this equation Q is the ionization rate obtained from formulae given by Jordan (1966).

As a result of the work of Gabriel and Jordan (1969a, b), it is now possible to infer the electron density in regions where ions in the helium-like isoelectronic sequence originate by comparing the strength of the $1\,^1S_0-2\,^3S_1$ and $1\,^1S_0-2\,^3P_1$ transitions. Figure 12 shows that this ratio changes during a flare for the lines of Si XIII. Before the flare, the forbidden line ($1\,^1S_0-2\,^3S_1$) is clearly more intense than the intercombination line ($1\,^1S_0-2\,^3S_1$) and consistent with an upper limit on the electron density of 1.2×10^{13} cm^{-3}. During the flare, the forbidden line declines in intensity relative to the $1\,^1S_0-2\,^3P_1$ transition. Using the most recent calculations by Freeman *et al.*

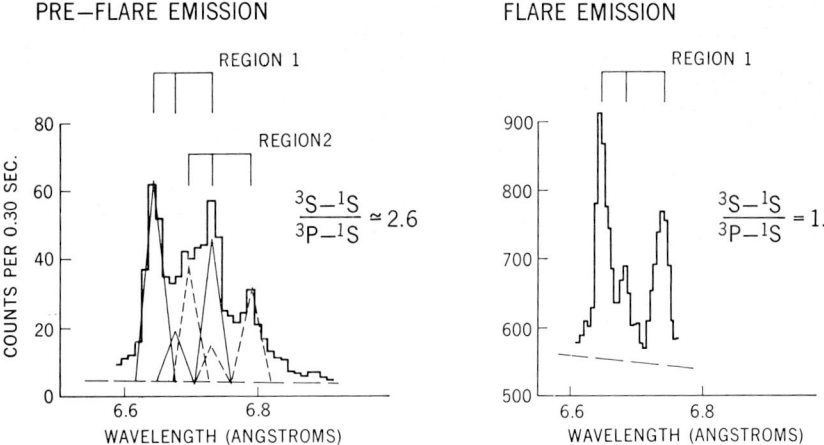

Fig. 12. Distribution of intensity in the resonance ($1\,^1S_0 - 2\,^1P_1$), intersystem ($1\,^1S_0 - 2\,^3P_1$) and forbidden ($1\,^1S_0 - 2\,^3S_1$) transitions of Si XIII. The distribution on the left is the result of emission from two sources on the Sun prior to a flare in one of these regions. During the flare the emission from region 1 is strongly enhanced and the ratio of forbidden to intersystem line strengths in this region is diminished.

(1970) and Gabriel and Jordan (1970), we infer that the electron density in the emitting region has increased to more than 1×10^{13} cm^{-3}. Assuming a typical emission measure of 2×10^{48} cm^{-3} (Culhane and Phillips, 1970), we conclude that the emitting volume is between 10^{22} and 10^{23} cm^3. Thus the emitting region must be highly localized (for instance a spherical volume of the order of 300 km in diameter) or be composed of fine thread-line or filamentary structures.

Direct evidence of the compactness of the soft X-ray emitting region associated with a flare of Importance 1N has been obtained by Vaiana *et al.* (1968) by photographically recording its X-ray image at the focus of a telescope employing glancing-incidence optics. This image, together with an Hα photograph obtained by the

Boulder Observatory of ESSA, is shown in Figure 13. Although the two images resemble one another in areas of weak X-ray emission, there appears to be an intense loop-like structure in X-rays which has no strong counterpart in Hα. This loop, which appears to connect several brighter portions of the Hα flare, is about 1 arc min (42000 km) in length and a few arc sec in diameter; this diameter is most probably an upper limit imposed by the limitations of the instrumentation.

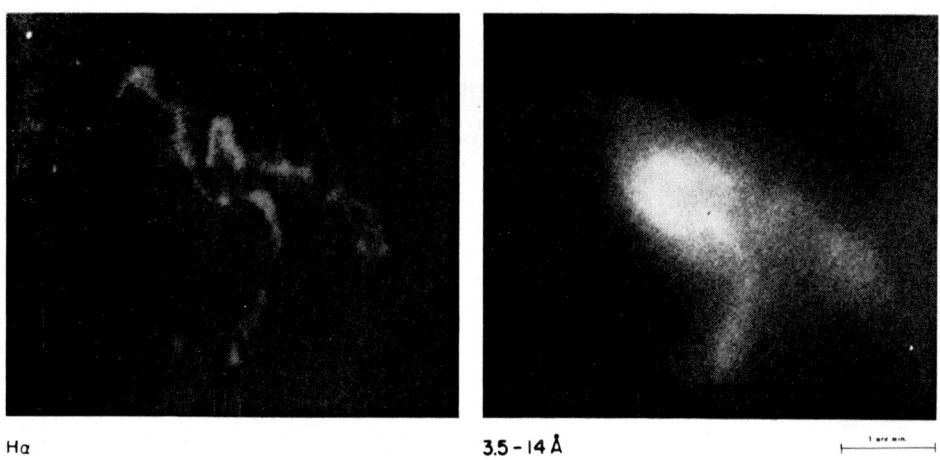

Fig. 13. Comparison of soft X-ray flare emission (3.5 Å–14 Å filter passband) with an H-alpha photograph recorded by the Boulder Observatory of ESSA. Courtesy of G. S. Vaiana, American Science and Engineering.

Additional indirect evidence for extreme localization of the hot plasma associated with a solar flare was obtained by the OSO-5 satellite on the occasion of the solar eclipse of March 7, 1970. As the moon passed in front of the Sun's disk, it occulted and later deocculted a solar flare which was in progress at that time (approximately 16:20 UT). The motion of the Moon's limb at occultation was almost perpendicular to its motion at deoccultation so that two nearly orthogonal one-dimensional spatial scans having a resolution of a fraction of an arc sec (approximately 400 km) were recorded. The X-ray emission recorded at 1.86 Å and in the 2 Å–8 Å band was found to have a diffuse component within which was imbedded a hotter nucleus, longer in one dimension than the other. The boundary of this core was very sharp, such that within 0.3 sec of time, corresponding to an apparent lunar motion of 400 km across the solar disk, the integrated X-ray emission from the Sun increased by nearly 50% as the core of the flare was deocculted. The observations in this case can be accounted for by a loop of X-ray emission which, at least in one dimension, had a sharply defined boundary. This boundary was apparently stable in space since it was observed well into the declining phase of the flare event.

Returning to the direct observation of Vaiana et al., if we assume the density of 10^{13} cm^{-3} derived from the line ratios we find that the actual diameter of the X-ray emitting filament is only 16 km in diameter, far smaller than the resolution of the optical system. This would only be the 'core' of the filament since the evidence we have discussed also indicates the presence of a larger diffuse volume which is also emitting soft X-rays. These sources lie in the lower corona or at the corona-chromospheric interface at a distance of 5000–20000 km above the photosphere. Such regions should have high continuous emission in the visible spectrum produced by Thomson scattering of the photospheric continuum by free electrons. Indeed, such emission has been frequently observed, and densities of 10^{11} cm^{-3} inferred on the assumption that the electron cloud is spherical with a diameter of 20000 km (Zirin, 1964). If, instead, the source were filamentary and occupied only a fraction of this volume, the electron density could be increased to those values inferred from the X-ray line ratios. Although the source region is small, it is not formed instantaneously at the beginning of a flare. The buildup to maximum emission measure takes place in a time span of 30 sec to 20 min or more, at the time when the impulsive microwave emission is observed. The characteristic rise and decay times of the various emissions are related: a short-lived microwave burst will be followed by short-lived soft X-ray emission; a complex microwave burst having a duration of many minutes and probably composed of many short bursts of radiation (Frost, 1970) will often be followed by a soft X-ray event of long duration. The magnetic field configuration and electron density appear ultimately to control both the release and dissipation of the energy in the flare event.

In conclusion, we can attempt to outline the physical characteristics of the hot coronal plasma frequently associated with chromospheric flares from current observations:

1. *Electron Temperature:* From the soft X-ray continuum spectrum between 1 Å and 8 Å,

$$T_e = 20 - 40 \times 10^6 \text{ K}.$$

2. *Ion Temperature:* From the observation of hydrogen and helium-like ions in argon, calcium and iron,

$$T_i = 20 - 40 \times 10^6 \text{ K}.$$

From the observation that Fe xxv emission tracks the soft X-ray continuum emission during the rising portion of the X-ray event, $T_i = T_e$.

3. *Emission Measure:* From observation of the soft X-ray continuum (Culhane and Phillips, 1970),

$$N_e^2 V \approx 2 \times 10^{48} \text{ cm}^{-3}.$$

From the intensity of the resonance transition $(1\,^1S_0 - 2\,^1P_0)$ in Fe xxv,

$$N_e^2 V = 2 \times 10^{48} \text{ cm}^{-3}$$

for a large soft X-ray event (Neupert et al., 1969).

5. *Electron Density:* From the ratio of forbidden and intersystem line intensities in the spectrum of ions of the helium isoelectronic sequence,

$$N_e \approx 10^{12} - 10^{13} \text{ cm}^{-3}.$$

From the ionization rate from Fe XXIV to Fe XXV, using an Fe XXIV lifetime of 1 sec,

$$N_e > \frac{1}{(1)(6.5 \times 10^{12})} = 1.6 \times 10^{11} \text{ cm}^{-3}.$$

6. *Volume of the Core of X-Ray Emission:* From direct imaging of X-rays,

$$V \leq 2 \text{ sec} \times 2 \text{ sec} \times 1 \text{ min} = 10^{25} \text{ cm}^3.$$

On the other hand, given a density of 10^{13} cm^{-3},

$$\frac{N_e^2 V}{N_e^2} \approx \frac{2 \times 10^{48}}{10^{26}} = 2 \times 10^{23} \text{ cm}^3.$$

7. *Kinetic Energy in Hot Plasma:*

$$3N_e V k T_e = 3 \times 10^{13} \times 1.4 \times 10^{-16} \times 3 \times 10^7 \times 2 \times 10^{22}$$
$$= 2.6 \times 10^{27} \text{ erg}.$$

8. *Magnetic Field:* From the observation that the hot plasma appears to remain localized,

$$\frac{B^2}{8\pi} > N_e k T_e$$

$$B \approx 1700 \text{ G}.$$

9. *Change in Magnetic Field to yield Energy Required to Heat the Plasma:*

$$\Delta\left(\frac{B^2}{8\pi}\right) \simeq 3N_e k T_e$$

$$\Delta B \approx 200 \text{ G}.$$

10. *Electron Gas Pressure:*

$$N_e T_e = (10^{13})(3 \times 10^7)$$
$$= 3 \times 10^{20} \text{ dyne cm}^{-2}$$

By comparison, for active regions,

$$N_e T_e \leq (3 \times 10^{11})(4 \times 10^6)$$
$$= 12 \times 10^{17} \text{ dyne cm}^{-2}$$

and for the corona away from active regions,

$$N_e T_e \approx (3 \times 10^8)(2 \times 10^6)$$
$$= 6 \times 10^{14} \text{ dyne cm}^{-2}.$$

We have therefore arrived at a point where we can describe the general physical conditions which must exist in the region producing flare-associated X-rays. However, these facts lead to new problems which now must be considered: Why are some physical conditions, such as the particle density and magnetic field apparently similar to those considered reasonable for regions of Hα emission, while other properties, such as the electron temperature and the size (smaller in X-rays than in Hα) are much different? How can regions of hot X-ray emitting plasma exist in close proximity to Hα regions. Then, again, why are some large chromospheric flares weak X-ray emitters while strong X-ray events are sometimes recorded for which no observable Hα activity exists (perhaps because the region is beyond the solar limb). We look forward to the acquisition of better observations and a more complete understanding of atomic processes which will guide us to answers to such questions.

References

Behring, W. E., Neupert, W. M., and Lindsay, J. C.: 1963, *Space Res.* **3**, 814.
Billings, D. E.: 1966, *A Guide to the Solar Corona*, Academic Press, New York.
Blake, R. L., Chubb, T. A., Friedman, H., and Unzicker, A. E.: 1965, *Astrophys. J.* **142**, 1.
Boardman, W. J. and Billings, D. E.: 1969, *Astrophys. J.* **156**, 731.
Culhane, J. L. and Phillips, K. J. H.: 1970, *Solar Phys.* **11**, 117.
Culhane, J. L., Vesecky, J. F., and Phillips, K. J. H.: 1971, 'The Cooling of Flare Produced Plasmas in the Solar Corona', *Solar Phys.*, in press.
Evans, K. and Pounds, K.: 1968, *Astrophys. J.* **152**, 319.
Freeman, F. F., Gabriel, A. H., Jones, B. B., and Jordan, C.: 1971, 'Helium-like Ion Forbidden Line Emission and Solar Active Regions', *Proc. Roy. Soc.*, to be published.
Frost, K. J.: 1969, *Astrophysics J. Letters* **158**, L159.
Gabriel, A. H. and Jordan, C.: 1969a, *Nature* **221**, 947.
Gabriel, A. H. and Jordan, C.: 1969b, *Monthly Notices Roy. Astron. Soc.* **145**, 241.
Gabriel, A. H. and Jordan, C.: 1970, *Phys. Letters* **32A**, 166.
House, L. L.: 1964, *Astrophys. J. Suppl. Ser.* **8**, 307.
Jordan, C.: 1966, *Monthly Notices Roy. Astron. Soc.* **132**, 515.
Jordan, C.: 1969, *Monthly Notices Roy. Astron. Soc.* **142**, 501.
Meekins, J. K., Kreplin, R. W., Chubb, T. A., and Friedman, H.: 1968, *Science* **162**, 891.
Neupert, W. M.: 1967, *Solar Phys.* **2**, 294.
Neupert, W. M., Gates, W., Swartz, M., and Young, R.: 1967, *Astrophys. J. Letters* **149**, L79.
Neupert, W. M., Swartz, M., White, W. A., and Young, R. M.: 1969, *Solar Flares and Space Research*, North-Holland Publ. Co., Amsterdam, p. 95.
Neupert, W. M. and Swartz, M.: 1970, *Astrophys. J. Letters* **160**, L189.
Rugge, H. and Walker, A. B. C.: 1968, *Space Res.* **8**, 439.
Tousey, R.: 1967: *Astrophys. J.* **149**, 239.
Vaiana, G. S., Reidy, W. P., Zehnpfennig, T., Van Speybroeck, L., and Giaconni, R.: 1968, *Science* **161**, 564.
Van Speybroeck, L. P., Krieger, A. S., and Vaiana, G. S.: 1970, *Nature* **227**, 818.
Walker, A. B. C. and Rugge, H. R.: 1970, *Astron. Astrophys.* **5**, 4.
Zirin, H.: 1964, *Astrophys. J.* **140**, 1216.
Zirin, H.: 1966, *The Solar Atmosphere*, Blaisdell Publishing Co., Waltham, Mass.

17. A NOTE ON A RECENT IDENTIFICATION OF THE SOLAR FLARE 1.9 Å LINE FEATURE

K. J. H. PHILLIPS

Mullard Space Science Laboratory, University College London, England

Bragg crystal spectrometer techniques have enabled studies to be made of the solar soft X-ray spectrum with good wavelength resolution. A number of investigators have now achieved results in various wavelength bands over the range 1 to 30 Å, for example. The spectra of solar active regions, which have typical electron temperatures (T_e) of 3 or 4×10^6 K and electron densities (N_e) of 10^{11} cm^{-3}, show intense lines due to the ions Fe XVII and XVIII, Ne IX, O VIII, and N VII, for example. Before 1967, crystal spectrometer experiments were generally rocket-borne, but the advent of some of the more recent OSO satellites enabled solar flare spectra to be studied in detail.

Some of the first results from OSO-3 and OSO-4 indicated the presence of extremely high ions of Fe and other metals. However, two groups gave different identifications, both based on precise wavelength measurements, of an intense feature at 1.9 Å due to iron. The group at Goddard Space Flight Centre (Neupert *et al.*, 1967) found that the peak emission of the feature coincided with the permitted and intercombination resonance lines of the ion Fe XXV, which is helium-like, whereas a group at the U.S. Naval Research Laboratory (Meekins *et al.*, 1968) identified the feature with inner-shell, Kα type transitions in ions Fe VII through Fe XXIII. Such transitions occur when an electron in the K-shell of an atom is removed by electron bombardment, and the vacancy filled by an electron previously in the L-shell. Meekins *et al.* attributed the observed skew shape of the line feature on its red side to these transitions occurring in the lower-stage ions.

The situation seems to have been resolved by some high resolution studies of the feature with another Goddard experiment aboard OSO-5 (Neupert and Swartz, 1970), where the line is registered in third and fourth orders of diffraction. With a wavelength resolution of 0.001 Å, the feature is observed to consist of a number of components which Neupert and Swartz attribute to the ions Fe XXV, XXIV, XXIII. A long-wavelength tail present is thought to be due to inner-shell transitions of lower ions of Fe. Table I summarizes the identifications of Neupert and Swartz.

A proportional counter spectrometer aboard OSO-4, constructed jointly by University College London and the University of Leicester, has been giving good data for some while, some of which is still being processed and analyzed. The instrument, which has crude wavelength but good time resolution, is sensitive in the region 1–18 Å and 44–56 Å. Experimental details are described elsewhere (Culhane *et al.*, 1969). During some intense X-ray flares, the 1.9 Å feature is clearly discernible, and a method has been developed (Culhane and Phillips, 1970) which extracts from count histograms an electron temperature T_e and $N_e^2 V$ ($V=$ source volume), from an ex-

TABLE I

Ion	Transition	Wavelength (Å)
Fe xxv	$1s^2\,{}^1S_0 \leftarrow 1s2p\,{}^1P_1^0$	1.850
Fe xxv	$1s^2\,{}^1S_0 \leftarrow 1s2p\,{}^3P^0$	1.855
Fe xxiv	$1s^2\,2s\,{}^2S \leftarrow 1s2s2p\,{}^2P^0$	1.860
Fe xxiv	$1s^2\,2p^2P^0 \leftarrow 1s2p^2\,{}^2D$ }	1.865
Fe xxv	$1s^2\,{}^1S_0 \leftarrow 1s2s\,{}^3S_1$ }	
Fe xxiii	$1s^2\,2s^2 \leftarrow 1s2s^2\,2p$	1.868
Fe xxiv	$1s^2\,2s\,{}^2S \leftarrow 1s2s2p\,{}^2P^0$ }	1.875
Fe xxiv	$1s^2\,2p\,{}^2P^0 \leftarrow 1s2p^2\,{}^4P$ }	

amination of a line-free part of the continuum, and a value for the flux in the iron line feature. The values of T_e and $N_e^2 V$ are obtained by comparing the observed slope of the spectrum with free-free and free-bound continua which have been calculated for coronal-type plasmas (Culhane, 1969).

Two X-ray events observed by the OSO-4 instrument have been examined in detail, and the results summarized in Table II. The observed iron like fluxes are compared with those calculated for the resonance line transition, $1s^2\,{}^1S_0 \leftarrow 1s2p\,{}^1P_1^0$, occurring

TABLE II

Observed and calculated 1.9 Å line fluxes

Event date	Peak temperature (10^6 K)	1.9 Å iron line flux	
		Calculated for permitted $1s^2 \leftarrow 1s2p$ transition (erg cm^{-2} sec^{-1})	Observed value at event peak (ergs cm^{-2} sec^{-1})
Dec. 21, 1967	16.0	1.2×10^{-5}	$(1.0 \pm 0.3) \times 10^{-4}$
Jan. 15, 1968	15.8	7.1×10^{-6}	$(0.7 \pm 0.2) \times 10^{-4}$

in Fe xxv. The formula of Pottasch (1964) was used, and it was assumed that T_e and $N_e^2 V$, derived from the continuum measurements, applied also to the line-emitting region. A coronal iron abundance of $N(\text{Fe})/N(H) = 5.7 \times 10^{-5}$ (Jordan, 1966; Pottasch, 1967), and recent ionization equilibrium calculations of Jordan (1970), were used.

The order of magnitude difference between the observed and calculated values can readily be explained by the new OSO-5 crystal spectrometer observations. Neupert and Swartz estimate that no more than 53% of the total iron line flux is due to the Fe xxv transitions, and it would seem that about 25% is due to the permitted $1s^2\,{}^1S_0 \leftarrow 1s2p\,{}^1P_1^0$ transition. This would reduce the error to a factor of about 2, and this is of the same order as the experimental accuracy of the observed fluxes.

The OSO-4 proportional counter observations would therefore appear to support the recent identifications of Neupert and Swartz. Moreover, for the peak of these two events at least, the assumption of ionization equilibrium does not appear to lead to significant errors.

Acknowledgements

I am grateful for numerous helpful discussions with my project supervisor, Dr J. L. Culhane, and also with Drs C. Jordan and A. Gabriel of the Astrophysics Research Unit, Culham Laboratory, U.K.

References

Culhane, J. L.: 1969, *Monthly Notices Roy. Astron. Soc.* **144**, 375.
Culhane, J. L., Sanford, P. W., Shaw, M. L., Phillips, K. J. H., Willmore, A. P., Bowen, P. J., Pounds, K. A., and Smith, D. G.: 1969, *Monthly Notices Roy. Astron. Soc.* **145**, 435.
Culhane, J. L. and Phillips, K. J. H.: 1970, *Astrophys. J.* **160**, 309.
Jordan, C.: 1966, *Monthly Notices Roy. Astron. Soc.* **132**, 463.
Jordan, C.: 1970, *Monthly Notices Roy. Astron. Soc.* **148**, 17.
Meekins, J. F., Kreplin, R. W., Chubb, T. A., and Friedman, H.: 1968, *Science* **162**, 891.
Neupert, W. M., Gates, W., Swartz, M., and Young, R.: 1967, *Astrophys. J. Letters* **149**, L79.
Neupert, W. M. and Swartz, M.: 1970, *Astrophys. J. Letters* **160**, L189.
Pottasch, S. R.: 1964, *Space Sci. Rev.* **3**, 816.
Pottasch, S. R.: 1967, *Bull. Astron. Inst. Neth.* **19**, 113.

18. RECENT INVESTIGATIONS ABOUT SOLAR X-RAYS EMITTED BY THE SOLAR CORONA BY MEANS OF SOLRAD SATELLITES

M. LANDINI and B. C. MONSIGNORI FOSSI

Arcetri Astrophysical Observatory, Florence, Italy

1. Introduction

The SOLRAD satellites, which have been instrumented by the Naval Research Laboratory, have been scheduled for monitoring the solar radiation in the soft X-ray region by means of broad band photometers. The nine SOLRAD satellites launched during the last ten years have produced an extensive amount of data which cover a large interval of the present solar cycle and the decreasing phase of the last one. The data have been recorded at Arcetri Observatory since 1964.

The photometers used on the satellites are Geiger Müller counter and ionization chambers whose spectral sensitivity is selected by proper choose of filling gases and of window materials. Mica-beryllium windows with argon, krypton and bromine filling gases limit the range of sensitivity to the region 0.5–3 Å; mica-beryllium windows counters filled with neon, argon, bromine cover the region between 1 and 8 Å; aluminium window counters filled with nitrogen are sensitive to the 8–16 Å interval; Mylar and aluminium window counters filled with carbon tetrachloride are sensitive to the 1–20 Å band; Mylar window counters filled with nitrogen select the region 44–60 Å. Additional counters have been used sometime to investigate UV regions around H Ly-α.

Because of the broad band sensitivity the measured signal cannot be converted to energy flux without assumptions on the spectral distribution within the spectral region of sensitivity.

Usually the data have been reduced using a 'grey body' distribution with $T= =2\times10^6$ K in the 1–20 Å band and $T=0.5\times10^6$ K in the 44–60 Å band (Kreplin, 1961).

In this assumption changes of the spectral distribution are not taken into account and any variation of the signal is attributed to variation of the 'grey body' emission.

Changes of spectral energy distribution indeed occur, and the conventional 'grey body' assumption is rather rough approximation to the true spectral energy distribution.

However it is possible to investigate changes in the spectral energy distribution when two or more photometers measure the same spectral range with different efficiency.

It has been possible to do such an investigation using the data of the last two satellites (SOLRAD 8 and SOLRAD 9) because each spectral range 0.5–3 Å, 1–8 Å and 8–20 Å was covered by one pair of photometers with different efficiencies which in many cases gave both on scale signals.

For the purpose of this investigation it has been necessary to compute the spectral energy distribution of the solar corona for different temperature.

The emission of the coronal plasma has been computed by several authors, but most of them do not take in account the processes of dielectronic recombination in computing the ionization balance (Elwert, 1954; Mandel'stam, 1965). Recently dielectronic recombination has been included for computation of continuum emission in the 1–30 Å (Culhane, 1969) continuum and line emission using photospheric abundances in the 1–30 Å (Beigman and Vainshtein, 1969) continuum and line emission from 1 to 100 Å for temperature ranging from 1 to 10×10^6 K using coronal abundances deduced by Pottasch (1967) (Landini and Monsignori Fossi, 1970).

2. The Spectral Distribution of the Energy

a. IONIZATION BALANCE

All processes which contribute to the emission of a thermal plasma depend on the abundance of the elements in the ionization stages present at the different temperatures. These abundances can be computed assuming equilibrium conditions produced by balance between the collisional ionization together with the autoionization and the radiative and dielectronic recombination. All these processes are effective for temperature between 1 and 10×10^6 K and have been taken into account in recent computation of ionization balance developed by Jordan (Jordan, 1969; Jordan, 1970).

b. CONTINUUM EMISSION

The continuum radiation in the coronal plasma is produced by free–free and free–bound processes.

The free–free emission per cm^2 sec in a $d\lambda$ interval of wavelengths at distance a from the Sun is given by

$$dE = 2.65 \times 10^{-27} \frac{T^{-1/2} \lambda^{-2}}{4\pi a^2} 10^{-5040 \frac{W}{T}} d\lambda \int_V N_e^2 \, dV \qquad (1)$$

W is the energy (eV) of one photon of wavelength λ, T is the temperature measured in K, λ is the wavelength measured in cm and a in cm. $\int_V N_e^2 \, dV$ is the 'emission measure' in cm^{-3}.

The free-bound emission in the same units as above is given by

$$dE = \frac{4.83 \times 10^{-22}}{4\pi a^2} \frac{n_z}{n_E} \frac{n_E}{n_H} \frac{z^4}{n^3} \left(\frac{Z}{2n^2}\right) T^{-3/2} \lambda^{-2} \cdot 10^{5040 \frac{I_{zn}-W}{T}} d\lambda \int_V N_e^2 \, dV \qquad (2)$$

where the ratio n_z/n_E is the abundance of the Z stage of the element E referred to the total abundance of element, n_E/n_H is the abundance of element E referred to the hydrogen, n is the principal quantum number of the level at which recombination occurs, $\bar{z}/2n^2$ is the incomplete fraction of shell n, I_{n_z} is the ionization potential of the nth level of ion Z.

C. LINE EMISSION

Line emission for dipole transition to the ground level using the Gaunt factors computed by Van Regemorter (Van Regemorter, 1962) is given by

$$dE = \frac{4 \times 10^{-16}}{4\pi a^2} f_{nn'} \frac{n_z}{n_E} \frac{n_E}{n_H} T^{-1/2} 10^{-5040 \frac{W_{nn'}}{T}} \int_V N_e^2 \, dV. \qquad (3)$$

where $f_{nn'}$ is the oscillator strength for the transition nn' in the z ion with $W_{nn'}$ excitation energy.

A sample of the computed results is given in Figure 1 where the energy distribution for $T = 1.5 \times 10^6$ K and $\int_V N_e^2 \, dV = 3.8 \times 10^{49}$ cm^{-3} is plotted against the wavelength.

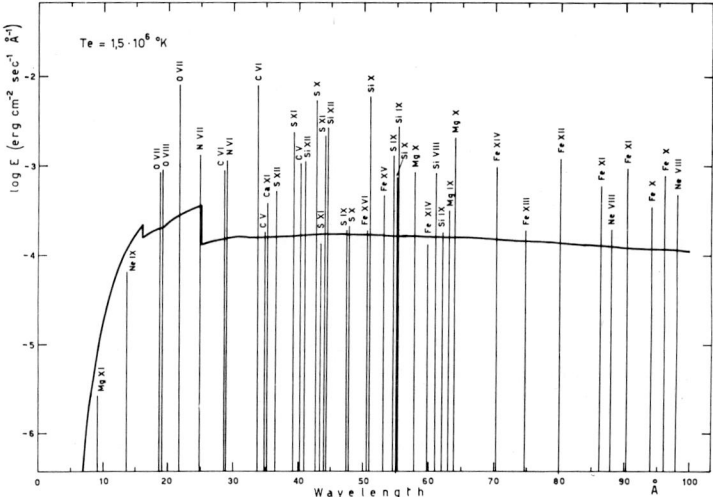

Fig. 1. Coronal emission in the range 1–100 Å for $T_e = 1.5 \times 10^6$ K and $\int_V N_e^2 \, dV = 3.8 \, 10^{49}$ cm^{-3}.

These values are plausible for quiet solar corona (Landini and Monsignori Fossi, 1970). The continuum radiation is mainly due to free–free processes for wavelengths greater than 30 Å; radiative recombination become effective for wavelength below 30 Å; the main contribution to free-bound emission comes from C v and C vi, O vii and O viii and N vii. For larger temperature contribution from Ne ix and x is important below 10 Å.

The line emission is an important feature of the spectrum: the most important lines are indicated in the figure by a vertical bar, whose amplitude represents the intensity of line. Below 25 Å O vii, O viii and N vii give the main contribution; with increasing temperatures strong emission lines of Fe XVII, Ne IX, Ne X, Mg XI, Si XIII appears to be important. For very high temperature ($T = 10^7$ K) almost all the line emission fall below 20 Å. As above it is possible to investigate changes in the spectral distributions when two photometers measure the same spectral range with different efficiencies. The signal of one photometer is proportional to the spectral distribution computed by means of Equation (1), (2), and (3) weighted through the efficiency function of the photometer.

In fact if an optically thin atmosphere is observed, the emission is proportional to the emission measure through a function of the temperature which depends on the spectral range used:

$$E(\lambda, T) \, d\lambda = f(\lambda, T) \, d\lambda \int_V N_e^2 \, dV$$

Because the efficiencies of Solrad photometers are known the theoretical signal of the chambers can be computed for different temperature and emission measure. Since the flux of energy is proportional to the emission measure:

$$\Phi = \text{const} \int_\lambda E(\lambda, T) \varepsilon(\lambda) \, d\lambda$$

the ratio of the signal of two photometers, having different efficiencies on the same spectral region, is function of temperature only. The measurement of experimental ratio allows the determination of temperature.

Once the temperature is known the signal of the 2 photometers gives the value of the emission measure.

This method of investigation has been applied to the pairs of photometers operating in the ranges 0.5–3 Å, 1–8 Å, 1–20 Å for data, selected from Solrad telemetry records, in the period March 1968–August 1969.

The results are shown in Figures 2, 3, 4. Figure 2 shows the result for the spectral region 0.5–3 Å: the signal ratio (scale on the left) or temperature (scale on the right) are plotted versus the logarithm of the signal of Geiger Müller counter.

Full lines represent the theoretical relationship between signal and temperature for constant emission measure. The experimental data appears to cover the region of T from 6–25×10^6 K with $\int_V N_e^2 \, dV$ between 10^{47}–10^{48} cm^{-3}. The large errors in temperature are due to the ionization photometer which was scheduled for extremely high activity and gave low signals when the GM counter was on scale.

The experimental points for temperature larger than 14×10^6 K are remarkable and suggest the presence of 2 types of very hot emitting regions on the sun.

This characteristic feature is present also in Figure 3 where the results for the spectral region 1–8 Å are shown.

Temperature scale and emission measure are indicated as in the previous figure. The experimental points appear to cover the interval of temperature from 3 to 20×10^6 K with emission measure between 10^{47} and 10^{49} cm^{-3}.

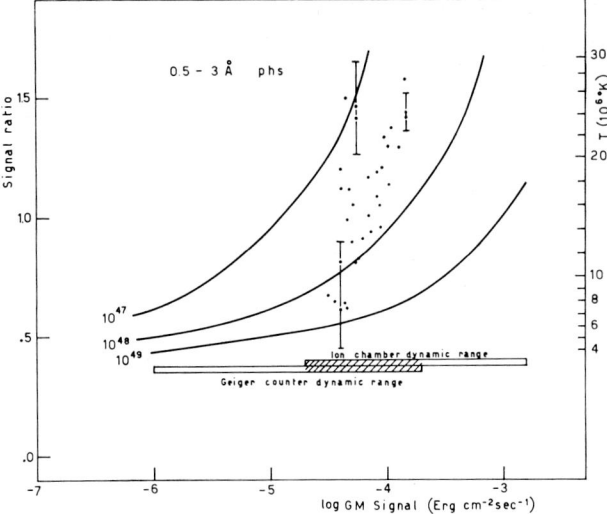

Fig. 2. Ratio of signals of ion chamber and Geiger counter operating in the 0.5–3 Å band versus the signal of the Geiger counter. The ratio is a measure of the electron temperature which is indicated in the right scale. Full lines represent the theoretical relation between temperature and Geiger Müller signal for 'emission measure' equal to 10^{47} cm^{-3}, 10^{48} cm^{-3} and 10^{49} cm^{-3}. In the lower part of the diagram the dynamic range of the photometers is shown; flux values where overlapping occurs show the range where the measurement is possible.

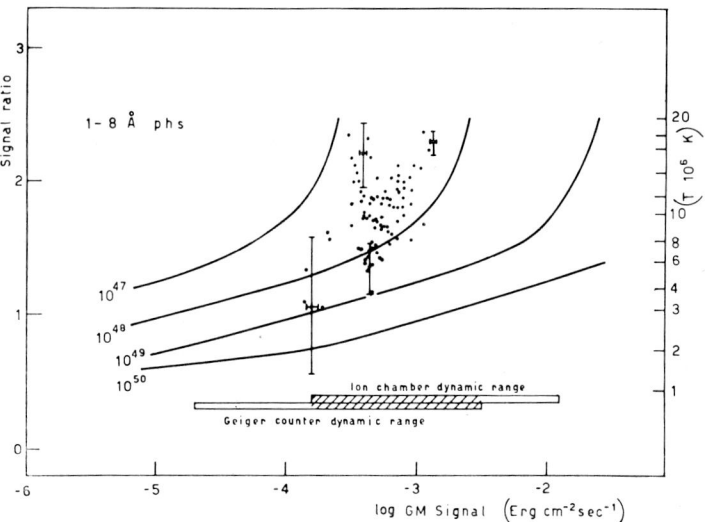

Fig. 3. The same information as in Figure 1 for photometers operating in the 1–8 Å band. Full lines represent the theoretical relationship between temperature and Geiger Müller signal for 'emission measure' equal to 10^{47} cm^{-3}, 10^{48} cm^{-3}, 10^{49} cm^{-3}, 10^{50} cm^{-3}.

The interpretation of 1–20 Å data has to be made very carefully because most of the emission falls in the region of low sensitivity of the photometers and a small variation in the efficiency highly affects the results. For this reason the results of the 1–20 Å photometers are not completely reliable. Anyhow in Figure 4 the results for the region 1–20 Å are shown. Most of the data show temperature between 1.2 and 3×10^6 K and emission measure between 3×10^{49} and 10^{51} cm^{-3}.

This emission measure appears to be very high compared with Pottasch or Goldberg measurements and with Solrad 8 measurements selected from observations in the period March to May 1966.

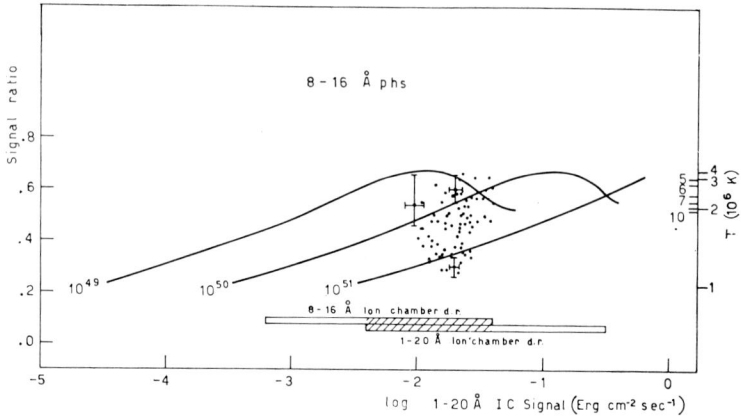

Fig. 4. The same information as in Figure 1 for photometers operating in the 1–20 Å band. Full lines represent the theoretical relationship between temperature and 1–20 Å ion chamber signal for 'emission measure' equal to 10^{49} cm^{-3}, 10^{50} cm^{-3}, 10^{51} cm^{-3}.

However, as the data of Solrad 9 concern a period of high activity, a value of the emission measure larger than for Solrad 8 is expected.

In the previous figures the temperature scale on the right hand and the full lines allow identification of temperature and emission measure of each experimental observation. The log of emission measure of each observation versus the temperature is plotted in Figure 5.

Data deduced by Solrad 8 and 9 are shown with the measurements of Goldberg (1967) Pottasch (1967) and, for very high activity, Hudson (1970).

Two sets of points with different emission measure for the same temperature are present for $T > 10^7$ K. These data appears to be related to the evolution of the flare activity: in fact all observations in 0.5–3 Å band of the lower branch fall before the peak of an X-ray burst and all the observations of the upper branch fall either after or just upon the peak of the burst.

Anyhow the attempt to study these effects by means of quasi thermal hypothesis may be wrong for high temperature.

3. Model of Active Regions

In Figure 5, for a 'mean active region', a relationship between the 'emission measure' and the temperature may be represented by the dashed line. In this assumption the hypothesis is made that the "mean active region" exhibits temperature as high as $7-8 \times 10^6$ K and that below 3×10^6 K the dispersion of data is due to the fact that a different number of condensations may be present on the disk at the same time.

An attempt to study a model for the 'mean active region' is made assuming that:

(1) the region is magnetically confined and equipartion between kinetic and magnetic energy holds

$$\frac{B^2}{8\pi} = 3kN_e T \qquad (1)$$

(2) the density distribution is spherical, according to existing models (Saito and Billings, 1964; Nishi and Nakagomi, 1962, Waldmeier, 1963) and follows the shape of the lines of force at least in the outer region.

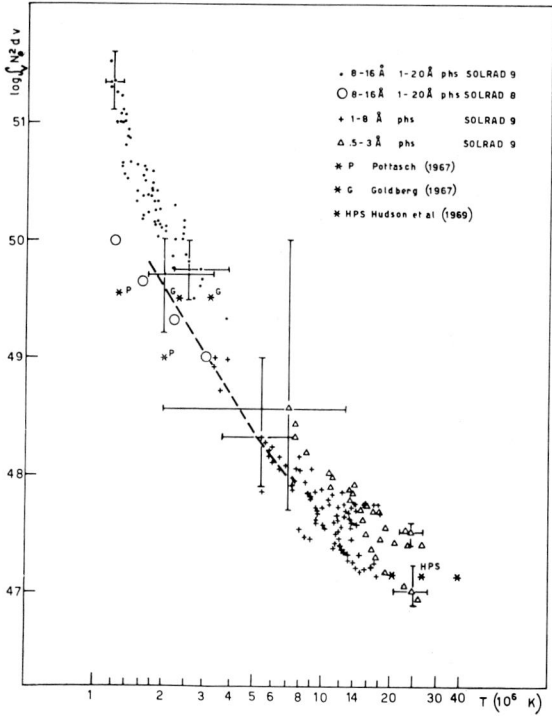

Fig. 5. Emission measure versus temperature of the emitting regions observed by the Solrad 9 photometers. Comparison with results of Pottasch (P), Goldberg (G) and Hudson *et al.* (HPS) is shown. Very large errors for Te between 4 and 8×10^6 K are due to the low signals of the ion chambers operating in the range 0.5–3 Å and 1–8 Å.

The dashed line represents the assumed 'emission measure' versus temperature relationship for a large condensation.

(3) the dashed line equation in Figure 5 is

$$T^\alpha \int_V N_e^2 \, dV = \text{constant}$$

and by logarithmic differentiation with assumption of spherical density distribution we obtain:

$$\frac{dT}{dh} \frac{1}{T} = \frac{2\pi h^2 N_e^2 \, dh}{\alpha \int_V N_e^2 \, dV} \quad (2)$$

where α is the slope of the line.

(4) hydrostatic equilibrium holds between gas plus magnetic pressure and tension of magnetic lines of force plus gravity, which along the vertical axis of the condensation gives

$$\frac{d}{dh}\left(2KN_e T + \frac{B^2}{8\pi}\right) = -\frac{B^2}{4\pi R} - 1.2 \, g \, m_H N_e \quad (3)$$

where R is the radius of curvature of magnetic lines of forces.

The 1.2 factor is due to the assumption that both helium and hydrogen are completely ionized.

Numerical integration of Equations (2) and (3) together with Assumption (1), has been performed using the Saito and Billings model for the outer condensation as a boundary condition. The results are shown in Figures 6 and 7. Figure 6 shows the

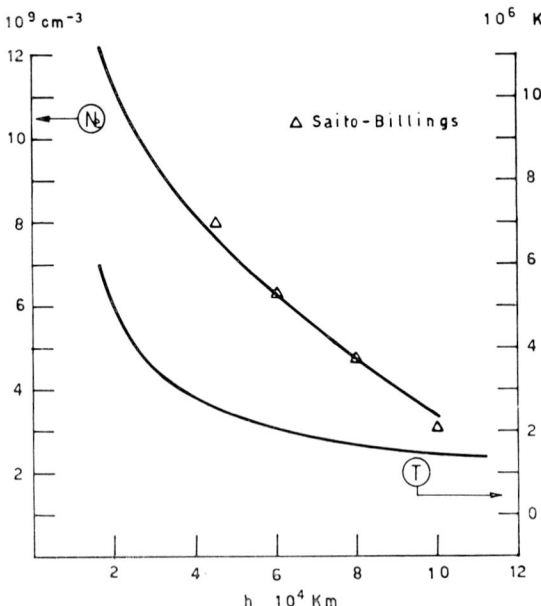

Fig. 6. Density and temperature distribution versus the altitude over the photosphere.

density and temperature distribution along the condensation; data from Saito and Billings used as boundary conditions are shown. The temperature remains below 2×10^6 K in the outer region of the condensation, while the gradient steeply increases going in the core.

In Figure 7 a schematic picture of the geometry of the 'mean condensation' is shown: the very hot core appears to extend from 15000 km to 25000 (over the photosphere) with density $\sim 10^{10}$ cm^{-3}; the temperature decreases outward until reaches 1.5×10^6 K at about 10^5 km above the photosphere.

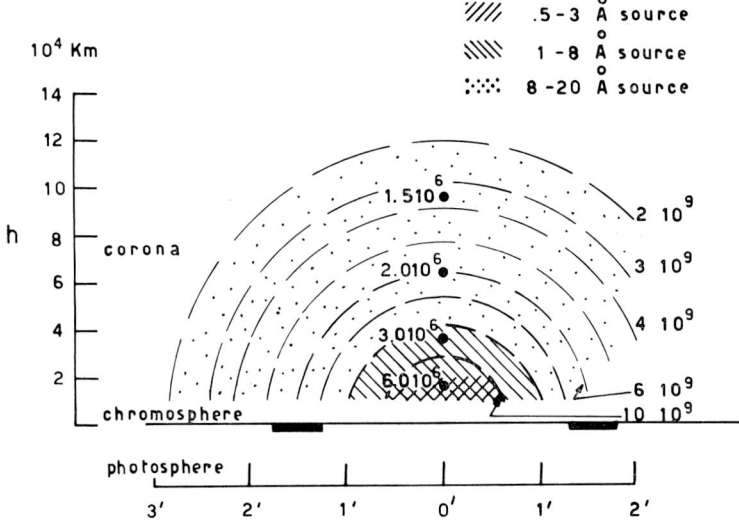

Fig. 7. Spherical model of a large condensation. Dashed lines indicate density distribution. Selected values of the temperature are shown along the vertical axis of the condensation.

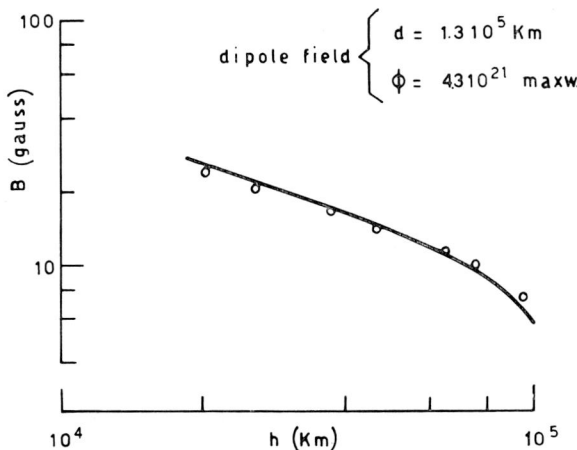

Fig. 8. Magnetic field along the vertical axis of the condensation (○); full line is the magnetic field for a dipole with $\Phi = 4.3 \times 10^{21}$ Mx and distance between the poles $d = 1.3 \times 10^5$ km.

The regions from which most of the emission in the 0.5–3 Å, 1–8 Å, 1–20 Å regions comes from, are also indicated, their approximate dimensions appear to be in the ratio 1:2:5; dimensions and ratio are not in disagreement with direct measurements (Negus and Glencross 1968; Reidy *et al.* 1968; Landini *et al.*, 1968).

By means of Assumption (1), the magnetic field along the vertical axis may be computed: it is plotted in Figure 8. It may be approximated by a dipole field with a magnetic flux of 4.3×10^{21} Mx and 1.3×10^5 km of distance between the poles.

References

Beigman, I. L., Vainshtein, L. A.: 1969, *Prep. Phys. Inst. Acad. Sci. USSR* **94**.
Culhane, F. L.: 1969, *Monthly Notices Roy. Astron Soc.* **144**, 365.
Elwert, G. Z.: 1954, *Naturforsch.* **(9a) 53**, 637.
Goldberg, L.: 1967, *Ann. Rev. Astron. Astroph.* **5**, 279.
Jordan, C.: 1969, *Monthly Notices Roy. Astron. Soc.* **142**, 501.
Jordan, C.: 1970, *Monthly Notices Roy. Astron. Soc.* **148**, 17.
Kreplin, R. W.: 1961, *Ann. Geophys.* **17**, 151.
Kreplin, R. W. and Horan, D. M.: 1969, Naval Research Laboratory Report 6800.
Landini, M., Russo, D., and Tagliaferri, G. L.: 1966, *Nature* **211**, 393.
Landini, M. and Monsignori Fossi, B. C.: 1970, *Astron. Astrophys.* **6**, 468.
Mandel'stam, S. L.: 1965, *Space Sci. Rev.* **5**, 587.
Negus, C. R. and Glencross, W. N.: 1968, *Nature* **220**, 48.
Nishi, K. and Nakagomi, J.: 1962, *Astron. Soc. Japan Publ.* **15**, 61.
Pottasch, S. R.: 1967, *Bull. Astron. Inst. Neth.* **19**, 113.
Reidy, W. P., Vaiana, G. S., Zehnpfenning, T., and Giacconi, R.: 1968, *Astrophys. J.* **151**, 353.
Saito, K. and Billings, D. E.: 1964, *Astrophys. J.* **140**, 760.
Waldmeier, M.: 1963, *Z. Astrophys.* **56**, 291.

19. RADIO EMISSION OF THE QUIET SUN

M. FELLI and G. TOFANI

Osservatorio Astrofisico di Arcetri, Firenze, Italy

1. Introduction

Since the first observations of the solar radio emission in the late 1940's, the existence of a clear correlation between the observed radio flux and the solar activity was found (Pawsey and Yabsley, 1949; Denisse, 1950). Several indicators of the solar activity were chosen to study this correlation, like the number of sunspots on the disk, the projected area of sunspots etc. and all were clearly indicating an increase of the radio emission as the optical activity was increasing. Extrapolating this correlation to the zero level of the optical activity, a lower value for the emitted radio flux was thus obtained. This extrapolated radio flux was attributed to a fictitious quiet sun, schematically represented as a disk devoided of all the active regions. In other words the radio emission from the sun was roughly separated into two components: the sporadic slowly varying component, due to the presence of centers of activity on the solar disk, and the quiet component, due to the thermal emission of a static, undisturbed atmosphere.

Total power measurements of the solar radio emission using small telescopes with large beams were first available. When telescopes with higher resolving power came into operation and maps of the sun were daily obtained with resolving powers of 4' or better, a statistical analysis of the two dimensional brightness temperature distribution on the disk became possible.

Using the same criteria previously indicated, i.e. attributing the quiet component to regions of the sun's maps devoided of sources, lower envelope curves were derived out of many scans. In this way a two dimensional picture of the sun at different wavelengths was obtained (Christiansen and Warbuton, 1953; Boischot, 1958; Bracewell and Swarup, 1961).

The problem of the theoretical prediction of the spectrum of the quiet sun and of its brightness temperature distribution has been considered since the early studies of solar radio astronomy (Martyn, 1946; Shklovskii, 1946; Ginzburg, 1946). The first model to be considered was one with a spherically symmetric structure. Then, assuming a distribution of electron density $N_e(h)$ and of electron temperature $T_e(h)$ as a function of the height h on the solar photosphere, the problem is that of solving the transfer equation:

$$T_b(\lambda, \varrho) = \int_0^\infty T_e(h) \exp\left[-\tau(\varrho, \lambda, h)\right] d\tau \qquad (1)$$

where T_b is the brightness or effective temperature, function of wavelength λ and radial distance $\varrho \, (= R/R_\odot = \sin \varphi)$, and where τ is the optical depth,

$$\tau(\varrho, \lambda, h) = \frac{1}{(1-\varrho^2)^{1/2}} \int_0^\infty k \, dh \qquad (2)$$

and the absorption coefficient k is:

$$k = 1.98 \times 10^{-23} \, g \, \frac{\lambda^2 N_e^2}{\mu T^{3/2}} \qquad (3)$$

where g is the Gaunt factor and μ is the refractive index.

In Equation (1), taking $\varrho = 0$ (the centre of the disk) and varying λ we can predict the quiet sun spectrum. Taking λ = constant and varying ϱ we can derive the brightness temperature distribution on the solar disk. When solving Equation (1) one must take into account the non rectilinear propagation of the rays through the solar atmosphere due to the variation of refractive index with height. This effect is much more pronounced at the longer wavelengths (Jaeger and Westfold, 1949; Smerd, 1950; Pawsey and Smerd, 1953; Bracewell and Preston, 1956). The negative gradient of electron density and the positive gradient of electron temperature existing when passing from the lower chromospheric levels to the upper corona, together with the spherical model, predict the variation of the height on the solar photosphere of the level at which $\tau(\lambda) = 1$, which corresponds to the effectively emitting layer at that wavelength. When one moves from the centre to the limb, close to the limb, the level at which $\tau(\lambda) = 1$ occurs at greater height from the solar photosphere, in regions of higher electron temperature. The simple spherical model therefore predicts the existence of a limb brightening at the edge of the solar radio disk. The phenomena should be particularly strong for wavelengths between 1 and 50 cm. In the meter wave range, the theory predicts a smooth equatorial profile without any enhancement at the limb.

As predicted by the theory two dimensional radio maps of the quiet sun have indicated the presence of a limb enhancement at the equator, while a limb darkening was found at the poles. We will discuss this later on, we want here to draw the attention to the very sharp predicted appearance of the limb brightening which cannot be resolved with conventional radio telescopes. In order to study its shape, especially in the cm range, eclipse observations have been used. At the phase of the four contacts a sharp decrease (or increase) of the total radio flux is expected if a very bright and very narrow region of the solar disk is being covered (or uncovered). Most of the observation on the shape of the limb brightening has been obtained in this way. The only cause of confusion is the presence of sources of the slowly varying component at the limb (Castelli and Aarons, 1965). This fact considerably reduces the reliability of the quiet sun eclipse observations.

2. Outline of the Problem

From this brief introduction it is clear that the basic problems for the understanding of the quiet sun radio emission are the following:

(i) Accurate determination of the quiet sun continuum radio spectrum.

(ii) High resolution radio maps of the quiet sun at different frequencies.

(iii) Accurate measurements of the equatorial and polar diameters.

(iv) Determination of the shape of the limb brightening with high resolving power.

Once these observational parameters are established, the problem is that of finding a model for the solar chromosphere and corona and a distribution of $T_e(h)$ and $N_e(h)$, which, once substituted in Equation (1), may lead to a good agreement with both the observed continuum spectrum and the two dimensional observed brightness temperature distribution at each wavelength. Caution should be taken, when determining these parameters, of the 11 years cycle of solar activity, since they might depend on the phase of the cycle.

We will now discuss each point separately.

3. The Continuum Spectrum

The plot of a number of observational results concerning the effective disk temperatures versus the frequency is given in Figure 1. (Shimabukuro and Stacey, 1968)*.

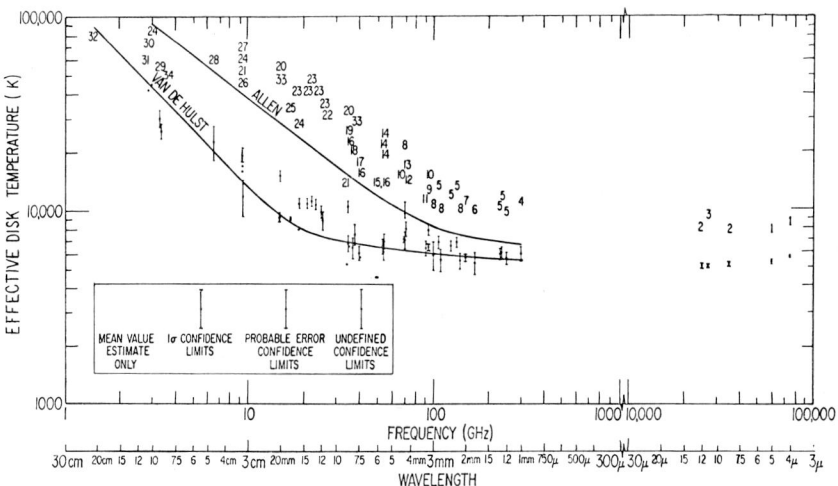

Fig. 1. Effective disk temperature versus frequency. References are listed in Table I.

The effective temperature can be also derived from the flux density, which is the observed parameter, from the relation:

$$T_b = \frac{S_\nu \lambda^2}{2K\Omega_s} \tag{4}$$

* Table I gives the references of the observations reported in Figure 1.

TABLE I
Central solar brightness temperature summary

Wavelength	Frequency (GHz)	Central Brightness Temperature (K)	Temperature Tolerance (K)	Figure 1 References	Source
4.0 μ	75000	5626	± 100	1	Murcray, Murcray, and Williams (1964)
5.0 μ	60000	5270	± 150	1	Murcray et al. (1964)
8.63 μ	35000	5160	± 40	2	Saiedy (1960)
11.1 μ	27000	5036	± 30	3	Saiedy and Goody (1959)
12.02 μ	25000	5050	± 80	2	Saiedy (1960)
1.0 mm	300.0	5900	± 500	4	Low and Davidson (1965)
1.2 mm	250.0	5600	± 400	5	Bastin et al. (1964)
1.3 mm	231.0	5900	± 400	5	Bastin et al. (1964)
1.3 mm	231.0	6000	± 500	5	Bastin et al. (1964)
1.8 mm	167.0	5300	± 700	6	Gorokhov et al. (1962)
2.0 mm	150.0	5670	± 230	7	Wort (1962)
2.15 mm	140.0	5433	± 500	8	Tolbert and Straiton (1961)
2.2 mm	136.0	6800	± 400	5	Bastin et al. (1964)
2.4 mm	125.0	6500	± 400	5	Bastin et al. (1964)
2.73 mm	110.0	5500	± 715	8	Tolbert and Straiton (1961)
2.8 mm	107.0	6800	± 500	5	Bastin et al. (1964)
3.0 mm	100.0	5870	± 950	8	Tolbert and Straiton (1961)
3.2 mm	94.0	6402	± 215 (16)	9	Simon (1965)
3.2 mm	94.0	7860	± 500	10	Tolbert, Straiton, and Walker (1962)
3.3 mm	91.0	6375	± 574 61	11	Rusch, Slobin, and Stelzreid (1966)
4.3 mm	70.0	7000	± 700	12	Coates (1958)
4.3 mm	70.0	9600	± 500	8	Tolbert and Straiton (1961)
4.3 mm	70.0	7100	± 200	10	Tolbert et al. (1962)
4.3 mm	70.0	8000	± 700	13	Kislyakov (1961)
5.50 mm	54.5	6950	± 600		
5.61 mm	53.50	6750	± 600		
5.62 mm	53.4	6900	± 600	14	Reber (1970)
5.62 mm	53.4	6100	± 600		
5.73 mm	52.4	6900	± 500		
6.0 mm	50.0	4500	—	15	Whitehurst et al. (1957)
6.0 mm	50.0	4500	—	15	Mitchell and Whitehurst (1958)
7.5 mm	40.0	6000	± 500	17	Whitehurst and Mitchell (1956)
7.5 mm	40.0	5700	—	16	Mitchell and Whitehurst (1958)
8.0 mm	37.5	6400	± 800	18	Salmonovich et al. (1959)
8.0 mm	37.5	7500	± 900	33	Salmonovich (1962)
8.5 mm	35.3	6500	—	16	Mitchell and Whitehurst (1958)
8.5 mm	35.3	6740	± 674	19	Hagen (1951)
8.6 mm	35.0	10420	± 730	20	Wulfsberg and Short (1965)
8.7 mm	34.5	5280	—	21	Aarons, et al. (1958)
1.18 cm	25.4	8870	± 980	22	Staelin et al. (1964)
1.18 cm	25.4	9800	± 700	23	Staelin et al. (1967)
1.28 cm	23.4	10700	± 700	23	Staelin et al. (1967)
1.35 cm	22.0	11000	± 700	23	Staelin et al. (1967)
1.43 cm	21.0	10800	± 700	23	Staelin et al. (1967)
1.58 cm	19.0	10800	± 700	23	Staelin et al. (1967)

Table I (continued)

Wavelength	Frequency (GHz)	Central Brightness Temperature °K	Temperature Tolerance °K	Figure 1 References	Source
1.6 cm	18.8	8000	—	24	Strezhneva, et al. (1958)
1.76 cm	17.0	9100	± 200	25	Tsuchiya and Nagane (1965)
2.0 cm	15.0	15100	±1057	20	Wulfsberg and Short (1965)
2.0 cm	15.0	9100	± 600	33	Buhl and Tlamicha (1968)
3.16 cm	9.5	11800	±2500	26	Veisig and Molchanov (1963)
3.2 cm	9.4	19300	±1351	27	Minnett and Labrum (1950)
3.2 cm	9.4	17000	—	24	Strezhneva et al. (1958)
3.2 cm	9.4	16000	—	21	Aarons et al. (1958)
4.6 cm	6.5	22800	±4560	28	Higgs and Broten (1966)
9.0 cm	3.3	30000	±3000	29	Swarup (1961)
9.1 cm	3.3	25000	—	34	Riddle (1969)
10.0 cm	3.0	45000	—	24	Strezhneva et al. (1958)
10.5 cm	2.9	42000	—	30	Hey and Hughes (1956)
10.7 cm	2.8	33000	—	31	Covington et al. (1955)
21.0 cm	1.4	47000	—	32	Christiansen and Warburton (1955)

where S_ν is the observed flux density
K is the Boltzmann constant
Ω_s is the solid angle of the source
λ is the wavelength of observation.

Assuming that the emission mechanism of the observed radiation is essentially due to bremsstrahlung process in a quiet plasma, the solar atmosphere can be approximated with a two layer atmosphere (Zheleznyakov, 1970), with a chromospheric temperature of the order of 3×10^4 K and a coronal temperature of 10^6 K. The clear cut boundary between the chromosphere and the corona is at an altitude of $h = 10^4$ km from the photosphere. The variation of the electron concentration $N_e(h)$ in the chromosphere at altitudes from 10^2 to 10^4 km is of the form:

$$N_e(h) = 5.7 \times 10^{11} e^{-7.7 \times 10^{-4}(h-500)}$$

where h is the height in km above the photosphere. In the corona the function $N_e(h)$ can be approximated by the Baumbach-Allen formula (Allen, 1947):

$$N_e(R) = 10^8 \left[1.55 (R/R_\odot)^{-6} + 2.99 (R/R_\odot)^{-16} \right]$$

For a more detailed analysis the spectrum can be divided into different parts. In the mm region, up to 2 mm, the temperature spectrum appears to be flat. If interpreted as a thermal emission, considering Equations (1) and (3), this behaviour can be explained with a rapid decrease of the electron density in the lower chromosphere from the photospheric level (Shimabukuro and Stacey, 1968). The region between 2 mm and 3 cm is the most complex. The observations are very scattered as shown in Figure 2. The dip in temperature present at about $\lambda = 5$ mm suggested an electron temperature inversion in the lower chromosphere. The scatter of the measured temperatures in this wavelength interval suggests also that the lower chromosphere is far from smoothly behaved. The need of a non-uniform model for the layers responsible of the emission will appear more strongly when discussing the limb problem at mm wavelengths.

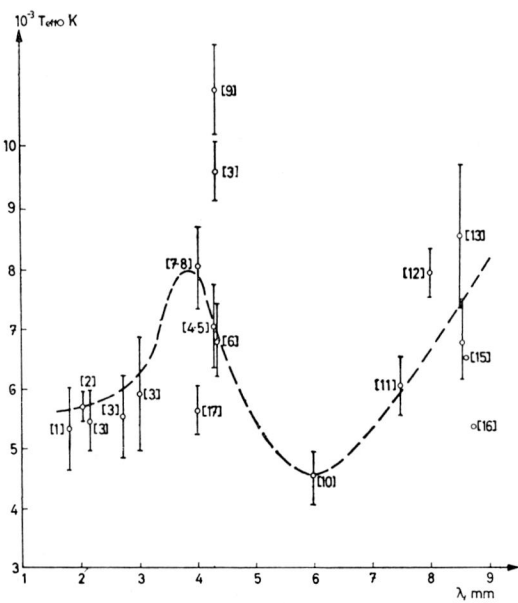

Fig. 2. Frequency spectrum of the quiet Sun in the millimetric range (after Zheleznyakov, 1970). The dotted line shows the most probable form of the frequency spectrum. The figures indicate the sources as follows: 1 – Gorokhov, Dryagin and Fedoseev (1962); 2 – Wort (1962); 3 – Tolbert and Straiton (1961); 4 – Coates (1958b); 5 – Coates (1958a); Edelson and Grant (1960); 7 – Kislyakov (1961a); 8 – Kislyakov (1961b); 9 – Straiton, Tolbert and Britt (1958); 10 – Whitehurst, Copeland and Mitchell (1957); 11 – Whitehurst and Mitchell (1956a and 1956b); 12 – Amenitskiy et al. (1958); 13 – Hagen (1949); 14 – Hagen (1951); 15 – Weaver, Mitchell and Whitehurst (1958); 16 – Aarons, Barrow and Castelli (1958), 17 – Coates (1957).

Assuming an electron density in the chromosphere of the form,

$$N_e(h) \simeq \exp\left[\alpha h^2 - \beta h + \gamma\right]$$

where $\alpha = 5.8 \times 10^{-18}$ cm^{-2}, $\beta = 1.28 \times 10^{-8}$ cm^{-1}, $\gamma = 27.24$ which takes into account the spicular nature of the chromosphere, Zheleznyakov fitted the observed spectrum shown in Figure 2 and obtained an electron temperature distribution with a marked dip near 3.3×10^3 km. In order to confirm the validity of this distribution, Zheleznyakov predicts the existence of a limb darkening in the waves near 5 mm, which has never been experimentally confirmed. Recent observations in this wavelength range (Reber, 1970) indicate a more uniform increment in the spectrum, thus eliminating the need to introduce a T_e inversion in the lower chromosphere.

In the region between 3 cm and 1.5 m, (see Figure 3) the temperature increases almost linearly, and it can be well approximated with the relation (Zheleznyakov, 1958):

$$\left[\frac{T_b}{K}\right] = 5 \times 10^3 \left[\frac{\lambda}{cm}\right]$$

As mentioned above this part of the spectrum can be fitted with a two layer atmosphere (Van de Hulst, 1953). Above 1.5 m the temperature spectrum decreases as shown in Figure 3. Recent measurements in period of maximum of solar activity

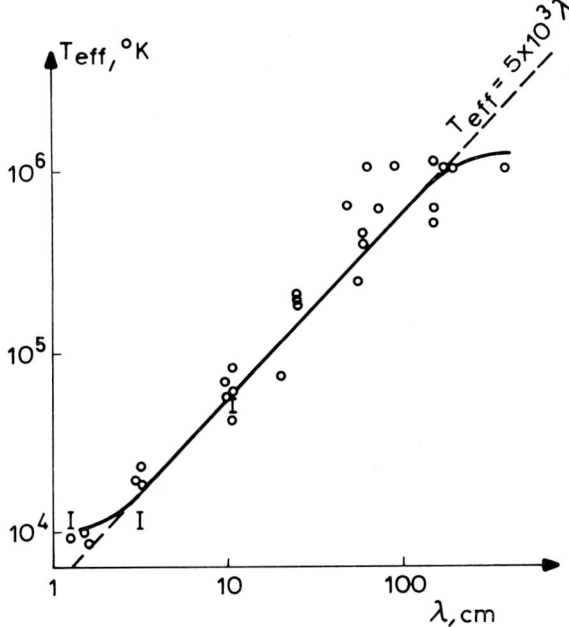

Fig. 3. Frequency spectrum of the quiet sun radio emission in the centimetric, decimetric and metric wavelengths (Zheleznyakov, 1970).

(August 1970), made by the Nançay group (Leblanc, 1970) with the Arecibo 1000 foot radiotelescope, have shown a very sharp decrease in the brightness temperature as the wavelength increases. The preliminary results are summarized in Table II. Unless we take into account the scattering of the radio waves on coronal inhomogeneities (Scheffler, 1958), then the only cause of this decrease can be the fall of the coronal kinetic temperature with altitude (Zheleznyakov, 1970).

TABLE II

Preliminary measurements of the quiet sun made in Arecibo by Leblanc (1970)

Frequency	Temperature T_{eff}	Radio sun mean sizes minutes of arc
29.3 MHz	$3.6\ 10^5$ K	$70' \pm 4'$
36.9 MHz	$5\ \ \ 10^5$ K	$53' \pm 5'$
60 MHz	$6\ \ \ 10^5$ K	$45' \pm 4'$

4. Brightness Distribution over the Solar Disk

The two dimensional brightness distribution of the quiet sun, together with the problem of the limb brightening is very strongly dependent on the wavelength of observation. The mm, cm, and m wave emission will be discussed separately, the separation being also related to the distinct layers at which the emission occurs.

4.1. Observations at mm waves

Table III lists all the observations carried up to now, which were aimed to study the quiet sun brightness distribution. Due to the weakness of the limb brightening in this wavelength range, there is still some ambiguity in the observations and in their interpretation. Some general remarks can be drawn in any case.

TABLE III

Observations of the temperature brightness distribution at mm wavelengths

Wavelength	Reference	Results
1.0 mm	Kundu (1970)	Limb darkening (?)
1.2 mm	Noyes et al. (1968)	Limb brightening
1.2 mm	Newstead (1969)	Limb brightening
3.3 mm	Shimabukuro and Stacey (1968) and (1970)	Limb brightening
3.5 mm	Kundu (1970)	Limb brightening (?)
4.3 mm	Coates (1958)	No evidence of limb brightening
4.3 mm	Tlamicha (1969)	No evidence of limb brightening
8.6 mm	Coates (1958)	Limb brightening
9.0 mm	Kundu (1970)	Limb brightening (?)

(1) A limb brightening seems to be present in all the more accurate observations, obtained either in eclipse or by averaging scans of the solar disk taken over a long interval of time. The limb brightening is concentrated in a very narrow spike. For all the observations, exception made for that by Coates at 8.6 mm (Coates, 1958b), the centre to limb distribution is flat up to $\varrho = 0.8$ ($\varrho = R/R_\odot = \sin\phi$).

In the E-W profile found at 8.6 mm there is a small dip just before the enhancement present on the limb. An analogous dip in temperature at $\varrho = 0.3$ is visible in the E-W distribution derived by Shimabukuro and Stacey (1968) at 3.3 mm (Figure 4). The dip is correlated with a region devoid of magnetic field. Simon and Zirin (1969) discussing recent high resolution observations in the mm wave range, point out that the observed limb brightening is always less than the value predicted using a spherically symmetric model (Figure 5). The obvious conclusion derived is that the model atmosphere from which the theoretical profiles were derived must be revised and that the introduction of inhomogeneities (spicules) is essential.

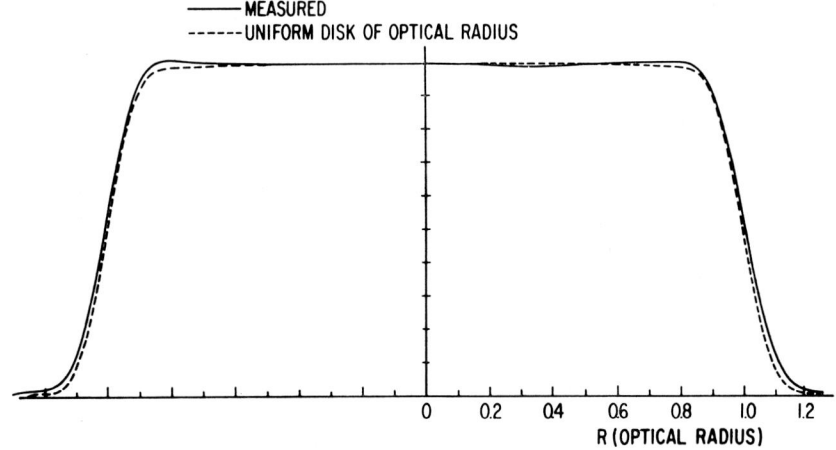

Fig. 4. Center to limb equatorial distribution at $\lambda = 3.3$ mm obtained by Shimabukuro (1968).

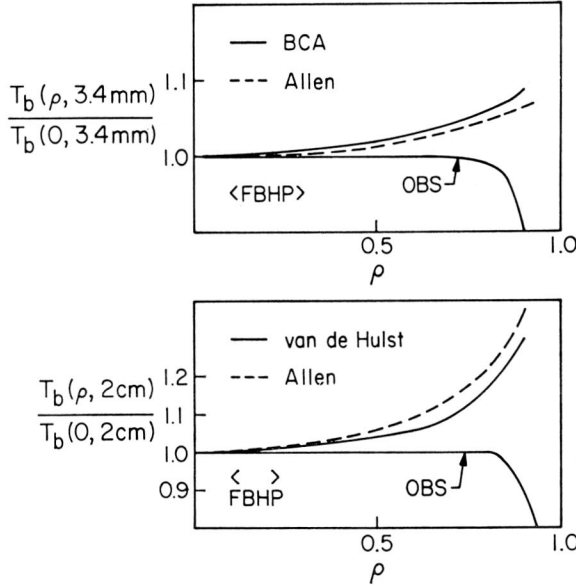

Fig. 5. Comparison of theoretical and observed distributions made by Simon and Zirin (1969). The observed 3.4 mm profile is the average polar scan derived by Shimabukuro (1970); the 2 cm profile has been observed by Tlamicha (1969).

(ii) Only one measurement at 3.3 mm of the N–S distribution has been obtained by Shimabukuro (1970). When the smoothing effect of the beam is taken into account, he does not find any significant difference between the N–S and the E–W profiles, both indicating the presence of limb brightening at the edge of the disk. The measurements are consistent with a solar brightness curve that is flat to about $\varrho = 0.8$ with a rapid increase to a peak value of about 1.3 at the limb (Figure 6). Buhl and Tlamicha

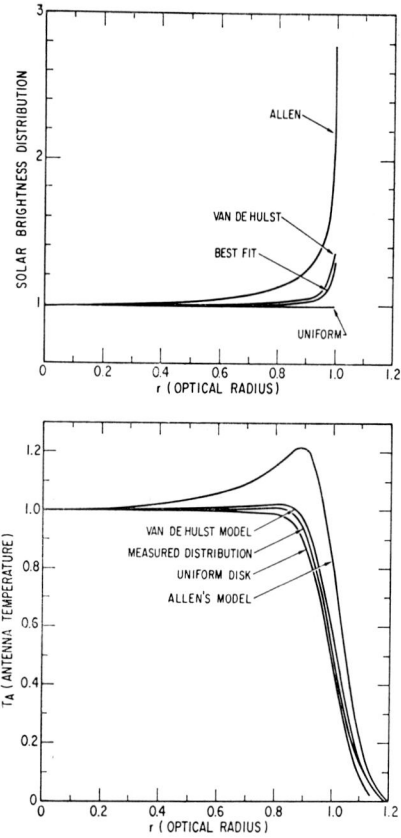

Fig. 6. Convolved profiles of different brightness distribution obtained by Shimabukuro (1970).

(1970) have recently mapped the sun at 3.5 mm with the 36 ft of the NRAO. Their observations show clear evidence of a darkening at the poles, while nothing can be said on the equatorial distribution, due to the presence of strong sources of the S component. The amount of the darkening at the poles is 200 K with respect to a brightness temperature of 6×10^3 K for the centre of the disk. In any case, it is difficult to conclude whether this polar darkening is typical of the quiet sun distribution or whether it is simply a lack of sources in the polar regions (a well known optical phenomenon).

Noyes *et al.* (1968) and Newstead (1969) find at 1.2 mm a value of 1.11 for the ratio of the mean T_b over the disk to the central T_b. They explain the brightness distribution and the limb brightening by using a spherically symmetric model for the solar chromosphere. A recent work carried out by Kundu (1970), using the 36 ft radio-telescope of NRAO, was aimed to study the temperature brightness distribution at mm wavelengths. The profiles, as shown in Figure 7, are the minimum averaged

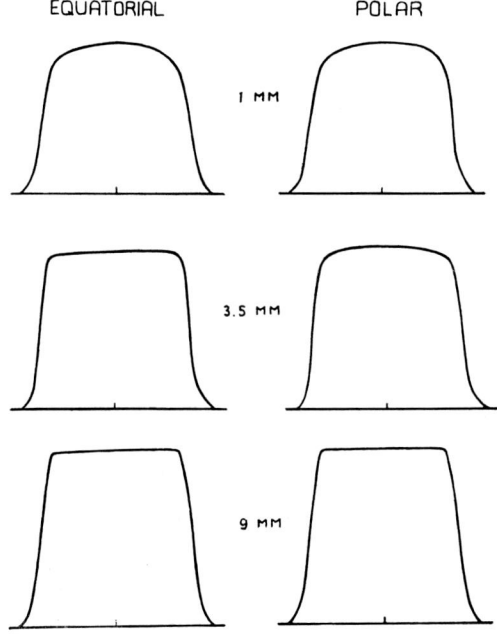

Fig. 7. Observed polar and equatorial brightness distributions at 1.0, 3.5 and 9 mm wavelengths (Kundu, 1970).

values of many scans taken in the equatorial and polar directions at 1.0, 3.5, 9.0 mm. These profiles are not deconvoluted from the antenna beam but a clear indication of limb darkening at 1 mm is apparent, while in the two others frequencies a limb brightening seems to be possible in the equatorial directions. A spherically symmetric model can be applied to the 3.3 mm distribution of Shimabukuro (1970) when considering a distribution for $N_e(h)$ and $T_e(h)$ as given by van de Hulst model (1953), provided lower temperatures ($T_e = 6 \times 10^3$ K) and an electron density with a smaller scale heigth are assumed. The chromosphere becomes optically thick for the 3.3 mm radiation over a small temperature range with the bulk of the emission coming from a region approximately 1500–3500 km above the photosphere as shown in Figure 8.

A spherically symmetric model cannot explain the 8.6 mm results obtained by Coates (1958, b) especially for the dip in temperature present near the limb and for the very sharp limb brightening. A symmetric model predicts in fact a gradual rise in

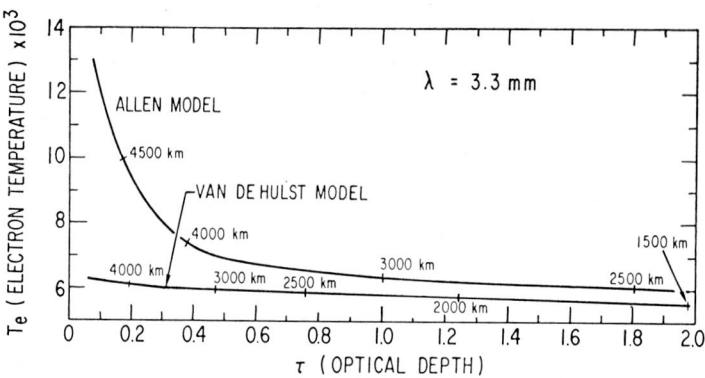

Fig. 8. Electron temperature vs. optical depth at $\lambda = 3.3$ mm for Allen and Van de Hulst models (Shimabukuro, 1970).

brightness temperature from the centre to the limb. The observed profile can be best interpreted assuming an inhomogeneous chromosphere where spicules are imbedded in a hotter and less dense plasma. Such a model, first considered by Hagen (1956) and subsequently Coates (1958, c) and Athay (1961) has been developed in more details by Molchanov (1964). As shown in Figure 9, the central part of the disk has a temperature which results from a mean value of the temperature of the two components (spicules and interspicular plasma). Close to the limb, the spicules inclined with respect to the line of sight, shield the hotter region thus giving an average lower temperature. At the limb itself, the radiation is emitted by the plasma region and is not intercepted by any of the filaments. Thus in this region, just as in the case of the spherically symmetric atmosphere, the temperature increases at the edge and forms a narrow bright limb.

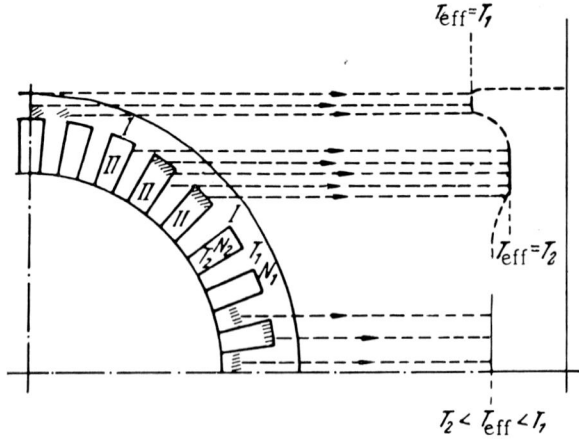

Fig. 9. A possible model of the solar lower atmosphere, with spicules (region II) and coronal plasma (region I), and the relative center to limb brightness distribution (Molchanov, 1964).

At wavelengths shorter than 8 mm, the observations do not show any dip near the limb. Assuming the same spicular model previously mentioned, this result can be explained by the fact that the two media (spicules and interspicular plasma) become so transparent that the main part of the radiation for all ray directions is emitted near the photosphere, that is in a layer whose temperature is low at all points of the disk.

On this hypothesis, decreasing the wavelength, the observed brightness distributions will have less importance on the comparison between the model with homogeneous atmosphere and the spicular model.

4.2. Observations at cm and dm waves

Table IV summarizes the observations made in this wavelength range. The symmetric spherical model together with the $N_e(h)$ and $T_e(h)$ as given by Smerd (1950) predict in this frequency range a strong limb brightening near the edge of the disk. The observed brightness profiles also show the presence of a limb brightening which does not however agree with the predicted center to limb profile. The observed maximum of temperature is always located inside the solar disk and not outside as the theory predicts. This maximum tends to shift towards the centre as the wavelength increases. An other peculiar aspect of the observations is the existence of a clear asymmetry between the E–W and N–S profiles, namely a limb darkening is present in the polar profiles. The situation is better shown in Figure 10 where theoretical and observed profiles at 9.1 cm are compared (Riddle, 1969). The amount of observed limb brightening is always less than the one predicted, and the profile appears to be

TABLE IV

Observations of the temperature brightness distribution at cm wavelengths

Wavelength	Reference	Results
2 cm	Tlamicha (1969)	No evidence of limb brightening
3 cm	Drago et al. (1964)	E–W Limb brightening
3.1 cm	Felli, Tofani (1970)	E–W Limb brightening
3.2 cm	Alon et al. (1953)	Limb brightening
7.5 cm	Kakinuma (1955)	Limb brightening
9.1 cm	Swarup (1961)	Limb brightening and polar darkening
9.1 cm	Riddle (1969)	Lower limb brightening than Swarup (1961)
9.4 cm	Haddock (1957)	Limb brightening
10.3 cm	Covington et al. (1955)	Limb brightening and polar darkening
21 cm	Christiansen and Warburton (1953)	Limb brightening and polar darkening
21 cm	Labrun (1960)	Limb brightening and polar darkening
60 cm	Stanier (1950)	No limb brightening
60 cm	Swarup and Parhtasarathy (1955)	Limb brightening
60 cm	O'Brien and Tandberg-Hansen (1955)	Limb brightening
88 cm	Firor (1959)	No limb brightening

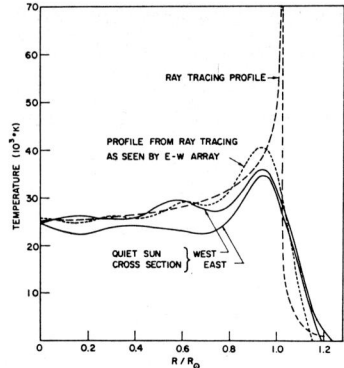

Equatorial cross-section of the quiet sun map and the comparable result of ray tracing.

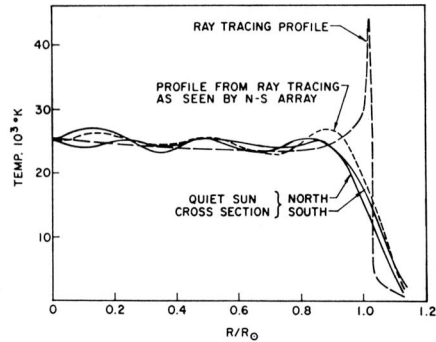

Polar cross-section of the quiet sun map and the comparable result of ray tracing.

Fig. 10. Equatorial and polar cross-sections of the quiet sun map at $\lambda = 9.1$ cm and the comparable result of ray tracing obtained by Riddle (1969).

flat up to $\varrho = 0.8$. To explain this behaviour the spicular nature of the lower solar corona in this wavelength range has been postulated.

The displacement of the limb brightening towards the center can be partially due to the aerial smoothing when the antenna beam is greater in size than the brightness peak (as shown by Smerd and Wild, 1957) but this effect cannot explain the constant shift toward the center as the wavelength increases. This effect has been interpreted (Molchanov, 1964) considering the negative temperature gradient in the outer coronal layers. In fact for the shorter waves the gradient of T_e is positive throughout the emitting region and the brightness maximum of the disk is situated outside the photosphere, for the longer waves the temperature gradient in all the emitting layers is negative and a limb darkening is observed. In the intermediate layers both of these effects take place and the T_b maximum is apparently displaced towards the centre of the disk.

The hypothesis of a spherically symmetric model is also questioned from the existence of the E–W, N–S asymmetry. This can be only explained considering a more realistic model of the corona which considers the variations of $N_e(h)$ from the equatorial to the polar regions, as also shown from optical observations. The variation of N_e

derived from the optical measurements between the equatorial and the polar regions is about 1.4–2.2, especially pronounced at the minimum of solar activity. Since these variations occur mainly in the corona the effect will be stronger in the dm and m range; in the cm range it will be less pronounced in proportion to the contribution of the coronal radiation at these wavelengths.

4.3. Observations at m Waves

The brightness distribution at m waves as resulting from the observations before 1964 has been reviewed by Zheleznyakov (1970) Figure 11 represents an interpretation of the observed profiles $T_b(r/R_\odot)$ in the metric band on the basis of a corona model with an irregular distribution of the electron concentration (Scheffler, 1958). Solid curves represent the theoretical distribution of radio brightness in a Baumbach-Allen isothermal model with a regular $N_e(r/R_\odot)$ curve; dashed lines represent the O'Brien's (1953) experimental results normalized so that the radio emission flux from the whole sun is the same as the value obtained from the theory; dashes and dots is the radio brightness distribution over the disk, found with allowance made for the scattering of radio waves on coronal inhomogeneities.

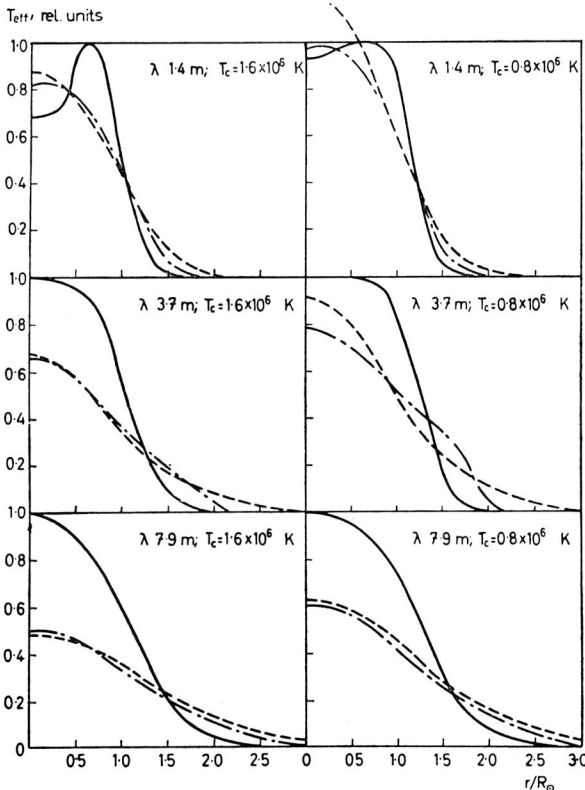

Fig. 11. Interpretation of the observed functions of the effective disk temperature in the metric band (Zheleznyakov, 1970).

Nevertheless the experimental results of O'Brien are not in good agreement with more recent observations, especially when the radio diameters are considered.

In Table II the radio diameters at 29.3, 36.9 and 60 MHz are given, indicating the strong increase in the radio diameter with the wavelength.

At these frequencies the levels at which $\tau=1$ are so high that the filaments do not have any significant influence on the brightness temperature distribution.

The experimental results at m waves are strongly affected by the presence of noise storms and most of all by the S component sources, whose presence can be eliminated with difficulty.

One of the most reliable discussion of the quiet sun in the m wavelength range is the recent analysis made by Leblanc and Le Sequeren (1969) of interferometric observations at 169 MHz taken during the last solar cycle. The great number of daily observations (~ 900) decreases to a minimum the error introduced by the presence of the S component, the presence of sources can be eliminated with an error of 10% for the flux.

The results are summarized in Table V. From the ratio of polar to equatorial diameter an ellipticity of 0.84 is found. From the measured flux and using an elliptical model for the corona, the electron temperatures during the maximum and minimum

TABLE V

169 MHz measurements (Leblanc and Le Squeren, 1969)

	Period of minimum of the solar cycle	Period of maximum of the solar cycle
Equatorial dimension	$38' \pm 1'$	$47' \pm 2'$
Polar dimension	$32' \pm 3'$	$36'$ [a]
Flux density	$6\ 10^{-22}$ W m^{-2} Hz^{-1}	$12,5\ 10^{-22}$ W m^{-2} Hz^{-1}
Temperature of the quiet Sun	$1.1\ 10^6$ K	$1.8\ 10^6$ K [a]
Temperature of the quiet Sun with a SVC centre	$1.5\ 10^6$ K	$2.2\ 10^6$ K [a]
Electron density	$2\ 10^{4+4.32\,a}$	$0.86\ 10^{6+2.64\,a}$

[a] Value deducted from calculations of Leblanc and LeSqueren (1969).

of the solar cycle were derived. The electron densities $N_e(R)$ were derived by solving the equation of transfer assuming an elliptical corona model. The corrected $N_e(R)$ function is the one which produces the best agreement in the E–W and N–S profiles.

A preliminary study of the two dimensional picture of the radio Sun at 80 MHz made by the Culgoora radioheliograph (Sheridan, 1970) has shown limb darkening and a shape which to the first order appears nearly circular; the radius is about 1.45 $R_\odot$ which corresponds to the 80 MHz plasma level, assuming an electron density plasma model as given by Newkirk (1961). The lack of any appreciable ellipticity is explained as due to the near maximum phase of the solar cycle. Figure 12 shows a good agreement between the 80 MHz radio map of the quiet sun and the corresponding polar plot of the K corona.

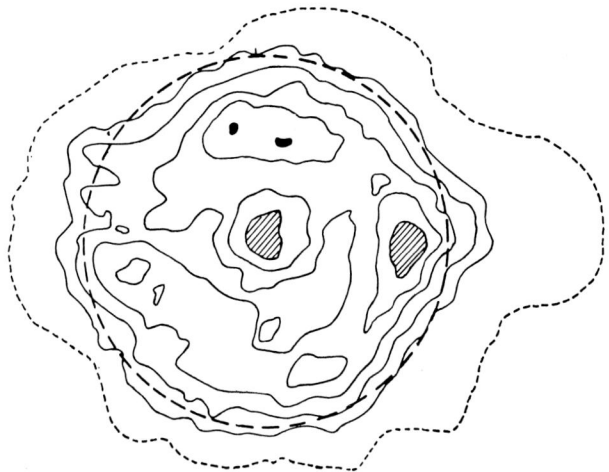

Fig. 12. 80 MHz solar image obtained by Sheridan (1970), surrounded by a polar plot of the K-corona, in which the radial distance is a measure of the square root of the intensity of the light scattered by electrons at a height of 0.5 $R_\odot$ above the photosphere (the dashed line is slightly lower at 0.45 $R_\odot$).

5. Conclusions

Some general considerations can be drawn from the previous exposition of the observational material concerning the radio quiet sun and its interpretation. Two different aspects of the quiet sun appear.

The first one is the concept used up to now, i.e. the statistical minimum value derived out of many daily observations of the two dimensional brightness temperature distribution. This statistical quiet sun seems to correspond to a smoothed picture of an irregular structure and provides us with the large scale characteristics of this fictitious body. The nature of the small scale structures has been inferred from some phenomena, like the limb brightening, but it is far from being clearly understood. A statistical approach cannot be used to study this fine structure since it would smooth and mask the real nature of the spicular structure.

The second aspect is therefore a quiet sun as a time varying structure on a very small scale and can be studied only by using very high resolution radio telescopes, observing regions of the disk devoid of active regions.

Recent observations (Simon *et al.*, 1970) in which high resolution has been achieved by means of big radio telescopes during the last solar eclipse suggest that in quiet regions the radio chromospheric network is present.

These preliminary results (shown in Figure 13) plot the normalized flux at the limb of the sun given at two different frequencies. The authors infer that these small scale fluctuations are fairly well correlated with the Hα filaments.

It is then clear that with increasing resolving powers and sensitivities the quiet sun will appear as an irregular structure which is far from the uniform quiet atmosphere of the early spherical models.

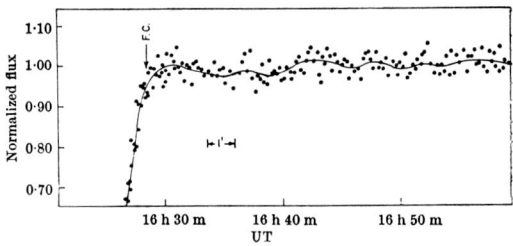

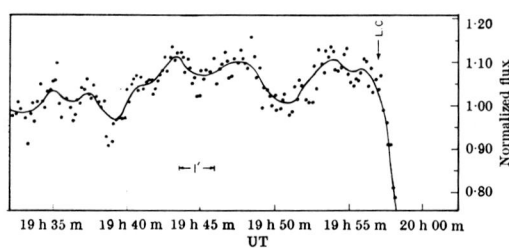

Fig. 13. Solar eclipse of March 7, 1970. (a) First contact data obtained at 3.4 mm with the 15 foot telescope of the Aerospace Corporation. (b) Last contact obtained at 1.35 cm with the 120 foot Haystack telescope of MIT-Lincoln Laboratory. The time of the contacts (FC and LC) and one min of arc size are indicated (Simon *et al.*, 1970).

References

Aarons, J., Barron, W. R., and Castelli, J. P.: 1958, *Proc. IRE* **46**, 327.
Allen, C. W.: 1947, *Monthly Notices Roy. Astron. Soc.* **107**, 426.
Alon, I., Arsac, J., and Steinberg, J. L.: 1953, *Compt. Rend.* **244**, 1726.
Athay, R. G.: 1959, *Paris Symposium on Radio Astronomy* (ed. by Bracewell), p. 98.
Bastin, J. A., Gear, A. E., Jones, G. O., Smith, H. J., and Wright, P. J.: 1964, *Proc. Roy. Soc. London* [A] **278**, 543.
Boischot, A.: 1958, *Ann. Astrophys.* **13**, 181.
Bracewell, R. N. and Preston, G. W.: 1956, *Astrophys. J.* **123**, 14.
Bracewell, R. N. and Swarup, G.: 1961, *IRE Trans., AP* **AP9**, 22.
Buhl, D. and Tlamicha, A.: 1968, *Astrophys. J.* **153**, L189.
Buhl, D. and Tlamicha, A.: 1970, *Astron. Astrophys.* **5**, 102.
Castelli, J. and Aarons, J.: 1965, *Solar System Radio Astronomy* (ed. by Aarons), p. 49.
Christiansen, W. N. and Warburton, J. A.: 1953, *Australian J. Phys.* **6**, 262.
Christiansen, W. N. and Warburton, J. A.: 1955, *Australian J. Phys.* **8**, 474.
Coates, R. J.: 1958a, *Proc. IRE* **46**, 122.
Coates, R. J.: 1958b, *Astrophys. J.* **128**, 33.

Coates, R. J.: 1958c, *Astrophys. J.* **128**, 83.
Covington, A. E., Medd, W. J., Harvey, G. A., and Broten, N. W.: 1955, *J. Roy. Astron. Soc.* **49**, 235.
Denisse, J. F.: 1950, *Ann. Astrophys.* **13**, 181.
Drago, F., Noci, G., and Piatelli, M.; 1964, *Ann. Astrophys.* **27**, 708.
Felli, M. and Tofani, G.: 1970, *Solar Phys.* **13**, 194.
Firor, J. W.: 1959, *Paris Symposium on Radio Astronomy* (ed. by Bracewell), p. 106.
Ginzburg, V. L.: 1946, *Compt. Rend. Acad. Sci. USSR* **52**, 487.
Gorokhov, N. A., Dryagin, Yu. A., and Fedoseev, L. I.: 1962, *Izv. Vysshikh Uchebn. Zavedenii, Radiofiz.* **5**, 413.
Haddock, F. T.: 1957, *Radio Astronomy* (ed. by Van de Hulst), p. 273.
Hagen, J. P.: 1951, *Astrophys. J.* **113**, 547.
Hagen, J. P.: 1957, *Radio Astronomy* (ed. by Van de Hulst), p. 263.
Hey, J. S. and Hughes, V. A.: 1956, *Observatory* **76**, 226.
Higgs, A. A. and Broten, N. W.: 1966, *J. Roy. Astron. Soc.* **60**, 272.
Hulst, H. C., Van de: 1953, *The Sun* (ed. by Kuiper), University of Chicago Press.
Jaeger, J. C. and Westfold, K. C.: 1949, *Australian J. Sci. Res.* **A2**, 322.
Kakinuma, T.: 1955, *Proc. Res. Inst. Atmosph. Nagoya Univ.* **3**, 96.
Kislyakov, A. G.: 1961, *Izv. Vysshikh Uchebn. Zavedenii, Radiofiz.* **4**, 433.
Kundu, M.: 1970, private communication.
Labrun, N. R.: 1960, *Australian J. Phys.* **13**, 700.
Leblanc, Y.: 1970, private communication.
Leblanc, Y. and Le Squeren, A. M.: 1969, *Astron. Astrophys.* **1**, 239.
Low, F. J. and Davidson, A. W.: 1965, *Astrophys. J.* **143**, 1278.
Martyn, D. F.: 1946, *Nature* **158**, 632.
Minnett, H. C. and Labrum, N. R.: 1950, *Australian J. Phys.* **3**, 60.
Mitchell, F. H. and Whitehurst, R. N.: 1958, Final Tech. Rep. University of Alabama, Rad. Astron. Lab.
Molchanov, A. P.: 1964, *Physics of the Solar System* (ed. by Mikhailov), p. 206.
Murcray, F. H., Murcray, D. G. and Williams, W. J.: 1964, *Appl. Opt.* **3**, 1374.
Newkirk, G. A.: 1961, *Astrophys. J.* **133**, 983.
Newstead, R.: 1969, *Solar Phys.* **6**, 56.
Noyes, R. W., Beckers, J. M., and Low, F. J.: 1968, *Solar Phys.* **3**, 36.
O'Brien, P. A.: 1953, *Monthly Notices Roy. Astron. Soc.* **113**, 597.
O'Brien, P. A. and Tandberg-Hansen, E.: 1955, *Observatory* **75**, 11.
Pawsey, J. L. and Yabsley, D. E.: 1949, *Australian J. Sci. Res.* **A2**, 198.
Pawsey, J. L. and Smerd, S. F.: 1953, *The Sun* (ed. by Kuiper), p. 466.
Reber, E. E.: 1970, *Icarus* **12**, 348.
Riddle, A. C.: 1969, *Solar Phys.* **7**, 434.
Rusch, W. V., Slobin, S. D., and Stelzreid, C. T.: 1966, Final rept. USCEE 183, Univers. of Southern California.
Saiedy, F.: 1960, *Monthly Notices Roy. Astron. Soc.* **121**, 483.
Saiedy, F. and Goody, R. M.: 1959, *Monthly Notices Roy. Astron. Soc.* **119**, 213.
Salomonovich, A. E.: 1962, *Soviet Astron.–AJ* **6**, 202.
Salomonovich, A. E., Koshchenko, F. N., and Noskova, R. I.: 1959, *Soln. Dannye* **9**, 83.
Scheffler, H.: 1958, *Z. Astrophys.*, **45**, 113.
Sheridan, K. V.: 1970, *Proceedings of ASA* **1**, 305.
Shimabukuro, F. I. and Stacey, J. M.: 1968, *Astrophys. J.* **152**, 777.
Shimabukuro, F. I.: 1970, *Solar Phys.* **12**, 438.
Shklovskii, I. S.: 1946, *Astron. Z.* **23**, 333.
Simon, M.: 1965, *Astrophys. J.* **141**, 1513.
Simon, M. and Zirin, H.: 1969, *Solar Phys.* **9**, 317.
Simon, M., Buhl, D., Cogdell, J. R., Shimabukuro, F. I., and Zapata, C.: 1970, *Nature* **226**, 1154.
Smerd, S. F.: 1950, *Australian J. Sci. Res.* **A3**, 34.
Smerd, S. F. and Wild, J. P.: 1957, *Radio Astronomy* (ed. by Van de Hulst), p. 290.
Staelin, D. H., Barret, A. H., and Kusse, B. R.: 1964, *Astron. J.* **69**, 69.
Staelin, D. H., Gant, N., Law, S., and Sullivan, W. T.: 1967, Rept. QPR 84, MIT Res. Lab. for Electr.

Stanier, H. M.: 1950, *Nature* **165**, 354.
Strezhneva, K. M., Plechkov, V. M., and Starodubtsev, A. M.: 1958, *Soln. Dannye* **7**, 71.
Swarup, G.: 1961, Rept. 13, Stanford Univ., Rad. Sci. Lab.
Swarup, G. and Parhtasarathy, R.: 1955, *Australian J. Phys.* **8**, 487.
Tlamicha, A.: 1969, *Solar Phys.* **10**, 150.
Tolbert, C. W. and Straiton, A. W.: 1961, *Astrophys. J.* **134**, 91.
Tolbert, C. W., Straiton, A. W., and Walker, T. A.: 1962, Rept. 6–45, Univ. of Texas, Electr. Eng. Res. Lab.
Tsuchiya, A. and Nagane, K.: 1965, *Publ. Astron. Soc. Japan* **17**.
Veisig, G. S. and Molchanov, A. P.: 1963, *Soln. Dannye* **11**, 58.
Whitehurst, R. N., Copeland, J., and Mitchell, F. H.: 1957, *J. Appl. Phys.* **28**, 295.
Whitehurst, R. N. and Mitchell, F. H.: 1956, *Proc. IRE* **44**, 1879.
Wort, D. J. H.: 1962, *Nature* **195**, 1288.
Wulfsberg, K. N. and Short, J. A.: 1965, Rept. AFCRL 65-75, Air Force Cambridge Res. Lab.
Zheleznyakov, V. V.: 1958, *Uspekhi Fiz. Nauk.* **64**, 113.
Zheleznyakov, V. V.: 1970, *Radio Emission of the Sun and Planets*.

20. SOLAR BURSTS AT DECAMETER AND HECTOMETER WAVELENGTHS

M. R. KUNDU
Astronomy Program, University of Maryland, U.S.A.

1. Introduction

One of the most interesting problems of solar radio astronomy is the study of the generating mechanisms of radio bursts. The most relevant observations for such studies are the measurements of dynamic spectra and positions of burst sources as a function of frequency and time. At meter and decameter wavelengths, the phenomena of most interest, bursts and storms, vary rapidly from second to second, and therefore the measurements must be made with high time resolution over the entire frequency range of interest. At meter wavelengths, the most sophisticated instrument capable of such measurements is the Culgoora radio heliograph at 80 MHz, which produces two pictures per second, one in each polarization. At decameter wavelengths, there is no heliograph. However, there are a number of interferometers including one at the Clark Lake Radio Observatory. The Clark Lake system gives the one-dimensional position and angular size of emissive regions on the sun nearly simultaneously at all frequencies over the range 20–60 MHz for studies of evolution of these regions as a function of time. In this paper we discuss some positional data of bursts at decameter wavelengths. At hectometer wavelengths, most of the burst data are obtained from experiments conducted in space vehicles. The most successful of such space experiments is that done by the Radio Astronomy Explorer Satellite in the range 0.2 to 5 MHz, with a time resolution of $\frac{1}{2}$ sec. Some of the more important properties of the low frequency bursts are reviewed here.

2. Type III Bursts at Decameter Wavelengths

At decameter wavelengths, the most commonly occurring phenomena are the type III bursts. It is now generally recognized that each individual type III burst is radiation from streams of electrons moving outwards through the corona and exciting plasma oscillations of progressively diminishing frequency as the stream passes through the coronal plasma of diminishing electron density. Their properties at decameter wavelengths are similar to those at meter wavelengths – their rise to maximum brightness temperature ($\lesssim 10^{11}$ K) occurs in a period of $\gtrsim 1$ sec and their subsequent decay with an exponential time constant of ~ 1 sec is generally attributed to collisional damping of the plasma oscillations by the coronal gas. The sudden rise time suggests that we are dealing with an extremely rapid acceleration process with a time constant of < 1 sec. At decameter wavelengths, one sometimes observes bursts of extremely short duration that look like type III but they start and decay within 1 sec at 30 MHz. This implies a very rapid acceleration process, and a rapid damping mechanism. In a two-million degree corona, typical decay times due to collisional damping are about 1.5 sec and 5 sec at 60 and 30 MHz respectively (see for example, Kundu, 1965). It is not unlikely that these short bursts are produced by the same generating mechanism as the type I burst, although their bandwidth is much larger than the type I.

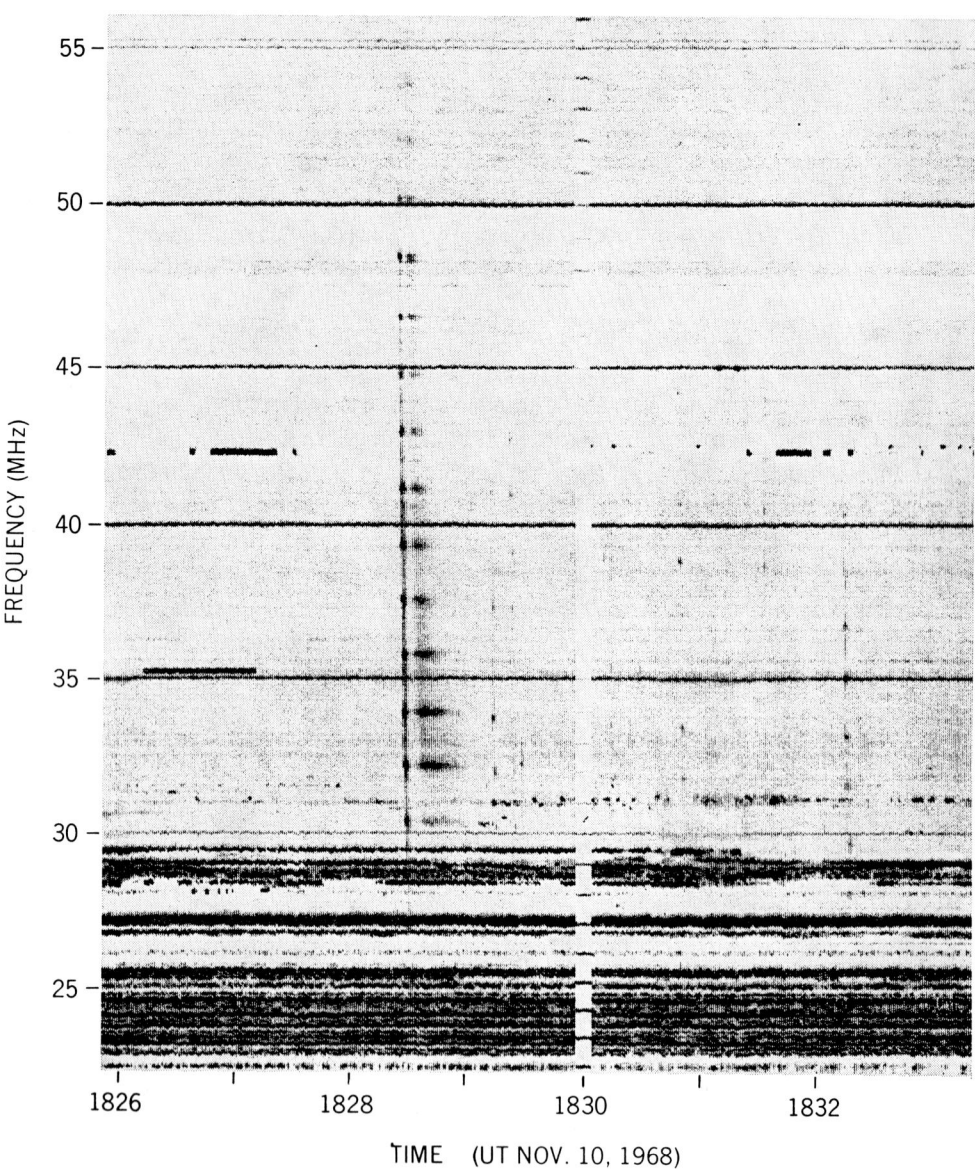

Fig. 1a. Interferometer record of a harmonic type III burst observed on Nov. 10, 1968. (From Kundu *et al.*, 1970.)

The positional data of type III bursts indicate a systematic dispersion with frequency, the lower frequencies occurring higher in the corona than the higher frequencies, consistent with the interpretation of type III bursts in terms of plasma radiation. When type III bursts occur in a group, the individual members of the group seem to originate at the same position, with some scatter probably attributable to instrumental errors. The 80 MHz heliograph observations of Culgoora indicate that sometimes different members of a group originate in different regions of the sun, but their distinct re-

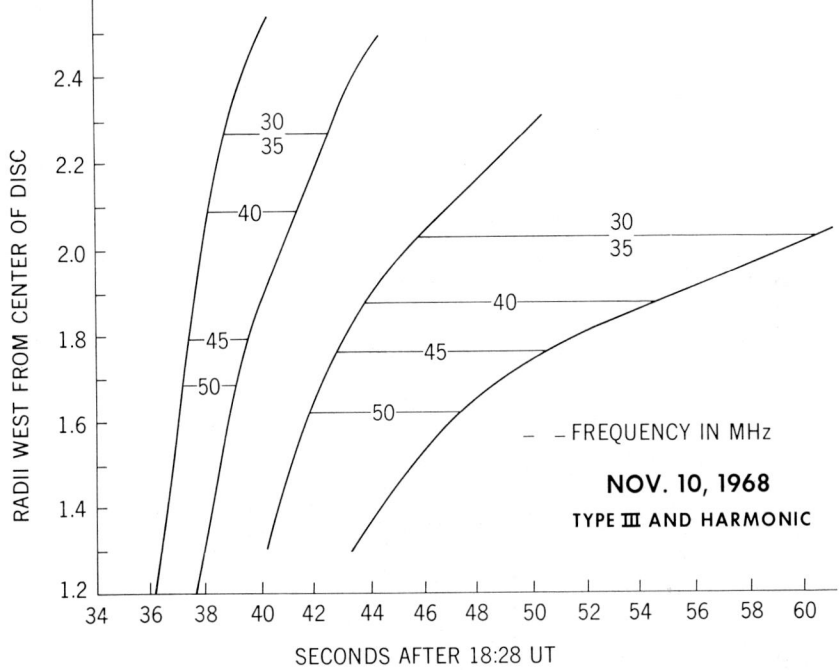

Fig. 1b. Positions of the type III and its harmonic as a function of frequency and time. (From Kundu et al., 1970.)

semblance to one another indicate that they may somehow be connected. The one-dimensional position measurements of Clark Lake show similar results at 20–60 MHz. From the position measurements of type III bursts occurring near the limb, and assuming that the disturbance (the stream of electrons) originating from the flare moves radially outwards producing the plasma oscillations responsible for type III bursts, one can compute the electron density as a function of height. The average electron density distribution determined in this manner is similar to that found in a coronal streamer. In harmonic type III bursts, one finds typically that the harmonic position is situated systematically lower than the fundamental (Figure 1). This result is consistent with the hypothesis of combination scattering by which the plasma waves are converted into electromagnetic radiation.

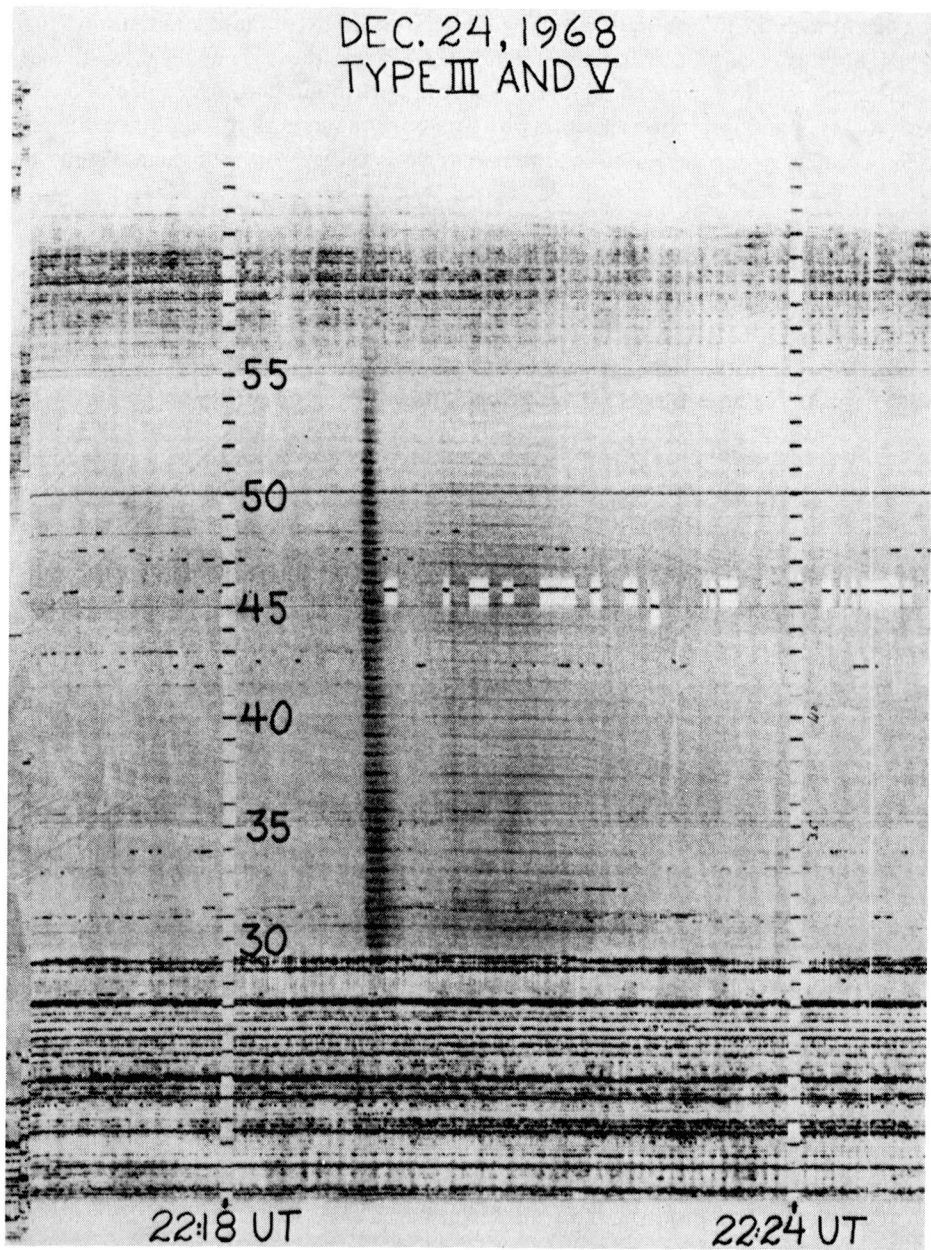

Fig. 2a. Interferometer record of a type III-type V burst observed on Dec. 24, 1968. (From Kundu *et al.*, 1970.)

The bursts show wide varieties of dynamic spectra. The simplest are sharp-featured and continue their curved sweep towards decreasing frequencies as far as the lowest frequencies observable, their duration increasing with decreasing frequency. It is conceivable that in such cases the electrons escape into the interplanetary medium. In other cases, the dynamic spectrum terminates at some frequency or it may continue with positive frequency drift, having the appearance of an inverted U. In such cases the electrons are believed to be magnetically guided back towards the sun. In still

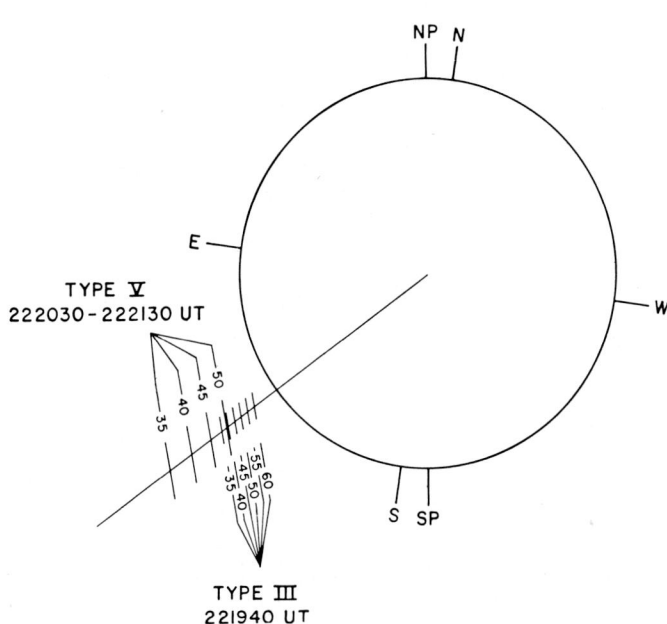

Fig. 2b. Positions of type III and associated type V as a function of frequency. The positions have been averaged over the durations of the type III and type V bursts. (From Kundu et al., 1970.)

further cases the bursts terminate in or are followed by a diffuse patch of low-frequency continuum radiation lasting for $\lesssim 1$ min – an event referred to as a type V burst. This behavior has been attributed to emission from electrons trapped between the mirror points of arched magnetic field lines. This interpretation has been supported by recent radioheliograph and interferometric observations which show the type V source has a much larger size than its parent type III source and is displaced in position by several arc minutes, although they both show similar dispersion in position with frequency (Figure 2).

At decameter wavelengths, one sometimes observes moving type III bursts, that is, the source position at a particular frequency changes systematically with time. A

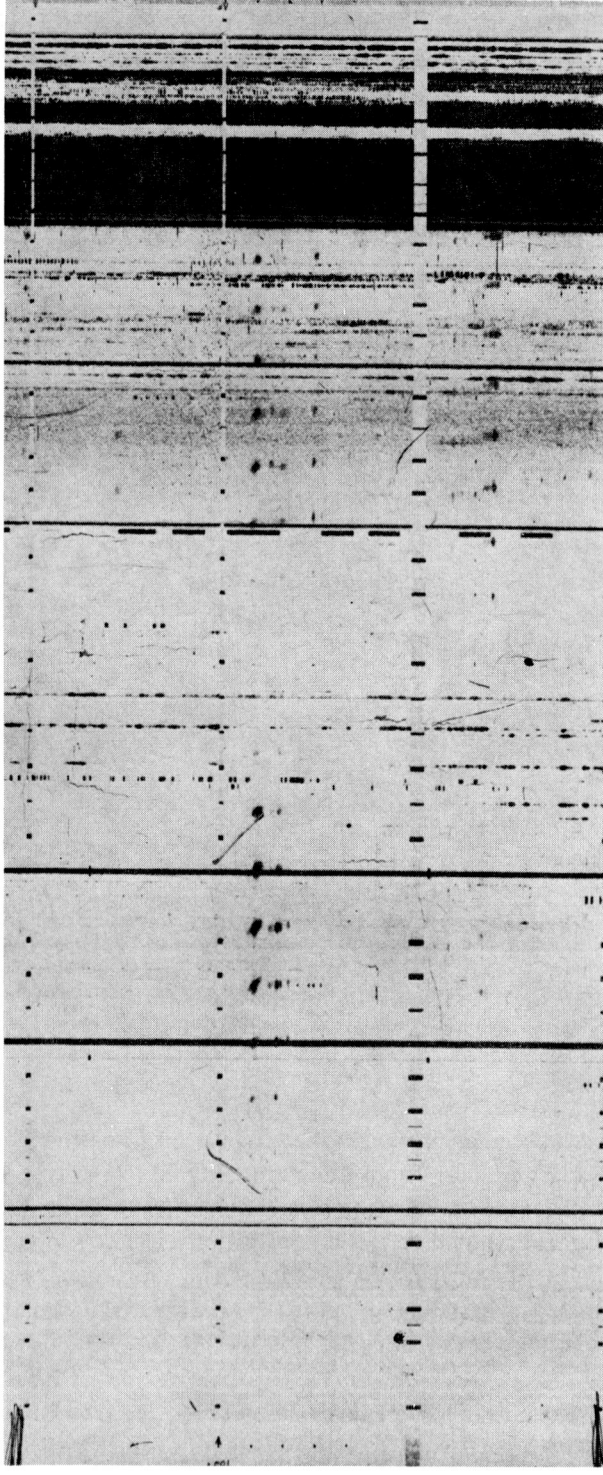

Fig. 3a. Interferometer record of a moving type III burst observed on May 26, 1969.

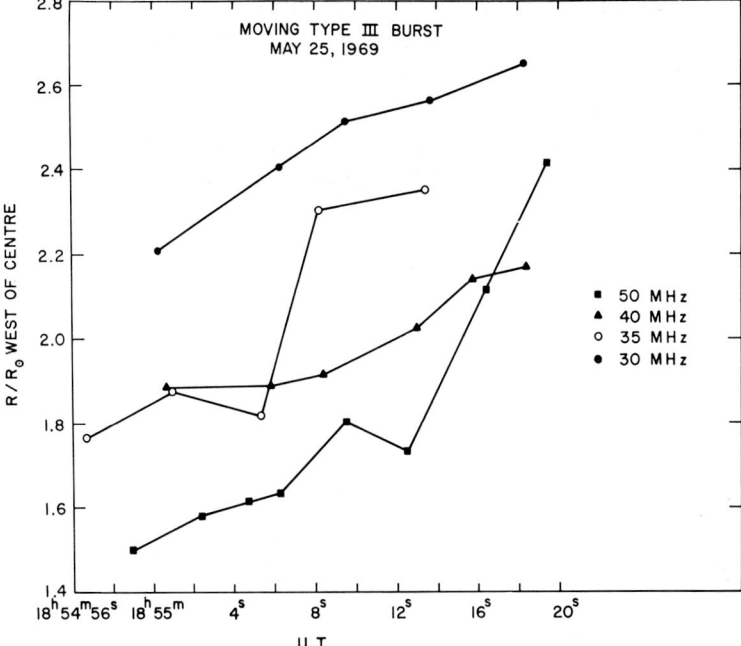

Fig. 3b. Position data of the burst.

typical example is shown in Figure 3. Such movements are possibly indications that the exciting particle stream travels tangentially to the contours of constant electron density.

The Culgoora heliograph observations have shown that successive brightenings often occur at two distant points of the sun with time separations of ~10 sec and it appears that such correlated activity may be similar in nature to a U-burst produced in wide magnetic arches. It is not easy to observe such bursts on one-dimensional interferometer records; however, the simultaneous existence of two systems of fringes over 20–60 MHz range in one type III event suggests that we might be dealing with a correlated event at decameter wavelengths. The resulting positional data (shown in Figure 4) indicate that at the high frequency end (near 50–55 MHz), the type III has two distinct sources separated by about 1 R_0, whereas at lower frequencies because of poorer resolution, the two sources are unresolved. Such correlated emissions appear to be similar in nature to those observed at meter wavelengths.

The identification of type III disturbances with fast electrons was made originally on the basis that no other type of disturbance with appropriate velocity was known which could generate radio waves; the tendency towards constant velocity supported this interpretation. The detection of bursts of electrons with energy in the range ~40 keV at the time of type III bursts has given further support to this interpretation.

Type III bursts often occur at the time of small chromospheric brightenings seen in

Hα and at the start of solar flares. They are often the only manifestation of instabilities in the upper corona, the type III bursts at decameter and hectometer wavelengths often not being associated with any Hα flares.

The number of fast particles in a type III disturbance has been estimated to be 10^{35} for a single burst (Wild, 1964), the uncertainty being several orders of magnitude. Upper limits to the linear dimensions of the particle stream are provided by angular

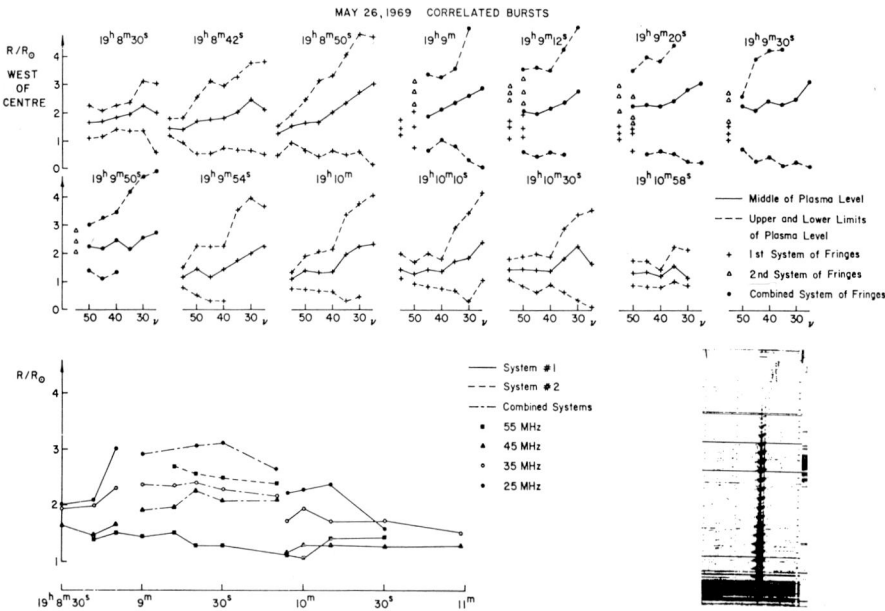

Fig. 4. Positional data of a 'correlated' type III event. The interferometer record is shown as an inset.

size measurements; the angular size increases sharply with decreasing frequency, an effect which may be due both to the 'fanning out' of the electron stream and to the scattering and refraction of the radio waves in the solar atmosphere.

Two important conclusions emerge from the study of type III bursts. The first is the unambiguous existence of open field lines which allow electrons to escape from their initiating instability to the interplanetary medium; the open field configurations are most likely to be along neutral sheets which themselves delineate coronal streamers. The neutral sheets may sometimes penetrate to near the base of the corona or they may form higher in the corona – e.g., as part of a helmeted structure above bipolar or multipolar field configurations. The second inference is that the bursts may be initiated by an instability capable of accelerating clouds of particles in a period not greater than ~ 1 sec. This instability may be the most basic one, since it may well provide the trigger by which the major energy of solar flares is released (Wild, 1969).

3. Type III Bursts at Hectometer Wavelengths

Our knowledge of the nature of solar bursts at hectometer wavelengths comes mainly from experiments conducted in space vehicles (Hartz, 1964; Graedel and Haddock, 1970; Fainberg and Stone, 1970). At these frequencies one often observes an extremely large number, tens of thousands, of sporadic bursts which in general can be classified as type III or fast-drift bursts. Such 'storms' of type III bursts are distinct from individual or groups of type III bursts because of their duration. Storms are observed over a half solar rotation, sometimes even up to two solar rotations. In contrast to isolated or groups of type III's the storm bursts rarely exceed 10 dB above the cosmic noise background. Figure 5 illustrates a section of type III storm data. The emissions at lower frequencies often appear as a slowly varying continuum, with occasional burst peaks recognizable as such.

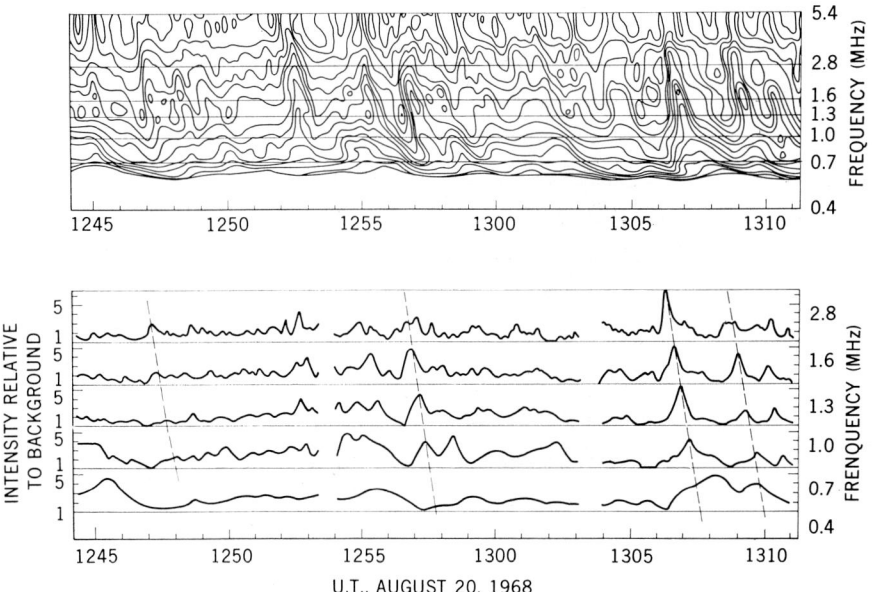

Fig. 5. Typical type III storm data obtained by RAE-satellite in the range 0.7 to 5.4 MHz. (From Fainberg and Stone, 1970.)

Concurrent data in the 20 to 60 MHz frequency range shows the presence of a continuum above the same active region that is presumably responsible for the hectometer type III storms. A decametric continuum, in contrast to decametric type IV, is characterized by a continuum background upon which are superimposed 'massive' numbers of type III events (Warwick 1965). Since the hectometric type III's are closely related to the occurrence of a decametric continuum, it seems reasonable that the two phenomena are closely related and that at still lower frequencies, decametric continuum may degenerate into a type III storm. However, it is still possible that a continuum component might exist along with the type III storm at hectometer wavelengths. The close connection between meter and decameter storms on one hand, and between decameter continuum and hectometer storms on the other, suggest that meter, decameter, and hectometer storms may all be produced by the same outward streaming electrons, and that under suitable conditions such energetic (~ 40 keV) electron streams are measured by space probes beyond the magnetosphere.

There is a systematic dependence of occurrence rate of hectometer type III on the heliographic longitude of an active region on the solar disc, the storm activity being maximum near CMP and minimum near the limb. Near the CMP the apparent drift rate is also maximum and as the active region approaches the limb, the average drift rate decreases.

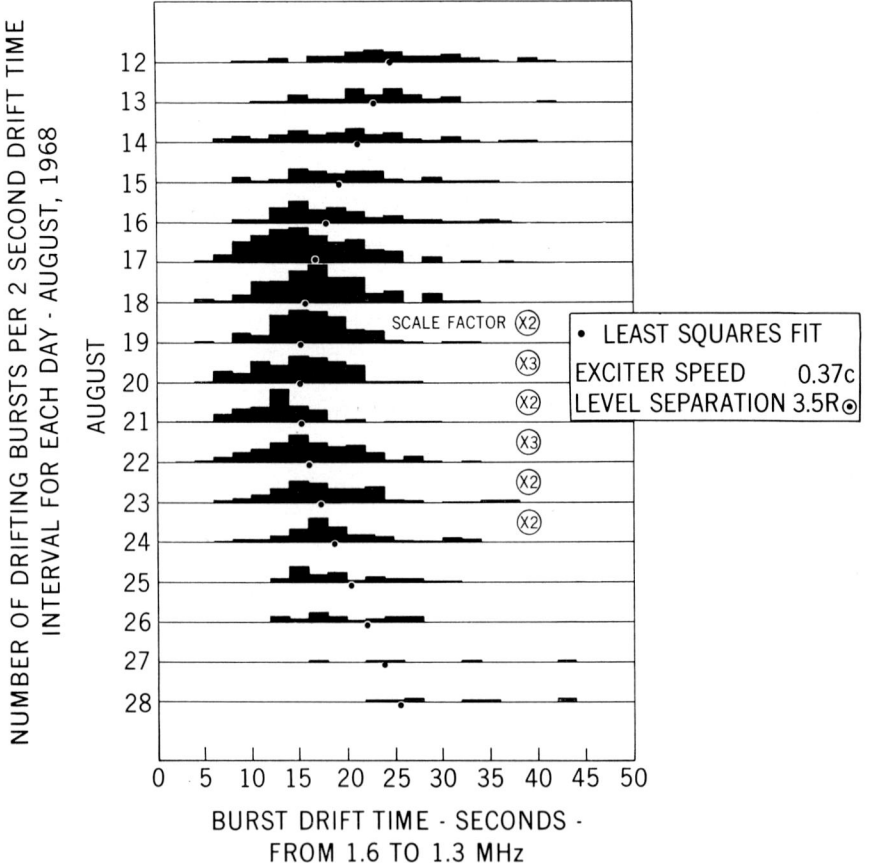

Fig. 6. Histogram of the number of bursts per two-second drift time interval vs. drift time for 24 h. groups of bursts drifting from 1.6 to 1.3 MHz. The circled figures are vertical scale factors relative to 1 for active days. The dots are the drift times yielded by a radially moving exciter at the midpoint of each day for the least squares fit of the model to these data. The number of bursts used was 1497. (From Fainberg and Stone, 1970.)

Figure 6 shows, according to Fainberg and Stone (1970), the distribution of drift times per 2-second drift interval over the range 1.6 to 1.3 MHz for each storm day. The observed drift rate depends on the time required for the exciter to travel between two plasma levels as well as on the time difference for the radio emission to travel from the two plasma levels to the observer. From these considerations, Fainberg and Stone made a least square analysis of 2500 drifting bursts and found that the average exciter speed between 2.8 and 0.7 MHz, corresponding to a total distance of about 19 R_0 (Alexander *et al* 1969) is 0.37 c, in reasonable agreement with the value of 0.33 to 0.4 c at 1.5 to 3 R_0. Haddock and Graedel (1970), on the other hand, using the density model of Hartz (1964) find an average velocity of about 0.11 c. It should be noted that, unlike Haddock and Graedel, the excess time of signal transmission through the corona was considered by Fainberg and Stone. The spread of drift seen in Figure 6 can be due to a variation of electron density gradient through which the exciters move or to a real spread of exciter speeds or both. If the spread is due to variations of the exciter speed, the exciter speed will then lie between 0.2 c and 0.6 c. However, it is more likely that the spread in drift rates is due to streamer density inhomogeneities.

The RAE results suggest that over a height range of 30 R_0 the exciter undergoes little, if any, deceleration. If the exciter consists of monoenergetic electrons, the speed of 0.37 c would correspond to electrons with kinetic energy of 40 keV. The low frequency determination of exciter speed obviously establishes a link between processes close to the sun and the detection of energetic electrons of solar origin near the earth. One would thus expect a good correlation between many hectometer type III bursts and the delayed occurrence of enhanced fluxes of $\lesssim 40$ keV electrons of solar origin measured by Lin and Anderson (1967). If this is really so, then the exciters could definitely be interpreted as packets of superthermal electrons of average energy $\lesssim 40$ keV which are able to propagate over distances of the order of 1 AU with little deceleration.

4. Type II Bursts

As we know, a type II burst usually consists of two main emission bands which drift from high to low frequencies at a rate some 200 times slower than for type III bursts and which persist for periods ranging from 3–30 minutes. The two bands are harmonically related and are identified with emission at the first and second harmonics of the plasma frequency; by the same kind of arguments as were used for type III bursts one infers that the type II burst is caused by a disturbance moving outwards through the corona with a velocity of $\sim 10^3$ km/sec. This disturbance is generally attributed to a collisionless magneto-hydrodynamic shock wave.

We know that the disturbances inferred from Hα observations (such as those of Lockheed) travel across the surface, whereas the type II shock waves travel radially outwards; however, the precise time-coincidence of both kinds of disturbances in some cases leave no doubt that the two observations refer to the same phenomenon. The angular dimension of type II bursts gives an estimate of the solid angle, Ω, into

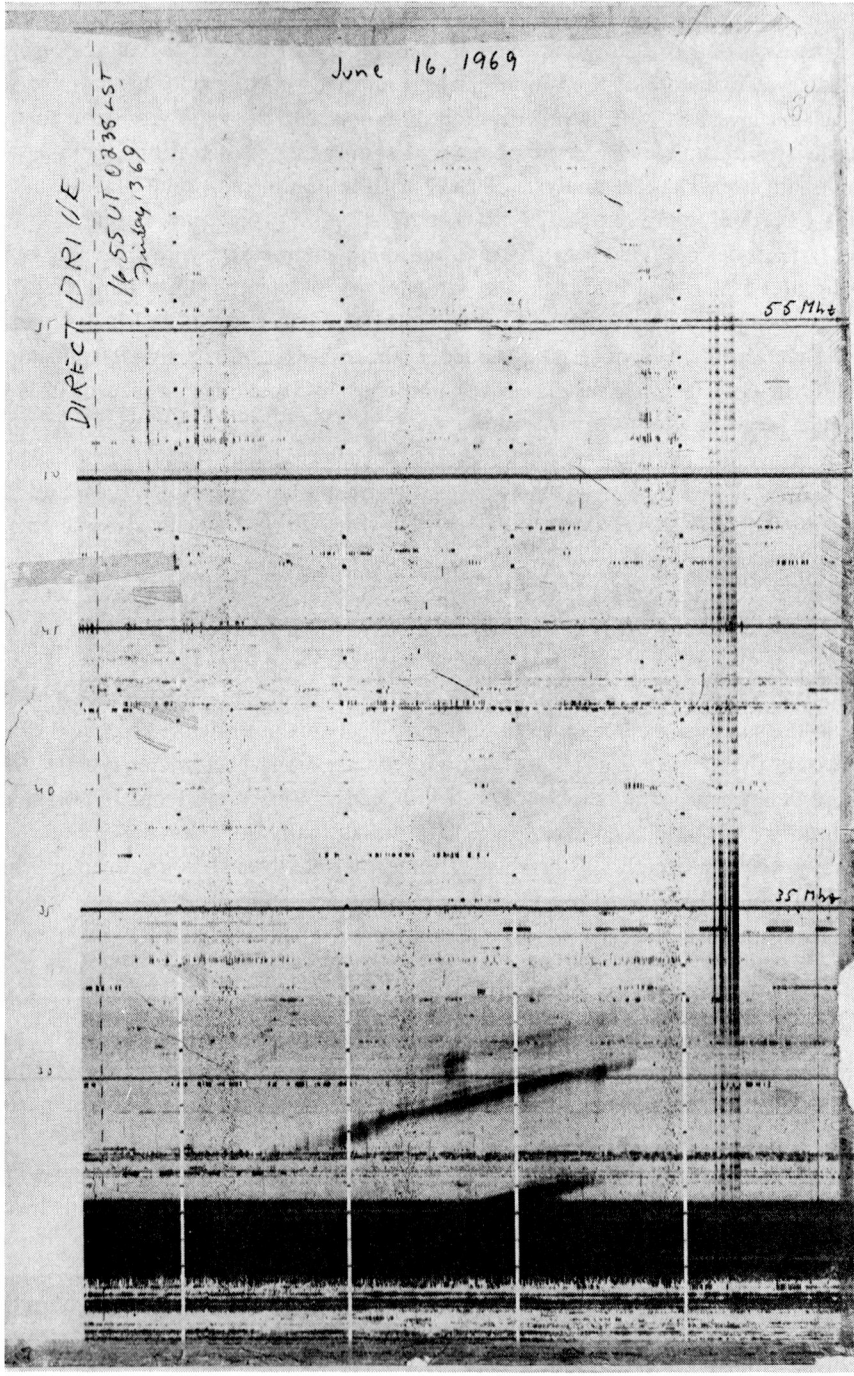

Fig. 7a. Interferometer record of a type II burst (followed by type IV) which occurred simultaneously at two positions.

which the disturbance propagates. The size measured by the 80 MHz heliograms is of the order of that of the optical disk. The 20–60 MHz Clark Lake data indicate even larger size. Since this size represents the projected area subtended by Ω at the height of the appropriate plasma levels, one can conclude that Ω is a substantial fraction of a hemisphere. We, therefore, envisage a quasi-spherical shock front expanding outwards from the center of the disturbance and exciting plasma oscillations of a given frequency from regions where the shock front crosses the appropriate plasma level. A spectacular evidence of such a spherical shock wave is found in the Clark Lake interferometer record of a type II event, shown in Figure 7. The type II event consists

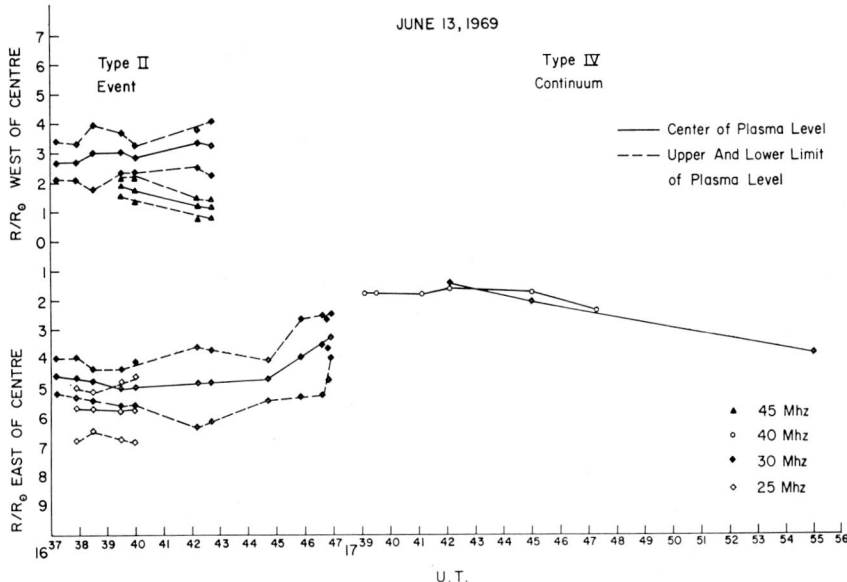

Fig. 7b. Positional data of the burst.

of two components originating from locations separated by about $7 R_0$. This event which occurred on June 13, 1969 at 1637 UT gives further evidence of interaction between distant active regions by a common disturbance originating at some intermediate point. The type II was followed by a moving type IV.

The physical processes which take place in the shock front of a type II disturbance are not fully understood but it is certain that type III-like instabilities can repeatedly take place at intervals of a few seconds in or near the front. These manifest themselves as small type III-like bursts which often grow out of the main emission bands of the type II and correspond to bursts of electrons which travel outwards from the shock front. Other parameters of interest to understanding a type II shock front are the existence of rapid movements in the distribution of brightness within the extended source regions of type II bursts (Wild, 1969) and also in the heights of type II sources.

There is strong evidence that the shock responsible for type II emanates from a region near the flare at about the time of the flash phase. As the shock travels outwards – covering great distances of up to $\sim 10^6$ km – it interacts with other features of the solar atmosphere.

An example of such interaction was revealed by the Culgoora observations (Wild, 1968) on February 25, 1968: shock waves from a flare on the sun's disk reached the 80 MHz plasma level ($\sim 4 \times 10^5$ km height) 11 min after flare onset; a few minutes later it caused a quiescent prominence on the limb of the sun, 10^6 km from the flare, to erupt and ascend. The prominence eruption was itself accompanied by a radio burst of type IV, high in the corona above it, and there is evidence that it was itself responsible for the further emission of shock waves. It appears that on encountering suitable magnetic configurations and physical conditions in its path the shock wave triggers new instabilities which release energy for further eruptions. A possible way in which this can happen is shown in Figure 8 (Wild, 1968).

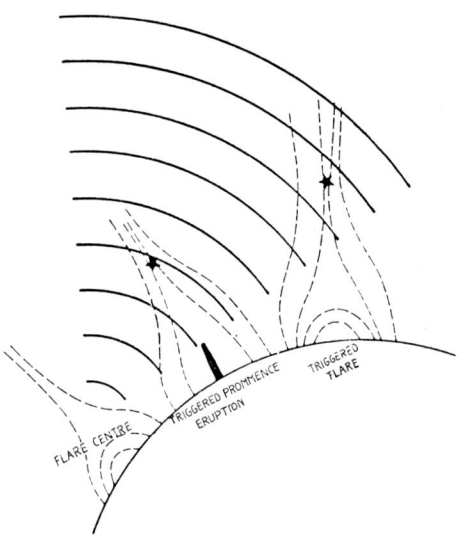

Fig. 8. Diagram showing the possible way in which the shock wave from a flare may trigger prominence eruptions and other flares by causing instabilities at neutral sheets high above them in the corona.

Type II bursts have been observed at frequencies as low as 5 MHz by the Sydney group. However, no type II burst has been observed at frequencies below 5 MHz at flux density levels of $\geqslant 10^{-17}$ W m^{-2} (c/s)$^{-1}$ by either the Michigan group (Haddock and Graedel 1970) or by the NASA-GSFC group. At frequencies of $\geqslant 4$ MHz one often observes only type III bursts concurrent with type II (accompanied by type IV) bursts at higher frequencies.

5. Type IV Bursts

During and following large flares, type IV continuum radiation is observed at decameter wavelengths – almost always associated with type IV at meter wavelengths. Even at decameter wavelengths, type IV is occasionally observed to occur in association with a type II which usually precedes it. Position determinations usually indicate a dispersion in frequency. Occasionally near the beginning of the type IV event, the radiation at all frequencies seems to come from the same region. This result is consistent with the commonly accepted interpretation of meter-wave type IV in the initial and late phases of the event. We shall illustrate the positional characteristics of the large events by describing several events.

The event of March 12, 1969: This event (Figure 9) is interesting in many respects. A flare occurred at 1735 UT near N10 W80, Imp. 2B. The flare reached its maximum at 1742 UT with erupting filaments. The prominence reached a height of 0.5 R_0 around 1743.5 UT when the radio event started with a type II. No harmonic has been reported for this type II, although there is a suggestion on our records that a harmonic might have been present. The type II source position (Figure 9) shows the characteristic dispersion in frequency; however, it moves to a lower position toward the end of the burst (around 1754 UT). If the harmonic is present, then the lower position at later times at the same frequency is quite consistent with the interpretation that the harmonic is due to combination scattering (Smerd *et al.*, 1962). In the absence of a harmonic, one has to interpret this movement as tangential to the surfaces of constant

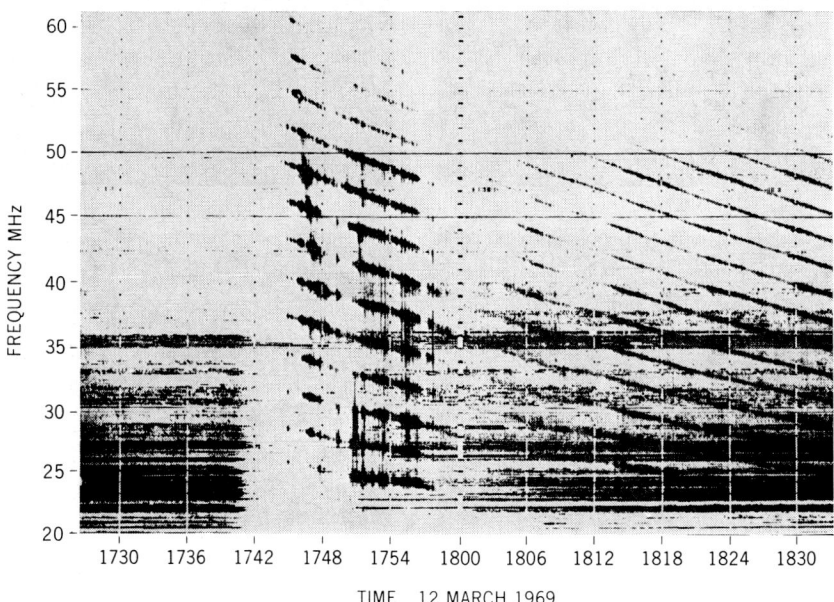

Fig. 9a. Interferometer record of a type II-type IV event observed on March 12, 1969.

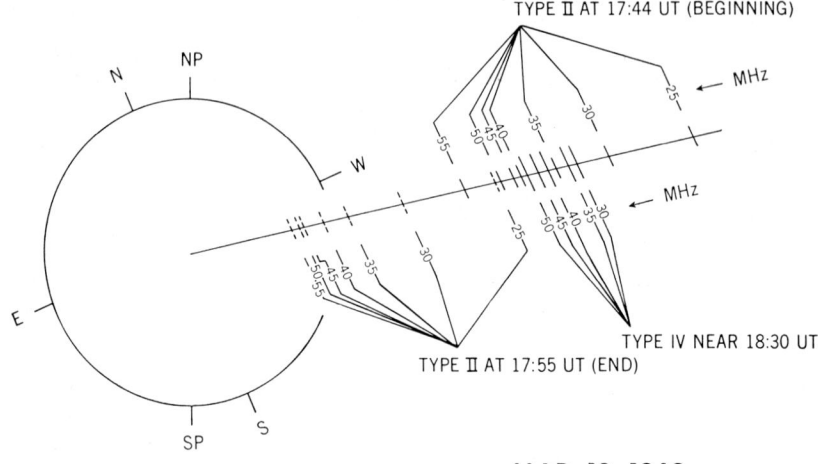

MAR. 12, 1969
TYPE II FOLLOWED BY TYPE IV

Fig. 9b. Positions of type II and type IV bursts as a function of frequency at some representative times in course of the event.

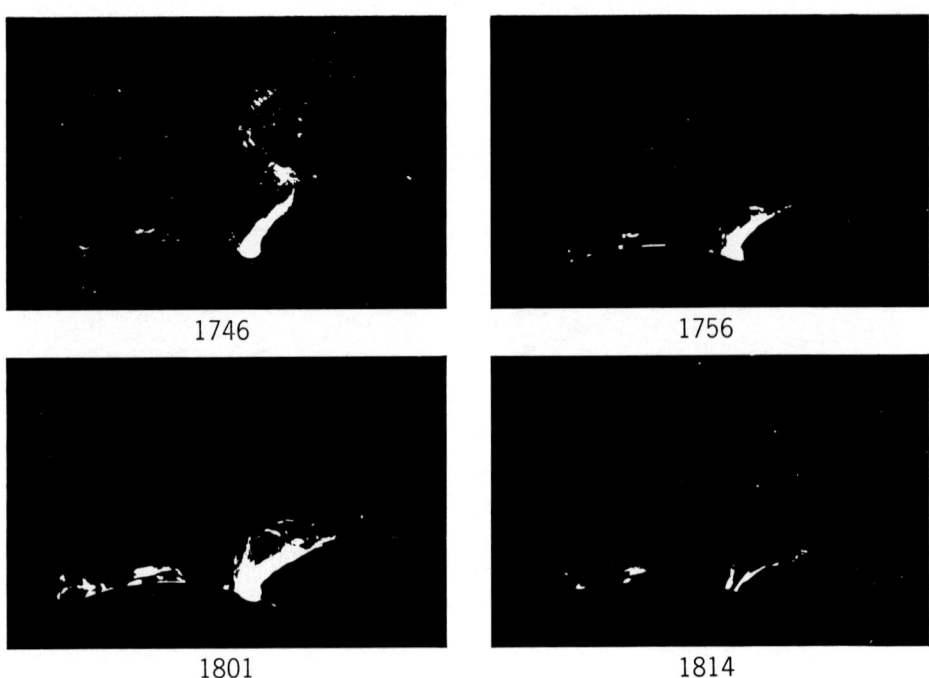

Fig. 9c. Hα coronagraph pictures of the prominence observed during the type II-type IV event of March 12, 1969 at some representative times. The times are in U.T. (Courtesy: Institute for Astronomy, University of Hawaii.) [From Kundu et al., 1970.]

electron density (Weiss, 1963). It is also possible that the type II burst was caused by multiple shocks, the later burst at lower altitude being the result of a different shock traveling on a path inclined to that of a first shock with slower speed. Indeed, the careful analysis by McCabe (private communication) of Hα prominences indicates the existence of multiple trajectories along which the prominence material traveled. The type IV emission at our wavelengths started at 1759 UT. The position determinations indicate a frequency dispersion which is similar to, but smaller than, that observed for type II. The initial 'synchrotron' source which would be indicated by all frequencies coming from the same region, does not seem to be present on our records. It is quite possible, however, that a synchrotron source exists simultaneously with, but is swamped by, the source of plasma radiation which is much stronger. At a later time, at $\sim$ 1830 UT, such a source tends to show up when the 'plasma' source is weak. The source of type IV remains more or less stable within a height range of 2.3 to 3.1 R_0. It is interesting to note that the prominence which started at 1744 UT developed into a looped structure at a height of approximately 1 R_0 where it was visible until about 2112 UT [Figure 9(c)].

The event of March 16, 1969: This event started with a strong group of type III's at 1926.5 UT and the type IV event started on our records at 2006 UT and lasted until 2129 UT. Optically, the event was associated with a flare which occurred before 2011 UT at N16 E07. The flare was of Imp 1F. The type III positions show the usual dispersion with frequency; the type IV source moves with a velocity of about 600 km/sec. The frequency dispersion of position seen in Figure 10 may be due partly to ionospheric refraction and partly to the fact that the source may have a plasma wave origin.

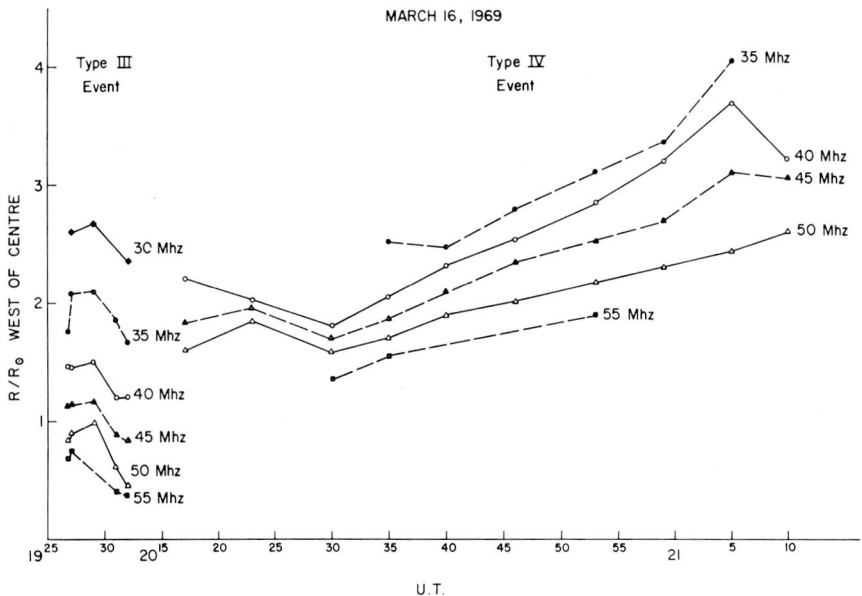

Fig. 10. Positional data of a type III-type IV event observed on March 16, 1969.

The event of June 11, 1969: This was a large type II-type IV event. The type II was not observed on our records. The type IV started at 1642.5 UT and lasted until about 2200 UT (Figure 11). The associated flare occurred about 1635 UT at S11 W23, and was of Imp B. One remarkable feature of the decametric type IV is that a second continuum source was ejected from the parent source with a high velocity of more than 4000 km/sec toward the east and ultimately reached a distance about 5 R_0 away in 15 min. The parent continuum source appeared to be of synchrotron origin in that emission at all frequencies had a common origin and the source had a slow motion of a few hundred kilometers per second.

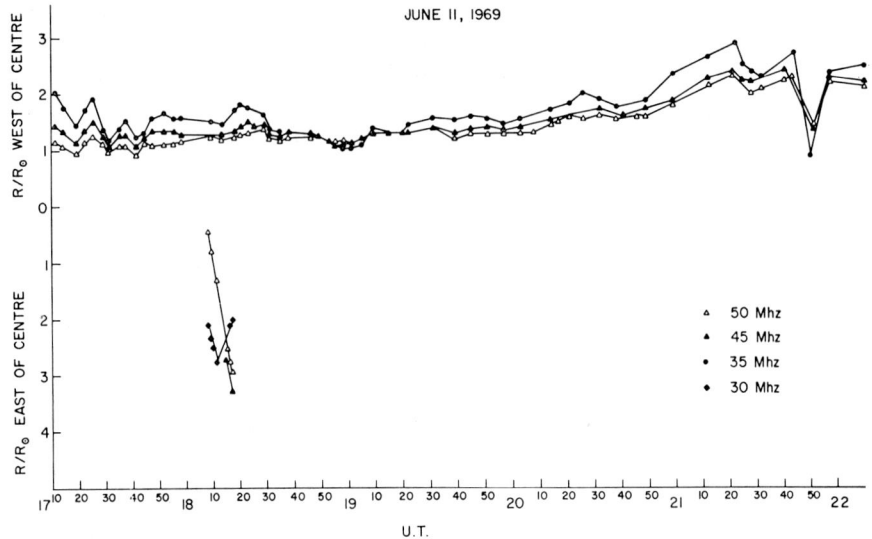

Fig. 11. Positional data of a type II-type event observed on June 11, 1969.

The event of June 14, 1969: This continuum event was associated with a flare of Imp. 2N, which occurred around 2038 UT at S12 W66. The continuum event was observed at both meter and decameter wavelengths; it started at 2119 UT and lasted until about 2300 UT. The source position moved to a higher level with a velocity of about 400 km/sec, at the same time showing frequency dispersion. Such dispersion of position with frequency is characteristic of 'stationary' type IV; consequently, it must be concluded that the disturbance producing plasma waves responsible for type IV moved tangentially to the constant electron density contours (Figure 12).

Although the characteristic dispersion of position with frequency is clearly seen in many large continuum events such as that of March 12, 1969, the positional measurements of large complex events are often difficult to interpret, particularly if there are two or more sources simultaneously present on the sun. Further, it is difficult to separate the frequency dispersion of position due to ionospheric refraction from the real effect. In general, in decameter-wave type IV associated with flares, we mostly see a smooth continuum; only occasionally we see structure resembling a continuous series of type III. Regarding the type IV power spectra at decameter wavelengths, we do not observe any decrease in intensity with decreasing frequency. On the contrary, sometimes we observe a continuum at 20–30 MHz but not at higher frequencies. This observation is not consistent with the Razin-Tsytovich Effect of the plasma on the synchrotron radiation. This effect predicts that if the plasma in which the synchrotron

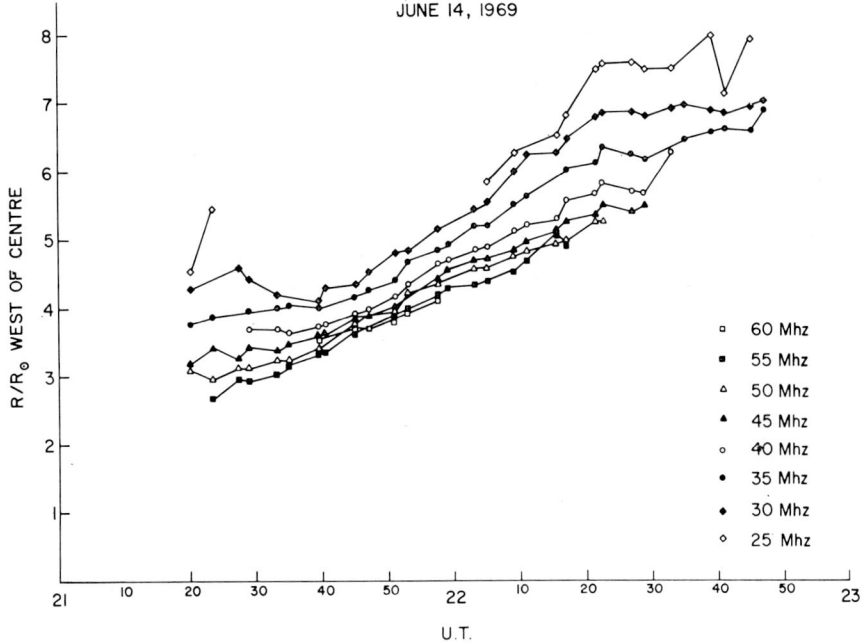

Fig. 12. Positional data of a type IV event observed on June 14, 1969.

radiation is generated is dense enough to influence the refractive index of the medium, then the intensity should decrease very steeply with decreasing frequency (Ramaty and Lingenfelter, 1967). However, this kind of power spectrum may be quite consistent with the origin of type IV in plasma radiation. Since we see the 'synchrotron part' of the source only occasionally, at the beginning of a type IV, we can conclude that the Razin effect probably plays an important role in suppressing the decameter wavelength type IV of synchrotron origin.

6. Flare Continuum

At meter-wave wavelengths many events that accompany or are triggered by flares are dominated by sources generated essentially by ejecta – e.g. the type II, III and moving IV bursts: one sees little or no sign of the brightening of the active region above the actual flare region, as is observed at centimeter wavelengths. However, in some high energy events a strong stationary continuum source is seen on 80 MHz radioheliograph records above the flare region. Such a source typically begins to brighten at the time of flare onset, intensifies to a maximum in ~ 10 m and gradually fades over a period of $\lesssim 1$ h. This kind of source is not to be confused with the storm continuum which characteristically begins ~ 1 h after flare onset, and has been referred to as 'flare continuum' by Wild (1970). Wild (1970) has suggested that the flare continuum is generated by energetic electrons trapped in closed magnetic structures above the flare and between the flare and other active regions.

7. Discussion

The positions of type III bursts at decameter wavelengths show the same characteristic dispersion with frequency as that observed at meter wavelengths. The second harmonic of type III is situated systematically at a lower level than the fundamental. Both these results are consistent with the currently accepted interpretation of type III generation, namely the excitation of plasma waves by electron streams and subsequent conversion of longitudinal plasma waves into transverse electromagnetic waves by scattering on density and charge fluctuations. When a type V burst follows a type III, it is found that the type V position is displaced from that of type III, but shows similar dispersion with frequency. This is consistent with the interpretation that, like type III, the type V is due to plasma radiation, but it is excited by an electron stream different from that producing the type III. It is possible that type V is caused by electrons trapped between mirror points of arched magnetic fields, whereas type III is caused by electron streams escaping along neutral sheets into the interplanetary medium (Weiss and Stewart, 1965; Zheleznyakov and Zaitsev, 1968). However, the higher position as well as higher dispersion of type V relative to those of type III found in one case is not necessarily consistent with the interpretation of Zheleznyakov and Zaitsev (1968) that the type V, like type III, is generated primarily at twice the electron plasma frequency. It is obvious that we need further positional data in order to better understand the generating mechanism of type V. At decameter wavelengths, certain type III sources show systematic motion, which implies that the disturbing agency moves tangentially to the constant electron density contours. Finally, the correlated type III activity first observed by the 80 MHz Culgoora heliograph seems to have been observed at least in one case at decameter wavelengths.

Two conclusions emerge from studies of large events in the 20–60 MHz range. First, the type IV burst at decameter wavelengths shows the same kind of complexity as that observed at meter wavelengths, probably being made of two parts, a 'synchrotron' one and a 'plasma wave' one. The first seems to be less frequent than on meter wavelengths, probably due to the Razin effect. The source of type IV at 20–60 HMz is, on the average, much higher in the solar atmosphere than at meter wavelengths, the height increasing with decreasing frequency. One can conclude that the source of decameter continuum is quite distinct from that of meter-wave continuum. This is consistent with the plasma origin of decameter continuum. Secondly, in large complex events when there are several sources at the same time on the sun, the one-dimensional positional measurements do not lead to easy and meaningful interpretation. The only way this difficulty can be overcome is by making fast two-dimensional maps of the sun at decameter wavelengths.

At hectometer wavelengths during active periods one observes a continuous series of type III's or type III storms, often associated with decametric continuum. The exciter speed in the upper corona, derived from hectometer drifting bursts, is essentially the same (about 0.37 c) as that obtained in the inner corona from observations at meter and decameter wavelengths. Type II or type IV bursts do not seem to be observed at hectometer wavelengths.

References

Alexander, J. K., Malitson, H. H., and Stone, R. G.: 1969, *Solar Phys.* **8**, 388.
Fainberg, J. and Stone, R. G.: 1970, *Solar Phys.* **15**, 222.
Fainberg, J. and Stone, R. G.: 1970, in preparation.
Haddock, F. T. and Graedel, T. E.: 1970, *Astrophys. J.* **160**, 293.
Hartz, T. R.: 1964, *Ann. Astrophys.* **27**, 831.
Kundu, M. R.: 1965, *Solar Radio Astronomy*, John Wiley-Interscience Publishers, p. 329.
Kundu, M. R., Erickson, W. C., Jackson, P. D., and Fainberg, J.: 1970, *Solar Phys.* **14**, 394.
Lin, R. P. and Anderson, K. A.: 1967, *Solar Phys.* **1**, 446.
Ramaty, R. and Lingenfelter, R. E.: 1967, *J. Geophys. Res.* **72**, 879.
Smerd, S. F., Wild, J. P., and Sheridan, K. V.: 1962, *Australian J. Phys.* **15**, 180.
Weiss, A. A.: 1963, *Australian J. Phys.* **16**, 526.
Weiss, A. A. and Stewart, R. T.: 1965, *Australian J. Phys.* **18**, 143.
Warwick, J. W.: 1965, *Solar System Radio Astronomy* (ed. by J. Aarons), Plenum Press, N.Y., p. 131.
Wild, J. P., Sheridan, K. V., and Neylan, A. A.: 1959, *Australian J. Phys.* **12**, 369.
Wild, J. P.: 1964, *Physics of Solar Flares* (ed. by W. N. Hess), NASA, p. 161.
Wild, J. P.: 1969, *Plasma Instabilities in Astrophysics* (ed. by D. Wentzel and D. Tidman), Gordon and Branch Publishers, p. 119.
Wild, J. P.: 1970, *Proc. Astron. Soc. Australia*, in press.
Wild, J. P.: 1968, *Proc. Astron. Soc. Australia* **1**, 137.
Zheleznyakov, V. V. and Zaitsev, V. V.: 1968, *Soviet Astron. – AJ* **12**, 14.

21. MODELS OF THE SOLAR TRANSITION REGION CHROMOSPHERE-CORONA

G. NOCI

Osservatorio Astrofisico di Arcetri, Firenze, Italy

1. Introduction

The models of the transition region chromosphere-corona developed in these last years, based essentially on the total intensities of resonance emission lines in the ultraviolet, are characterized by a very high temperature gradient. We will show that, in spite of a rather diffused opinion, a similar steep gradient is obtained if the radio data are the empirical basis of the model.

2. Radio Models

From the observed radio spectrum $T_b(f)$ the electron temperature T as function of the electron density N can be extracted. Indeed at the distance ϱ from the disc center, the equation holds:

$$T_b(f, \varrho) = \int_0^\infty T e^{-\sigma/f^2} \frac{d\sigma}{f^2} \tag{1}$$

where for sufficiently high frequencies ($f > 500$ MHz) and negligible magnetic fields (i.e. for refractive index equal to unity) the frequency-independent part of the optical depth is given by:

$$d\sigma = -\xi N^2 T^{-3/2} \sec\vartheta \, dh \tag{2}$$

where $h \sec\vartheta$ is a length along the line of sight ($\sin\vartheta = \varrho/R_\odot$) and ξ is a slowly varying factor. Equation (1) holds up to the point ϱ above the photospheric limb where the total optical depth of the emitting layer is still practically infinite. From Equation (1) $f^2 T_b(f)$ appears to be the Laplace transform of $T(\sigma)$ with respect to $x = f^{-2}$. By fitting the observed brightness temperatures with a function such that the inverse Laplace transform exists, it is possible to find $T(\sigma)$:

$$T(\sigma) = \frac{1}{2\pi i} \int_{c-i\infty}^{c+i\infty} \frac{T_b(x)}{x} e^{\sigma x} \, dx \tag{3}$$

the integration being on a straight line in the complex plane to the right of every singularity of $T_b(x)$.

Another equation is needed at this point to get separately $T(h)$ and $N(h)$, with the assumption of spherical symmetry. In principle such equations can still be given by radio observations using the center-limb profiles $T_b(f, \varrho)$. However this procedure is unreliable since the thickness of the transition region is by far smaller than the distance between two points resolvable with the best existing radiotelescopes. Only for an active region, using the greater resolution of an eclipse observation, such a procedure seems to have been successful (Christiansen *et al.*, 1960).

Quantities involving N and T, as functions of h, can be obtained from observations in other spectral regions, permitting the construction of a completely empirical model. Functions of this kind can also be provided by the theory: a semi-empirical model is obtained in this way.

3. Semi-Empirical Radio Models

In the latter case one can assume, for instance, that hydrostatic equilibrium holds. In a completely ionized hydrogenic atmosphere, this gives:

$$dp = -gmN\,dh, \tag{4}$$

where g is the gravity, m the proton mass and p the pressure:

$$p = 2KNT, \tag{5}$$

K being the Boltzmann constant.

Equation (4) together with Equations (2) and (3) allows the determination of both $N(h)$ and $T(h)$. This method was followed by Koeckelenbergh (1951), who however used an expression for $T_b(f, 0)$ open to criticism, being in it T_b boundless for $f \to 0$ (Chiuderi *et al.*, 1970). The author of the present paper together with two colleagues used the same method, without being aware of Koeckelenberg's work, for an active region (Chiuderi *et al.*, 1970). They fitted the observations (Chiuderi Drago, 1970) by the formula (Figure 1):

$$T_b(x) = T_0 + \frac{T_1 x}{x + a} \quad (x = f^{-2}) \tag{6}$$

where T_0 and $T_1 + T_0 = T_c$ are the limiting photospheric and coronal electron temperatures, respectively ($T_0 = 6.44 \times 10^3$ K, $T_c = 3.09 \times 10^6$ K). Equations (3) and (6) give:

$$T(\sigma) = T_0 + T_1 e^{-a\sigma}. \tag{7}$$

σ can be eliminated from Equations (2) and (7), giving:

$$\frac{T^{3/2}}{T - T_0}\frac{dT}{dh} = a \sec\vartheta \xi N^2. \tag{8}$$

Equations (8), (4) and (5) yield

$$\frac{dT}{dh} = \frac{2gm}{K}\frac{T - T_0}{T^{7/2}}\{R(T_c) - R(T)\}, \tag{9}$$

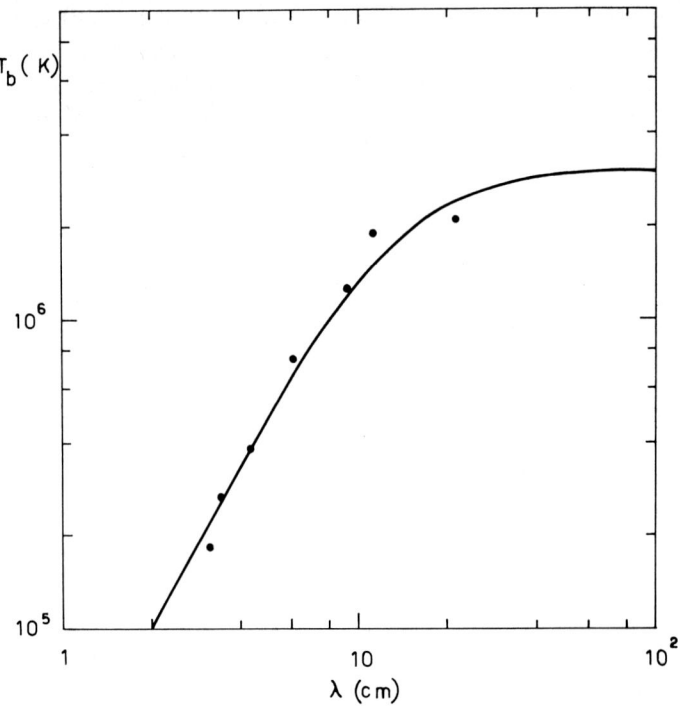

Fig. 1. Spectrum of a solar active region. Points: observations; line: $T_b = T_0 + T_1/(1 + a f^2)$, see text. (From Chiuderi-Drago, 1970.)

where

$$R(T) = T_0^{5/2} \left\{ \frac{1}{5}\left(\frac{T}{T_0}\right)^{5/2} + \frac{1}{3}\left(\frac{T}{T_0}\right)^{3/2} + \left(\frac{T}{T_0}\right)^{1/2} - \frac{1}{2} \ln\left[\frac{\left(\frac{T}{T_0}\right)^{1/2} + 1}{\left(\frac{T}{T_0}\right)^{1/2} - 1}\right] \right\}. \tag{10}$$

A very interesting property is contained in Equation (9). If $T^{5/2} \ll T_c^{5/2}$, say if $T \leq T_c/2.5 \ (=1.2 \times 10^6 \text{ K})$, it is: $R(T_c) \gg R(T)$; assuming also $T \gg T_0$, say $T \geq 7 \times 10^4$ K, Equation (9) gives:

$$T^{5/2} \frac{dT}{dh} = \frac{2gm}{5K} T_c^{5/2}. \tag{11}$$

Since the conductive flux, in absence of magnetic fields, is given by $F_c = 5.9 \times 10^{-7} T^{5/2} \, dT/dh$ erg cm^{-2} sec^{-1}, this equation means that in the transition region $(7 \times 10^4 \text{ K} \leq T \leq 1.2 \times 10^6 \text{ K})$ the conductive flux is constant; a result which, being not contained at all in the hypotheses, must arise from the empirical Equation (6): the radio observations suggest the constancy of the conductive flux in the transition region. Its value is fixed in a very simple way by the coronal temperature T_c: for the studied active region the result is:

$$F_c = 1.35 \times 10^6 \text{ erg cm}^{-2} \text{ sec}^{-1},$$

a value 2.3 times higher than the corresponding value for the quiet Sun deduced from the UV emission lines by Athay (1966) and Dupree and Goldberg (1967), and about one half of the one found by Noyes *et al.* (1970) in the active regions. This high value of F_c ensures by itself the existence of a very strong temperature gradient. Anyhow this can be seen explicitly by the temperature profile obtained by the integration of Equation (9), shown in Figure 2, curve (a). As boundary condition we employed the temperature of the BCA photospheric model (Gingerich and de Jager, 1968) at $h = 2210$ km ($T = 9.5 \times 10^3$ K).

The existence of a weak magnetic field ($\lesssim 10$ gauss) does not change essentially the above results, neither for the steepness of the temperature gradient, nor for the constancy of the conductive flux, which, channeled along the lines of force, results slightly reduced in value (Chiuderi *et al.*, 1970).

We must recall that the described model refers to an active region; however there are good reasons to believe that the results for the quiet Sun are not too different.

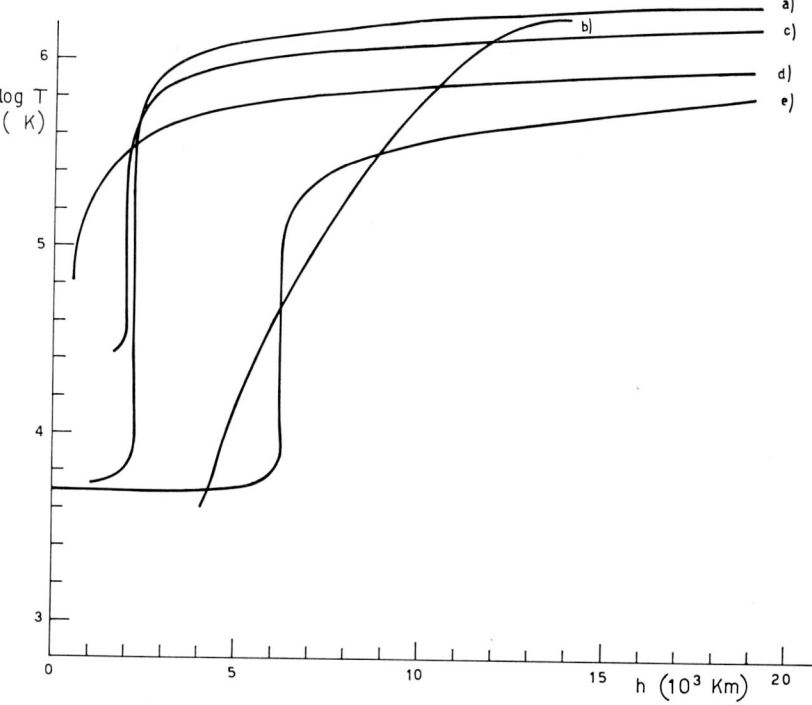

Fig. 2. Semi-empirical temperature profiles in the transition regions of the Sun: (a) Chiuderi *et al.*, 1970 (active region); (b) Oster, model II, 1956; (c) Dupree and Goldberg, 1967; (d) Koeckelenbergh, 1951; (e) Wolley and Allen, 1950. Athay's $T(h)$ (1966), which is given up to $h \approx 4000$ km, is indistinguishable from Dupree and Goldberg's $T(h)$, in the scale of the figure, above $T = 5 \times 10^4$ K.

In fact a remarkable result of Equation (11) is that in the region $10\,T_0 \leqslant T \leqslant T_c/2.5$ the temperature gradient does not depend on the parameter a of the brightness spectrum, but only on T_c/T. If therefore the spectrum of the quiet Sun can also be approximated by an expression like expression (6), it can be safely assumed that the temperature gradient in the quiet Sun differs from that in the considered active region solely because of the different coronal temperature T_c. Now, looking at Equation (11) we see that the value of the temperature gradient does not change varying T_c: the dT/dh curve simply shifts and contracts on the temperature scale, without variations in the ordinate dT/dh, when T_c changes; for instance the value of dT/dh corresponding to $T = 1.0 \times \times 10^6$ K if $T_c = 3 \times 10^6$ K is found at $T = 6.7 \times 10^5$ if $T_c = 2 \times 10^6$ K.

Concluding we can say that the radio data do not seem to indicate, by themselves, a mild temperature gradient in the transition region, but rather a steep one, like the ultraviolet resonance lines. It corresponds to a constant conductive flux which has the value

$$F_c = 7.84 \times 10^{-11} T_c^{5/2} \quad \text{(c.g.s.)} \tag{12}$$

according to Equation (11), in the region $10\,T_0 \leqslant T \leqslant T_c/2.5$.

To clarify why most of the models previous to the ultraviolet observations had slow temperature increases in the transition region, it will be useful to consider in some detail the most important ones. Therefore a review, limited to spherically symmetric models, is presented below.

4. Other Semi-Empirical Models

In the models of this type the observations used are not enough to determine $N(h)$ and $T(h)$ separately. These models, generally – but not, for instance, Oster's model – do not include direct observations of heights: the heights enter the model via the equation supplied by the theory.

Let us consider the most important:

(1) Wolley and Allen (1950) use two theoretical equations: they show that the electron pressure P_e is constant ($= P_0$) in the transition region, which, in a totally ionized layer of small thickness, agrees with the hydrostatic equilibrium condition; furthermore they assume the conductive flux equal to $c(h - h_0)^n$, where h_0 and are constants to be fixed by two boundary conditions together with P_0, and n an unknown exponent. The models which can be calculated by these two equations include n as a parameter, which is chosen by comparing the calculated radio spectrum with the observed one.

(2) Oster (1956) combines the hydrostatic equilibrium equation with values of the product NT interpolated between the chromospheric value of Böhm-Vitense (1955), corresponding to $h = 4000$ km, and the coronal one of Van de Hulst (1953), corresponding to $h = 20000$ km. The Böhm-Vitense model of the chromosphere is based on the optical eclipse observations of Cillié and Menzel (1935), who have very few observations corresponding to elevations as high as 4000 km. Furthermore the agree-

ment between the radiospectrum calculated with the model and the observed radiospectrum is rather poor, at least at the frequencies interesting the transition region.

(3) Athay (1966) expresses the total energy flux in a resonance ultraviolet line as $F = KA \langle dT/dh \rangle^{-1}_{average} \cdot \int_{AT} f_1(T) N^2 dT$, where K is a known constant depending on the line, A the abundance of the element, $f_1(T)$ a function given by ionization theory and collisional excitation formula, and ΔT refers to the layer where the line is emitted. The temperature of it is also given by the ionization theory. The above empirical relation, if the theoretical condition $P_e =$ constant is added, can be transformed into $\langle dT/dh \rangle^{-1} = F f_2(T)/A$, where $f_2(T)$ is known; at different temperatures, different ions of the same element exist, permitting the construction of the function $F f_2(T)$: there are as many functions as observed elements, which differ by the constants $1/A$ only, since the first member in the above equation is element independent. If one of the abundance is then assumed to be known, all the others and the curve $\langle dT/dh \rangle^{-1}$ as function of T can be obtained. Using for the silicon abundance the photospheric value, Athay finds in this way $T(h)$.

(4) Dupree and Goldberg (1967) proceed in a way similar to Athay, but include dielectronic recombination in the ionization theory. Furthermore they check the silicon abundance by comparing the radio spectrum inferred from the model with the observed one.

Temperature profiles of semi-empirical models are given in Figure 2.

5. Empirical Models

The empirical models considered here do not make theoretical assumptions, apart from the obvious ones and from the spherical symmetry hypothesis. While among the semi-empirical models described above Oster's model only includes direct determinations of heights, all the empirical models need them.

We give here a brief description of the most important ones:

(1) Piddington (1954) makes use of the equation $N^2 T^{-3/2} = \alpha e^{-\beta h}$ where α and β are known constants, extracted from an analysis of Wildt (1947) of the flash spectra of four eclipses, and of the equation $T_b = a + bf^{-1} + cf^{-2}$ for the radiospectrum, which is the same as that used by Koeckelenbergh (1951), criticized above. The first formula is used by Piddington up to $h = 16000$ km ($T = 8 \times 10^5$ K) although Wildt's analysis (based on the intensity gradient of the Balmer and Paschen series), which includes a single datum at 12000 km, gives a rather arbitrary interpolation formula above $h = 3000$ km.

(2) Ivanov-Kholodnyi and Nikol'skii (1961, 1963) first introduced the method based on the total intensity of solar ultraviolet lines, method later followed by other authors with some improvements, like the calculation of the abundances (Pottasch, 1963) and the explicit recognition of the information on the temperature gradient contained in the total intensities of the lines (Koyama, 1963; Athay, 1966). From an equation similar to the one written above when describing Athay's model, namely $F =$

$= kA\langle f_3(T)\rangle_{\text{average}} \int_{\Delta h} N^2 T^{-3/2} \, dh$, Ivanov-Kholodnyi and Nikol'skii, using Allen's (1955) abundances, get $\Delta \varphi = \int_{h(T_1)}^{h(T_1 + \Delta T)} T^{-3/2} N^2 \, dh$ and $\varphi = \int_{h(T)}^{\infty} T^{-3/2} N^2 \, dh$ as functions of T. (In these integrals T_1 and $T_1 + \Delta T$ are the temperatures characterizing the zone in which the ion emitting the considered line can exist.) Then, from the 1952 eclipse observations of the Balmer lines and continuum (Athay et al., 1955), which they interpolate from $h = 6000$ km to the coronal point at $h = 20000$ km, they get $\psi(h) = \int_h^{\infty} N^2 T^{-3/2} \, dh$. The equation $\varphi(T) = \psi(h)$ gives at this point $T(h)$; $N(h)$ follows simply. However the observations of the Balmer continuum, which yield $N^2 T^{-3/2}$, go so far as $h = 2400$ km, while Balmer line data, which extend up to $h = 6400$ km, give essentially only upper limits on the (negative) gradient of $N^2 T^{-3/2}$.

The authors include radio data in the model to separate the contribution of the active regions from that of the quiet sun.

(3) Pottasch (1964) uses the total intensities of the ultraviolet resonance lines similarly to the above authors but gets not only $\int_{\Delta h} N^2 \, dh$ as function of T but also the abundances relative to silicon, as later Athay (model 3 of Section 4). As function of height he takes the electron densities $N(h)$ deduced from the 1952 eclipse observations by Athay et al. (1955), and by Shklovskii and Kononovich (1958) up to $h \simeq 6000$ km. The $N(h)$ curve is then extrapolated as far as the coronal density corresponding to $h \simeq 30000$ km.

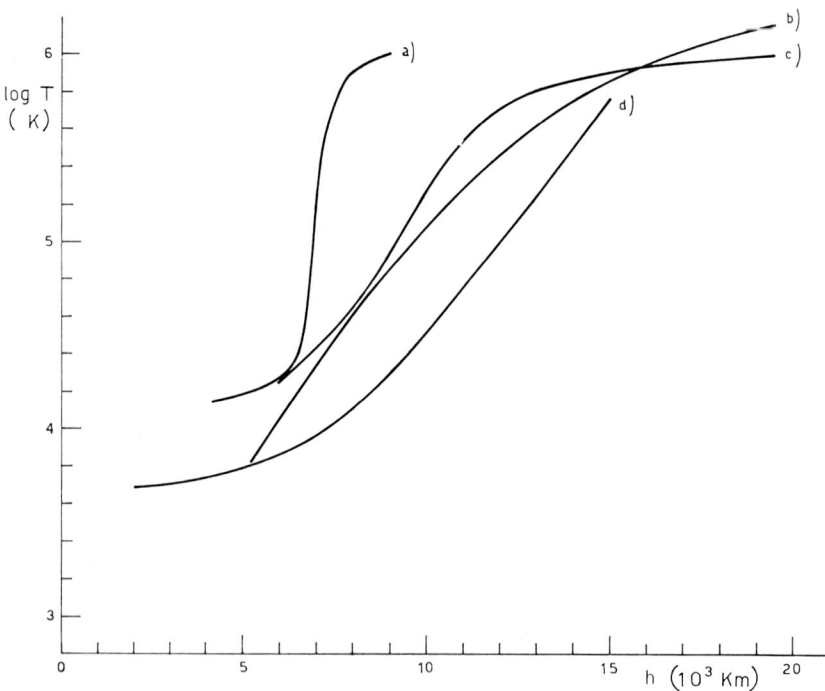

Fig. 3. Empirical temperature profiles in the transition region of the Sun: (a) Pottasch, 1963; (b) Ivanov-Kholodnyi and Nikol'skii, 1963; (c) Koyama, solar minimum, 1963; (d) Piddington, 1954.

Radio data are used to determine the silicon abundance, in a way similar to Dupree and Goldberg.

(4) Koyama (1963) also uses the total intensities in the extreme ultraviolet lines to get $d(\log T)/dh = N^2 T^{-3/2}/P(T)$ in a way essentially similar to that later followed by Athay. Then he takes $N^2 T^{-3/2}$ as function of h from Athay *et al.* (1955) and extrapolates this function to get the coronal point at $h = 20000$ km.

Radio data are used for the same purpose as in Ivanov-Kholodnii and Nikol'skii's model.

Temperature profiles of empirical models are given in Figure 3.

6. Conclusions

The analysis carried out in Sections 4 and 5 shows that all the models including direct measurements of heights had to resort to interpolation in the heights interval ~6000–20000 km, corresponding, for those models, to the region of temperature increase. This means that the shape of the temperature profile in the transition region was essentially arbitrary in those models: with the partial exception of the one due to Pottasch, they all show mild temperature gradients (Figure 2b and Figure 3).

Furthermore, as said above, the eclipse data more safely interpretable in terms of a simple function of N and T are the observations of the continuum at the Balmer limit (emitted flux proportional to $N^2 T^{-3/2}$), which has been detected up to $h \approx 2500$ km only. On the other hand the interpretation of data above $h \approx 1500$ km are not straightforward because of the break of spherical symmetry at these heights. For instance, low emission gradients observed in some lines during eclipses are attributed to cold parts of spicules by Zirin and Dietz (1963). Moreover the analysis of the line data requires the knowledge of the deviations from LTE populations, and the relative theory was not satisfactory until rather recently. As a matter of fact the improvement of the theory and the ultraviolet data led to a substantial change of the previous picture of the high chromosphere, decreasing considerably in it the electron density and fixing the upper point, where the temperature begins to rise sharply towards coronal values, at about $h = 2000$ km (see for instance Athay, 1965).

The consequence is that also the condition at the lower boundary was incorrect in the old models; a result that affects models which, like the one of Wolley and Allen, do not rely upon arbitrary interpolations in the transition region.

Therefore, to build correctly a model of the transition region one must reject 'observed' height distributions, which are not, in fact, supported by the observations, and use values at the chromospheric boundary that the most reliable data imply. This is what is done in the recent models, as that described in Section 3 and those at the numbers 3 and 4 of Section 4.

After this discussion and inspection of Figure 2 one would like to conclude that a very narrow transition region of a few hundred kilometers, occurring below

$h \approx 3000$ km, is strongly supported both by radio and ultraviolet data and does not conflict with visible eclipse observations.

Unfortunately one can't leave the argument without saying that this optimistic picture partly contrasts with very recent results of the Culham group (Burton *et al.*, 1970). Their essentially empirical model combines the total intensities of the ultraviolet lines with disc-limb ratios to find $T(h)$. Now, while the optically thin lines do give a very steep temperature gradient, in the analysis of the Culham group, the optically thick lines indicate that cool material ($\sim 10^4$ K) extends up to $\sim 10^4$ km above the transition region. This result is interpreted as an indication of the absence of spherical symmetry in the transition region.

References

Allen, C. W.: 1955, *Astrophysical Quantities*, Univ. of London Press.
Athay, R. G.: 1965, *Astrophys. J.* **142**, 755.
Athay, R. G.: 1966, *Astrophys. J.* **145**, 784.
Athay, R. G., Menzel, D. H., Pecker, J.-C., and Thomas, R. N.: 1955, *Astrophys. J. Suppl.* **1**, 505.
Böhm-Vitense, E.: 1955, *Z. Astrophys.* **36**, 145.
Burton, W. M., Jordan, C., Ridgely, A., and Wilson, R.: 1970, *Phil. Trans. Roy. Soc. London*, in press.
Chiuderi, C., Chiuderi-Drago, F., and Noci, G.: 1970, *Solar Phys.*, **17**, 369.
Chiuderi-Drago, F.: 1970, *Solar Phys.* **13**, 357.
Christiansen, W. N., Mathewson, D. S., Pawsey, J. L., Smerd, S. F., Boischot, A., Denisse, J. F., Simon, P., Kakinuma, T., Dodson-Prince, H., and Firor, J.: 1960, *Ann. Astrophys.* **23**, 75.
Cillié, G. G., and Menzel, D. H.: 1935, *Harvard Circ.* 410.
Dupree, A. K., and Goldberg, L.: 1967, *Solar Phys.* **1**, 229.
Gingerich, O. and de Jager, C.: 1968, *Solar Phys.* **3**, 5.
Ivanov-Kholodnyi, G. S., and Nikol'skii, G. M.: 1961, *Soviet Astron. – AJ* **5**, 31.
Ivanov-Kholodnyi, G. S., and Nikol'skii, G. M.: 1963, *Soviet Astron. – AJ* **6**, 609.
Koeckelenbergh, A.: 1951, *Bull. Acad. Roy. Belgique* **37**, 252.
Koyama, S.: 1963, *Publ. Astron. Soc. Japan* **15**, 15.
Noyes, R. W., Withbroe, G. L., and Kirshner, R. P.: 1970, *Solar Phys.* **11**, 388.
Oster, L.: 1956, *Z. Astrophys.* **40**, 28.
Piddington, J. H.: 1954, *Astrophys. J.* **119**, 531.
Pottasch, S. R.: 1963; *Astrophys, J.* **137**, 945.
Pottasch, S. R.: 1964, *Space Sci. Rev.* **3**, 816.
Shklovskii, I. S. and Kononovich, E. V.: 1958, *Soviet Astron. – AJ* **2**, 32.
Van de Hulst, H. C.: 1953, *The Sun* (ed. by G. Kuiper), Univ. of Chicago Press, p. 207.
Wildt, R.: 1947; *Astrophys. J.* **105**, 36.
Wolley, R. v. d. R. and Allen, G. W.: 1950, *Monthly Notices Roy. Astron. Soc.* **110**, 358.
Zirin, H. and Dietz, R. D.: 1963, *Astrophys. J.* **138**, 664.

22. STUDIES OF THE OUTER CORONA THROUGH SPACE RADIO ASTRONOMY

MICHAEL D. PAPAGIANNIS

Dept. of Astronomy, Boston University, Boston, Mass. 02215, U.S.A.

1. Introduction

Radio astronomy offers one of the most resourceful approaches to the study of the solar corona. Solar radio observations from the ground can be conducted only up to a frequency of about 20 MHz due to the effects of the terrestrial ionosphere, and very seldom they can be extended to frequencies below 10 MHz. The lowest penetrating frequency for solar radio waves is,

$$f_p = f_c \cos\theta = 9 \times 10^{-3} N_m^{1/2} \cos\theta \quad \text{MHz} \tag{1}$$

where N_m is the maximum electron density of the F2 layer of the ionosphere (usually of the order of $0.5-1 \times 10^6$ el/cm³) and θ is the zenith angle of the sun. Near the penetrating frequency, however, the signals suffer strong refraction effects and are badly distorted by ionospheric scintillations, which are produced by continuously changing irregularities in the electron density distribution of the ionosphere. As a result, useful ground observations start from at least twice the penetrating frequency, and solar radio astronomy is as a rule confined to frequencies higher than about 20 MHz.

An empirical expression, which is often used for the electron density profile of the quiet corona near the sun, is the Baumbach-Allen formula,

$$N = 10^8 \left(1.55\varrho^{-6} + 2.99\varrho^{-16}\right) \quad \text{el/cm}^3 \tag{2}$$

where $\varrho = r/R_\odot$ is the radial distance r from the center of the sun expressed in solar radii $R_\odot$. From the electron density N we can find the plasma frequency f_N through the expression,

$$f_N = AN^{1/2} \quad \text{MHz} \tag{3}$$

where $A = 9 \times 10^3$ if f_N is expressed in Hz and N in el/cm³. Thus from (2) and (3) we find that the plasma frequency of the quiet solar corona will be equal to about 20 MHz when $N \simeq 5 \times 10^6$ el/cm³, which occurs when $\varrho \simeq 1.8$, i.e. at $r \simeq 1.8 R_\odot$. The electron density profile in the coronal streamers, on the other hand, which appear over the active regions of the sun, is approximately 10 times higher than that of the quiet corona given by (2). In these streamers the plasma frequency will reach the value of 20 MHz at a radial distance of $r \simeq 2.6 R_\odot$.

Macris (ed.), Physics of the Solar Corona, 317–332. All Rights Reserved.
Copyright © 1971 by D. Reidel Publishing Company, Dordrecht-Holland.

Practically all radio studies of the outer corona are based on our observations of the type II slow drift radio bursts, and of the type III fast drift solar radio bursts. The type II bursts are produced by the shock front from a flare event as it advances with a velocity of about 10^8 cm/sec through the solar corona. The type III bursts on the other hand are caused by a burst of superthermal particles from the flare region streaming outwards in the corona with an average velocity of 0.1 to 0.6 the speed of light. The passage of these disturbances through the solar corona excites plasma oscillations which in turn produce radio waves at the local plasma frequency.

From the above it follows that even under very favorable conditions the study of the solar corona with ground based radio telescopes can be carried out only up to a distance of a few solar radii from the surface of the sun. To extend these studies into the outer corona we must conduct our radio observations beyond the barrier of the terrestrial ionosphere.

2. Space Radio Astronomy

The advent of the space age made it possible to send rockets and satellites, equipped with antennas and radiometers, above the maximum of the ionosphere. This has allowed us to extend the domain of radio astronomy to frequencies well below 1 MHz. Figure 1 shows a typical daytime profile of the ionosphere. In order to conduct observations at a given frequency f, the plasma frequency f_N of the surrounding ionosphere must be smaller than f so that the signals can reach the antenna. However, between

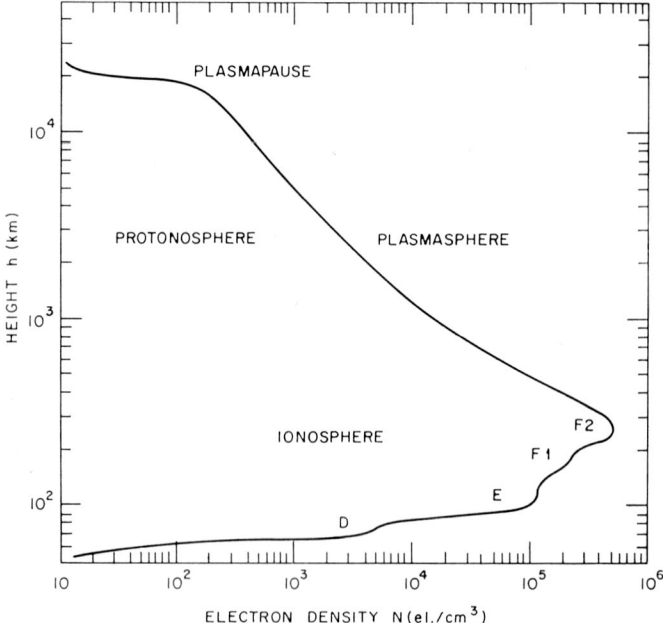

Fig. 1. A typical electron density profile of the ionosphere and the plasmasphere showing the daytime maximum at the F2-layer and the rapid decrease of the electron density at the plasmapause.

the local plasma frequency f_N and the upper hybrid frequency f_T,

$$f_T = (f_N^2 + f_H^2)^{1/2} \tag{4}$$

where f_H is the gyrofrequency of the earth's magnetic field, observations become practically impossible due to a resonance in the impedance of the antenna. Actually, due to focussing effects (Papagiannis and Huguenin, 1964) and problems with the antenna impedance (Stone *et al.*, 1966) useful observations can be conducted only when $f_N < f/2$ and preferably when $f_N < f/5$. For this reason in order to extend our observations below 0.5 MHz, we must fly our radiometers beyond the plasmapause (Figure 1) where the electron density drops rapidly from values above 100 el/cm^3 to values near or below 10 el/cm^3. At the equator the plasmapause occurs at a distance of about 5 earth radii from the center of the earth, and in a meridian plane it follows approximately the geometry of the L-shells of the earth's magnetic field. As a result, satellites on a high inclination orbit can reach low ambient electron densities at a smaller radial distance from the earth.

The Baumbach-Allen model gives a good fit to the solar corona only up to a distance of about 3 $R_\odot$. Beyond this distance we are essentially dealing with the solar wind which flows at supersonic velocities into the interplanetary space and therefore the electron density of the outer corona must vary approximately like r^{-2} in order to have the same number of solar wind particles crossing successive shells at different radial distances r. Blackwell and Petford (1966) on the basis of optical observations during a solar eclipse have derived the expression,

$$N = 1.46 \times 10^6 \varrho^{-2.3} \tag{5}$$

for the outer corona. According to this model the plasma frequency of the solar corona will reach a value of 1 MHz at a distance of about 8 solar radii, and in a coronal streamer, which has a plasma density approximately 10–20 times higher, the same plasma frequency will be reached at a distance of about 25 $R_\odot$. Thus radio-astronomy, which can extend solar radio observations to frequencies considerably below 1 MHz, offers a practically unique method for the study of the outer corona up to distances of 50 or more solar radii.

The angular resolution θ of an antenna is given essentially by the expression,

$$\theta = \frac{\lambda}{L} \tag{6}$$

where λ is the observing wavelength and L the length of the antenna. When L is shorter than about one wavelength, the antenna becomes almost omnidirectional and has very little power in resolving radiation coming from different regions of the sky. Most of the antennas flown on satellites for space radio astronomical observations are simple dipole antennas a few tens of meters from tip to tip. In the hectometer wavelength range ($100 < \lambda < 1000$ m) i.e. at frequencies between 3 MHz and 300 kHz, these antennas provide very little directivity and for this reason practically all observations of the galactic radio background in the hectometer range during the sixties (e.g., Huguenin and Papagiannis, 1965) were essentially measurements of the integrated (average) radio background of the entire sky.

In the range between 0.5 and 5 MHz the brightness temperature of the galactic background is of the order of 10^7 K and therefore the quiet corona with a temperature of the order of 10^6 can not be observed except with radio telescopes of very high angular resolution. Even non-directional antennas, however, in the same frequency range can detect a good number of type-III solar radio bursts, which with a flux density of about 10^{-17} Watts m^{-2} Hz^{-1} stand easily above the galactic background which produces a total flux density ($\sim 8\pi kT/\lambda^2$) of about 10^{-19} W m^{-2} Hz^{-1}.

Explorer 38, the so called Radio Astronomy Explorer I (RAE-1), which was launched on July 4, 1968, was the first major step to achieve a higher directivity for space radio observations. RAE-1 carried two oppositely directed V-type antennas, each leg of the V being 230 meters long. The two V-antennas which formed an X with a 60° acute angle, provided a gravity gradient stabilization for the satellite, which as a result had always one V-antenna pointing toward the earth and the other toward the local zenith of the celestial sphere. RAE-1 was equipped also with a 37 m dipole antenna which bisected the obtuse angle of the X and therefore was always in a horizontal position. Data on the galactic background with the short dibole antenna have already been reported (Alexander, Brown, Clark, Stone, and Weber, 1969) but the processing of the data from the V antennas has been slow due to heavy demands

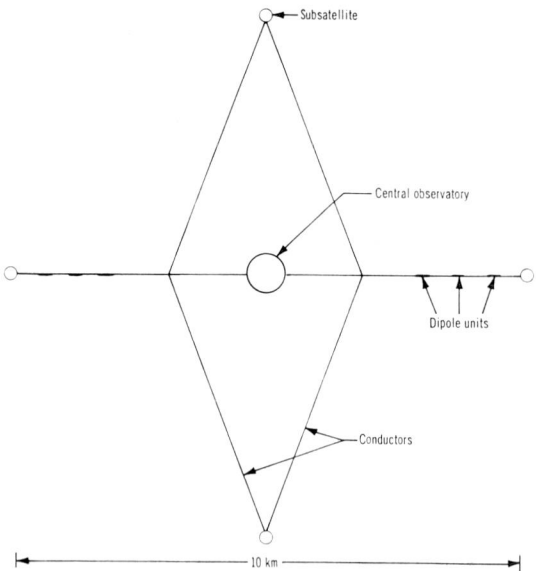

Fig. 2. A diagrammatic representation of the 10-kilometer rhombic antenna which has been studied under a NASA grant by the University of Michigan. When orbited into space this antenna will be able to provide a considerable degree of directivity for space radio astronomical observations in the hectometer range (Doyle, 1969).

on computer time. RAE-I had a 5,860 km nearly circular retrograde orbit at an inclination of 59° to the equator. At the high latitude points of the orbit the satellite went beyond the plasmapause and the local plasma frequency is considerably below the 0.2 MHz lower limit of the sweep frequency radiometer of RAE-1.

The second satellite in this series, RAE-II, is scheduled for launching at the end of 1972 and will probably be placed in a circumlunar orbit to free it from any problems with the ambient plasma density. An even more ambitious project, planned for the late seventies or early eighties, is the huge rhombic antenna, which has been studied by the University of Michigan under a grant from NASA. The diameter of this antenna, which is shown in Figure 2, will be 10 km and at 1 MHz it will be able to focus its beam on an area of approximately 80 square degrees, which is less than 0.002 of the entire celestial sphere, and is approximately the size of the radio corona of the sun at this frequency (Doyle, 1969).

3. Solar Radio Bursts in the Hectometer Range

Most authors agree (Takakura, 1967) that the instantaneous frequency of the type II and the type III solar radio bursts corresponds to the local plasma frequency at the point of the corona which is disturbed by the passage of the exciter. If this excitation was taking place in a homogeneous and spherically symmetric corona, the radio burst would be able to propagate only inside a very narrow cone in the forward direction. The type II and type III bursts, however, are most likely produced in the coronal streamers, which extend into the outer corona above the active regions, because these seem to be the paths which the exciters follow after they are ejected from the flare region. Since the plasma density of the streamers is approximately 10–20 times higher than the surrounding quiet corona, the frequency of the radio bursts is 3–4 times higher than the plasma frequency of the solar corona at the same distance, and thus the radiobursts can propagate in practically any direction of the solar corona which surrounds the coronal streamer.

Though this is a relatively well established theory, several authors have discussed the possibility that the bursts are emitted at a frequency which is higher than the local plasma frequency. Hartz (1969) has suggested as a good candidate the upper hybrid frequency f_T (4), or the sum of $f_T + f_N$. Kucker and Sudan (1969) have suggested the possibility that the emission occurs in one of the low harmonics of the cyclotron frequency that are higher than the local plasma frequency. Finally Slysh (1967b) has suggested that in the case of type III bursts, which are produced by particles with velocities V of the order of 0.2–0.5 c, the emission frequency f, is related to the local plasma frequency f_N, through the expression,

$$\frac{f}{f_N} = \frac{V}{4\pi} \left(\frac{m}{kT}\right)^{1/2} \qquad (7)$$

and therefore for $V = c/3$ the estimated electron density N' will be related to the actual electron density N by the expression,

$$\frac{N'}{N} = \left(\frac{f}{f_N}\right)^2 \simeq \frac{4 \times 10^6}{T} \qquad (8)$$

Actually in his paper Slysh (1967b) has an error of a factor of 10 in his equation, and for $V=c/2$ he obtains the relation $N'/N=0.9 \times 10^6/T$ instead of $9 \times 10^6/T$. With this numerical error, in order to accommodate his observations which show that at 200 kHz ($N' \simeq 500$ el/cm^3) the radio burst occurred near the earth's orbit, he had to postulate a very low temperature of almost 10^4 K at 1 AU. With the error corrected, one can obtain good agreement with the observed values of about 10 el/cm^3 at 1 AU with a much more realistic electron temperature of about 10^5 K, which makes the entire process quite plausible.

All three of the above mentioned processes, which suggest that the emission frequency is higher than the local plasma frequency, are very interesting but they definitely require much more theoretical and experimental work to substantiate them. Until such a time, it is safer to proceed with the presently accepted theory that the frequency of the bursts corresponds to the local plasma frequency of the solar corona.

As the disturbance, which is exciting the radio burst, advances further away from the sun the local plasma frequency decreases and the peak intensity of the burst drifts to lower frequencies. Type II and type III solar radio bursts start usually near the top of the meter range, i.e. near 300 MHz, and continue at least up to the limit of ground based radio observations. Space-borne radiometers have detected repeatedly type III solar radio bursts (Hartz, 1964; Hartz, 1969; Haddock and Graedel, 1970; Slysh, 1967a; Slysh, 1967b; Alexander, Malitson and Stone, 1967; Fainberg and Stone, 1970) in the hectometer range but no type II bursts have been observed up to now in this range. Neither have type IV bursts; and type V bursts, which represent the broad band continuous emission that often follows the type III bursts, have been only tentatively identified in the hectometer range.

By following the type III solar radio bursts into the hectometer range we can obtain information on the density and the temperature of the solar corona up to distances of 50 or more solar radii, which are almost impossible to obtain with any other method. Let the electron density profile of the solar corona be given by a general power low as in the Expressions of (2) and (5)

$$N(r) = N_0 \left(\frac{r}{R_0}\right)^{-a} \qquad (9)$$

and let V be the radial velocity of the advancing disturbance. Let us also consider the general case when the velocity vector V makes an angle θ with the vector pointing toward the earth. The frequencies produced at the points O and A of Figure 3, will arrive at the earth with a time delay dt. This delay will occur primarily because the exciter of the burst must travel the distance OA with a velocity V. From this time, however, we must subtract the time it takes for the radio burst produced at O to travel the distance $OB = OA \cos\theta$. For simplicity we can take the velocity of propa-

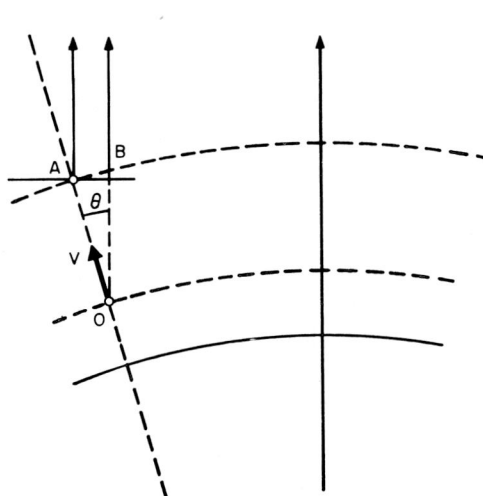

Fig. 3. A diagram showing the radial path followed by the exciter of a type III burst as it propagates with a velocity V through the solar corona, and the path followed by the radio burst produced at the point O which propagates toward the earth essentially with the velocity of light.

gation to be the speed of light. For higher accuracy one must include also the effects of the index of refraction, but the frequency of the burst produced in the coronal streamer is several times the plasma frequency of the corona and therefore the effect of the coronal plasma on the velocity of the radio waves will be negligible. Setting now $OA = dr$, we find that the time delay dt is,

$$dt = \frac{dr}{V} - \frac{dr \cos\theta}{c} = (1 - \beta \cos\theta)\frac{dr}{V} \qquad (10)$$

where $\beta = V/c$. The second term of this equation can be neglected for the type II bursts where $\beta \ll 1$, but for the type III bursts, where β is of the order 0.2–0.5 c, we must retain the entire expression of (10). The angle θ can usually be determined from the heliographic coordinates of the flare region which is associated with the radio burst. If the local plasma frequency at O and A differ by df_N, using (3), (9) and (10) we can compute the rate of the frequency drift at a given frequency f,

$$\begin{aligned}\left.\frac{df}{dt}\right|_f &= \frac{V}{(1-\beta\cos\theta)}\frac{df_N}{dr} = \frac{V}{(1-\beta\cos\theta)}\frac{dAN^{1/2}}{dr} = \frac{VAN^{-1/2}}{2(1-\beta\cos\theta)}\frac{dN}{dr} \\ &= \frac{VAN^{-1/2}}{2(1-\beta\cos\theta)}\left(-a\frac{N}{r}\right) = \frac{-aV}{2(1-\beta\cos\theta)}\frac{AN^{1/2}}{r} \\ &= -\frac{aVf_N}{2(1-\beta\cos\theta)r} = -\left[\frac{aV}{2(1-\beta\cos\theta)R_0(A^2N_0)^{1/a}}\right]f^{(2+a)/a}\end{aligned}$$

(11)

Thus by studying the drift rates of the type III bursts into the hectometer range of the radio spectrum with space-borne radio receivers, we can obtain extremely valuable information about the outer corona and the coronal streamers which, as we mentioned already, appear to be the paths the exciters follow as they advance through the outer corona. If we know the temperature, density and magnetic field of a streamer, we can compute the pressure inside the streamer and by equating it with the pressure of the surrounding corona we can find also the pressure of the quiet corona.

Equation (10), however, contains two unknowns, namely the velocity V of the disturbance at a given radial distance from the sun, and the electron density profile of the coronal streamer. At the present stage neither of these two quantities is accurately known and there are still doubts even on whether the speed of the energetic particles that produce the type III bursts remains constant, or decreases with distance from the sun. Thus (10) can yield the electron density profile of the streamer only if we assume a velocity for the exciter, and conversely the velocity of the exciter for an assumed profile of the streamer.

More observations will allow us to narrow the uncertainty range in all these parameters. Also observation with higher angular resolution, such as with more directional antennas or an interferometric arrangement with two satellites, one of them at large distances from the earth, will allow us to determine the radial distance of the center of the disturbance at any given instant of time and thus deduce directly the electron density profile of the outer corona and the velocity of the exciters which produce the radio bursts. A first, rather crude attempt in this direction was made by Slysh (1967b) using the radio receivers of a satellite in a lunar orbit.

4. The Temperature of the Outer Corona

Studies of the solar radio bursts in the hectometer range can provide also information about the temperature profile of the outer corona. This can be obtained from the rate of the exponential decay of the radio burst at a given frequency. Assuming that the decay of the disturbances in each point of the corona is due to collisional damping, we can relate the time constant τ_0 of the decay (the time interval for an e-fold decrease in the intensity of the burst), to the collision frequency of the coronal plasma. In a fully ionized plasma, like the solar corona, the only collisions that matter are those between the electrons and the ions (protons). Collisions between charged particles are the result of the interaction of their Coulomb forces which become effective at distances much larger than the radii of the particles. To a first approximation an electron will be substantially deflected from its trajectory, i.e. will collide with a proton, if at the distance of closest approach the Coulomb potential energy of the electron-proton pair will reach the kinetic energy of the electron. Thus we have,

$$\tfrac{1}{2}mV^2 = \tfrac{3}{2}kT = \frac{e^2}{r} \tag{12}$$

and therefore to a first approximation the collisional cross-section for Coulomb collisions is,

$$\sigma = \pi r^2 = \pi \left(\frac{2e^2}{3kT}\right)^2 \qquad (13)$$

and the collision frequency v (average number of collisions per second) will be,

$$v = NV\sigma = N \left(\frac{3kT}{m}\right)^{1/2} \pi \left(\frac{2e^2}{3kT}\right)^2 = CNT^{-3/2} \qquad (14)$$

where C is a numerical coefficient of the order of 50 if N is in el/cm^3 and T in K. Actually, when the Debye shielding of the ions is taken also into consideration, C is not a simple constant but it includes a logarithmic term of the temperatures and the density. Ginzburg (1961), gives for C the expressions,

$$C = 5.5 \ln\left(220 \frac{T}{N^{1/3}}\right) \quad \text{for} \quad T < 10^5 \text{ K} \qquad (15)$$

and

$$C = 5.5 \ln\left(10^4 \frac{T^{2/3}}{N^{1/3}}\right) \quad \text{for} \quad T > 10^5 \text{ K} \qquad (16)$$

The collisional damping of the disturbance at a given point of the corona follows an exponential law,

$$D = D_0 e^{-vt} \qquad (17)$$

where v is the collision frequency of Equation (14). Consequently the intensity of the radio burst at a given frequency, which is produced at the plasma frequency of the solar corona at this point, will decay to a value $D = D_0/e$ in a time interval,

$$\tau_0 = \frac{1}{v} = \frac{T^{3/2}}{CN} \qquad (18)$$

Since the frequency f of the radio burst is equal to the local plasma frequency f_N, introducing (3) in (18) and using an average value of $C \simeq 50$ we obtain the relation,

$$T \simeq 7 \times 10^{-5} \tau_0^{2/3} f^{4/3} \qquad (19)$$

where f is in Hz, τ_0 in seconds, and T in K. Actually, one should use the more exact expressions for C given in (15) and (16) to obtain a more accurate value for T. All of the temperature computations up to now (Hartz, 1969; Haddock and Graedel, 1969; Slysh, 1967a; Alexander, Malitson and Stone, 1969) have used values in the range $6-6.5 \times 10^{-5}$ for the numerical constant of (19). It is readily seen, however, that at large distances from the sun, where $T \simeq 10^5$ K and $N \simeq 10^4$ el/cm^3, the logarithmic term makes $C \simeq 75$ which gives $T \simeq 10^{-4} \tau_0^{2/3} f^{4/3}$. Thus the expression adapted by the above authors underestimates the temperature at large distances from the sun by a factor of approximately 1.5.

It should be emphasized, however, that plain collisional damping will be a very ineffective process at large distances from the sun where due to its extremely low plasma density the solar corona becomes essentially a collisionless plasma with a mean free path of the same order as its distance from the sun. Extensive studies of the solar wind, however, have shown that to a good approximation the solar wind can still be treated as a fluid. This occurs because the magnetic field of the interplanetary space replaces the collisions as the cohesive force of the fluid. For this reason at large distances from the sun one must replace collisional damping with the much more complex equations of magnetohydrodynamic damping.

5. Discussion of Experimental Results

Several groups have measured type III solar radio bursts with space-borne radiometers and have analyzed the data to obtain the properties of the outer corona and the velocities of the exciters. Table I summarizes the data available up to now,

TABLE I

Group	Satellite	Frequencies available
Hartz (1964)	Alouette I	0.5–12 MHz (sweep)
Hartz (1969)	Alouette II	0.1–15 MHz (sweep)
Slysh (1967a)	Venus 2	0.03, 0.2, 1 MHz (discr.)
Slysh (1967b)	Luna 11 and 12	0.03, 0.2, 1 MHz (discr.)
Alexander et al. (1969)	ATS-II	0.45, 0.7, 1.1, 2.6, 2.2, 3.0, MHz (discr.)
Haddock and Graedel (1970)	OGO-3	2–4 MHz (sweep)
Fainberg and Stone (1970a, b)	RAE-I	0.2–5.4 (sweep), 0.5, 0.7, 1.0, 1.3, 1.6, 2.8 (discr.)
Dunckel and Helliwell (1971)	OGO-3	20–100 kHz (sweep)
Haddock and Alvarez (1971)	OGO-5	3.5, 1.8, 0.9, 0.35, 0.2, 0.1, 0,05 MHz (discr.)

A typical case of the fast drift solar radio bursts in the hectometer range, as seen by both the sweep frequency and the discrete frequency radiometers of RAE-I, is shown in Figure 4. With the exception of Slysh, who observed only few events, all the other observers have studied large numbers of cases which allowed the use of statistical methods of computation. The average drift rate deduced by Hartz (1969) is shown in Figure 5. This figure includes also values of the drift rate obtained at higher frequencies by ground observations. It should be pointed out that this is the drift rate of the commencement of the disturbance and that the drift rate of the maximum of the burst as deduced by Alexander, Malitson and Stone (1969) is somewhat slower.

To change the drift rate to the electron density profile we must know the velocity of the exciter, or conervsely if we assume an electron density profile we can deduce the velocity of the exciter. Hartz (1969) using the Blackwell and Petford (3) model of the quiet corona enhanced by a factor of 15 for the coronal streamers, has obtained an average speed for the exciter of 0.35 C. An average value of $V=0.37$ c has been obtained by Fainberg and Stone (1970b) from a statistical analysis of about 2500 bursts observed with RAE-I. This large number of cases allowed them also to observe clearly the change of the drift rate with heliographic longitude. The cause of this change is the second term of (10) which becomes maximum near the center

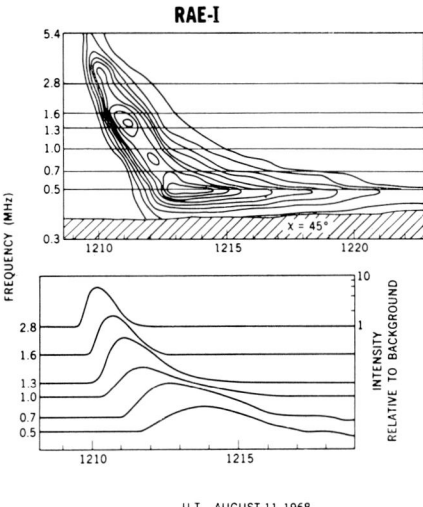

Fig. 4. A typical type III solar radio burst recorded in the hectometer range by both the sweep frequency (upper diagram) and the discrete frequency (lower diagram) receivers on board the RAE-1 satellite (Fainberg and Stone, 1970).

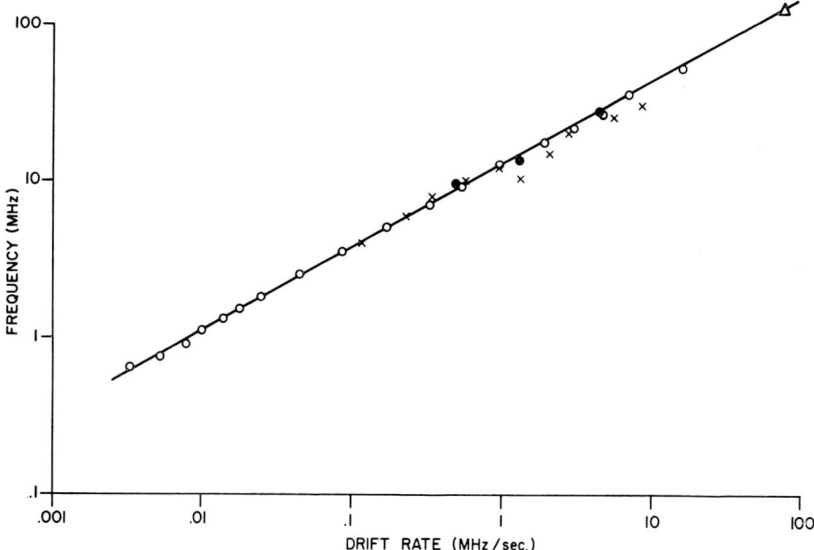

Fig. 5. The average drift rate in the MHz/sec of the commencement of the type III solar radio bursts over a wide range of frequencies as deduced from both satellite and ground based observations (Hartz, 1969).

of the solar disc ($\theta=0$) and tends to zero near the limb. The results for an active period in August of 1968 are shown in Figure 6. As it can be deduced also from these observations, the active region passed by the central meridian of the solar disc on August 20, 1968.

The spread in the drift rates seen in Figure 6 could be interpreted as a distribution of exciter velocities. It could, however, be and to a great extent is due to plasma density inhomogeneities in the streamer through which the individual exciters travel. The average velocity of 0.37 c corresponds to an electron kinetic energy of about 40 keV.

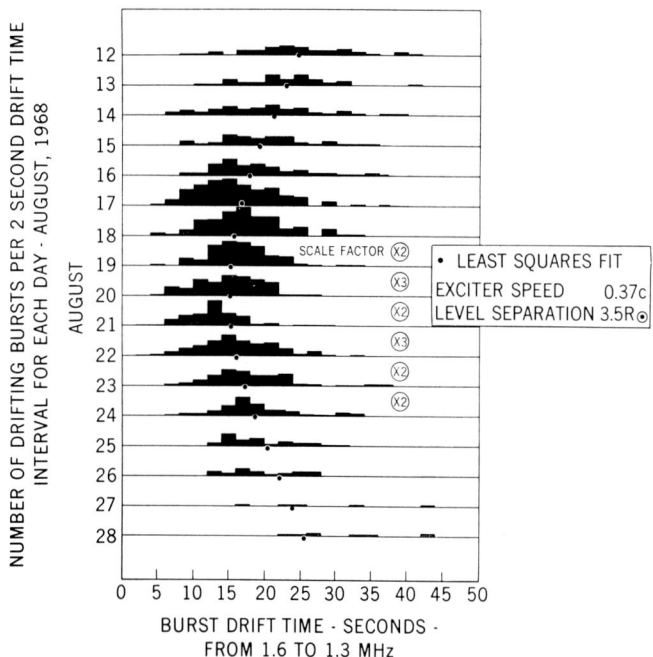

Fig. 6. The time interval in seconds which elapses as type III solar bursts drift from 1.6 to 1.3 MHz. The smallest delay occurs when the angle θ of Fig. 3 approaches zero, i.e., when the active region passes near the central meridian of the solar disc. A least squares fit to the data gives a value of 0.37 c for the speed of the exciter (Fainberg and Stone, 1970).

It is interesting that this is also the typical energy of streams of energetic electrons observed by spacecrafts beyond the magnetosphere (Lin and Anderson, 1967). It appears therefore that these packets of superthermal electrons, which excite the type III bursts, have an average energy of about 40 keV and propagate over distances of the order of 1 AU without significant deceleration.

Using this $V=0.37$ c and the observed drift rates, Fainberg and Stone (1970c) have obtained an average electron density profile for the streamers which is 15 times higher than the electron density profile of the outer corona deduced by Newkirk (1967). The stream densities in their profile (Figure 7) are by about a factor of 2 less than the

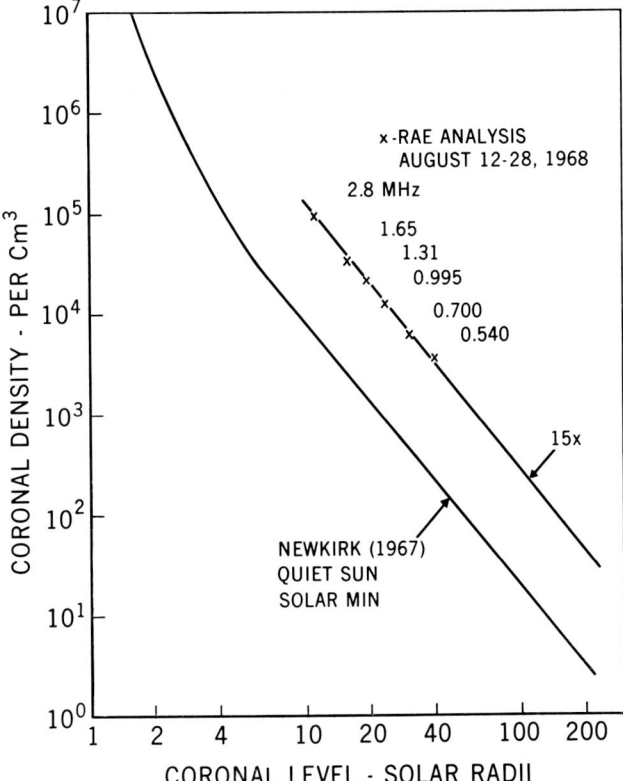

Fig. 7. The analysis of type III solar radio burst data obtained in the hectometer range with the RAE-1 satellite indicates that on the average the exciters of the bursts were propagating through coronal streamers with an electron density approximately 15 times higher than Newkirk's (1967) profile for the quiet corona (Fainberg and Stone, 1970).

ones derived by Hartz (1969), who obtains a density of about 8×10^3 el/cm^3 at 40 solar radii against 4×10^3 el/cm^3 of Fainberg and Stone.

An interesting result obtained also by Fainberg and Stone (1970) was that as the active region passes by the central meridian of the sun the minimum drift rate occurs half a day later at the lower frequencies. Since lower frequencies are produced at larger distances from the sun, this must be due to the backward curving of the streamers in their co-rotation with the sun. This is a well known effect (the garden-hose effect) of the interplanetary magnetic field which due to the radial velocity of the solar wind and the rotation of the sun assumes the form of an Archimedean spiral (Figure 8). The geometry of the Archimedean spiral is given by the equation,

$$\omega(r) = \frac{2\pi}{\tau V}(r - r_0) + \omega_0 \tag{20}$$

where τ is the period of rotation of the Sun, and V the velocity of the solar wind.

Thus by determining the lagging of the streamer at different frequencies, hence distances, we can use (20) to compute the velocity of the solar wind in this region.

Haddock and Alvarez (1971), have observed a large number of type III radio bursts with OGO-5, some of which could be detected even at 100 kHz and at least one at 50 kHz. From the times of the bursts at their 8 observing frequencies, and appropriate models of the interplanetary medium, they have deduced an increasing dominance of the second harmonic ($f = 2f_N$) of the type III bursts with decreasing frequency which approaches 100% at 100 kHz. Their measurements of drift rates indicate that the

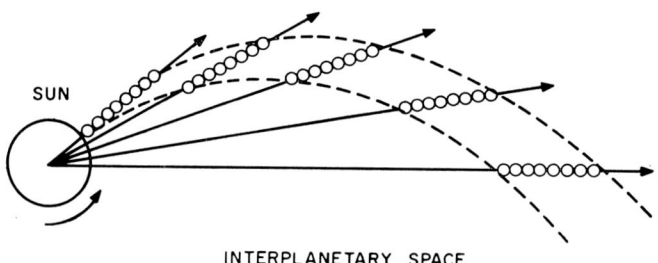

Fig. 8. The garden-hose effect produced by the radial outflow of the solar wind and the rotation of the sun about its axis. The solar magnetic field remains anchored on the sun but is also stretched out by the solar wind into the interplanetary space where it assumes the geometry of the Archimedean spiral.

straight line of Figure 5 continues at least up to 100 kHz, with df/dt remaining proportional to f^n, where $n \simeq 1.85$, which from (9) and (11) yields $a \simeq 2.35$.

Type III bursts in the kilometer range were observed also, somewhat unintentionally, by Dunckel and Helliwell (1971) with their VLF magnetic fields detectors on OGO-3. The bursts appear first at the high frequency end of their 20–100 kHz records and then slowly expand to lower frequencies with a cut-off around 35 kHz. At 100 kHz, type III bursts last usually for more than one hour.

Measurements of the temperature using the theory of collisional damping yield usually coronal temperatures which near the sun are in agreement with the accepted values of about $1-4 \times 10^6$ K. At longer distances from the sun, however, they tend to fall much faster with distance than expected. Thus Hartz (1969) finds that the temperature decreases to about 5×10^4 K at a frequency of about 0.6 MHz, where the e-fold decrease of the radio burst is of the order of minutes. Alexander, Malitson and Stone (1969) have computed a temperature of 5.4×10^4 K at 0.45 MHz with a decay constant of about 120 sec. Haddock and Graedel (1970) measured a time constant of 23 sec at 3.5 MHz which gives a temperature of 2.6×10^5 K at a distance of only 6 solar radii, and Slysh (1967a) obtained a temperature of 1.7×10^5 K at 1 MHz with a decay time of about 150 seconds. As we mentioned in Section 3, all these authors neglected the logarithmic term in the expression for the collisional frequency. This leads to an underestimation of the electron temperature by a factor of approximately 1.5 at a distance of about 50 solar radii where plasma frequencies of the order of 0.5 MHz are expected to occur. Even this correction however gives a temperature for the coronal streamers which is considerably colder than the ambient electron temperature. Spacecrafts observation beyond the magnetosphere of the earth find the electron temperature to be of the order of 10^5 K during quiet periods and considerably higher during periods of enhanced solar activity.

A suggestion that the streamers are denser and cooler than the ambient solar corona so as to balance the pressure inside and outside the streamer could account for some of this temperature deficiency but finds it difficult to account for all of it. Another possibility, of course, is that due to the very low collision frequency, at large distances from the sun, collisional damping is replaced by hydromagnetic damping and therefore the expressions used can not yield the correct value for the coronal temperature. This is a very interesting subject that requires further theoretical investigations.

The intensity of the type III solar bursts seems to increase with decreasing frequency. Hakura, Nishizaki and Tao (1969) using the Alouette II data find an increase of an order of magnitude in the flux density (W m^{-2} Hz^{-1}) of the type III bursts as the frequency decreases from 10 to 1 MHz. Haddock and Graedel (1970) find an average peak flux density of 10^{-16} W m^{-2} Hz^{-1} at 3.5 MHz. Finally, Fainberg and Stone (1970a) from the analysis of a very large number of type III bursts find that type III bursts occur over a wide range of intensities. Some of the most intense ones stand above the galactic radio background by as much as 60 db, while others are bearly detectable above the background noise level. It seems also that the number of type III bursts increases tremendously at the lower intensities. Thus while Haddock and Graedel (1970) observed type III bursts with an average flux density of 10^{-16} W m^{-2} Hz^{-1} at the rate of approximately one per 50 hours of observation, Fainberg and Stone (1970a) at the level of 10^{-18} W m^{-2} Hz^{-1}, i.e. less than 10 db above the galactic background, detected approximately one type III burst every 10 sec. Actually at even lower energy levels there appear to be so many type III bursts that due to their longer duration in the hectometer range they all overlap forming a hectometric continuum.

Fainberg and Stone (1970a, b) call this a type III solar radio burst storm at low frequencies, and by following the active center on the sun they have found that these storms can last for at least half a solar rotation. An interesting result of their data was that the occurrence of type III bursts reached a maximum when the active region passes near the center of the solar disc, an observation which can prove very useful in the study of the emission pattern of the type III solar radio bursts.

An interesting result is also that the intensity of a type III burst might have ups and downs with frequency. Thus Hartz (1969) reports that though most of the type III bursts, observed at the low frequency end, had a counterpart at frequencies above 10 MHz observed from the ground, there were also several cases when bursts were observed in the one frequency range and not in the other. This probably is the result of electron density inhomogeneities along the coronal streamers which affect the intensity and the propagation characteristics of the bursts.

6. Summary

Observations of type III solar radio bursts with spaceborne radio receivers in the hectometer range have provided valuable data about the densities and temperatures of coronal streamers. They have also helped to narrow down the uncertainties in the velocity of the packets of superthermal electrons which excite the type III radio bursts as they speed through the outer corona along a coronal streamer. They have also stimulated many theoretical questions (relation of the emission frequency to the local plasma frequency, the effectiveness of collisional damping at larger distances from the sun, etc.) which need further study. The general picture that seems to emerge is that in an active region of the sun there is a continuous outburst of small packets of superthermal electrons which produce a very large number of weak type III bursts. Occasionally there are also larger outbursts of energetic electron, usually related to the eruption of a flare. These packets of energetic electrons follow the curved paths of coronal streamers which start over the active regions and extend into the interplanetary space to at least 50 solar radii. These streamers seem to have a higher density but a lower temperature than the ambient corona.

In summary this is a very exciting new field, where much more remains to be learned. As more observations become available we will be able to answer many of the theoretical questions and we will be able to build a much better picture of the outer corona and of the coronal streamers as they unfold in the interplanetary space between the sun and the earth.

References

Alexander, J. K., Brown, L. W., Clark, T. A., Stone, R. G.. and Weber, R. R.: 1969, *Astrophys. J.* **157**, L163.
Alexander, J. K., Malitson, H. H., and Stone, R. G.: 1969, *Solar Physics* **8**, 388.
Blackwell, D. E. and Petford, A. D.: 1966, *Monthly Notices Roy. Astron. Soc.* **131**, 399.
Doyle, R. O. (ed.): 1969, 'A Long-Range Program in Space Astronomy', NASA SP-213, Washington, D.C.
Dunckel, N. and Helliwell, R. A.: 1971, URSI Spring Meeting, Washington, D.C.
Fainberg, J. and Stone, R. G.: 1970a, *Solar Phys.* **15**, 222.
Fainberg, J. and Stone, R. G.: 1970b, *Solar Phys.* **15**, 433.
Fainberg, J. and Stone, R. G.: 1970c, NASA Goddard S.F.C., Report No. X-693-70-389.
Ginzburg, V. L.: 1961, *Propagation of Electromagnetic Waves in Plasmas*, Gordon and Breach, New York, N.Y.
Haddock, F. T. and Graedel, T. E.: 1970, *Astrophys. J.* **160**, 293.
Haddock, F. and Alvarez, H.: 1971, URSI Spring Meeting, Washington, D.C.
Hakura, Y., Nishizaki, R., and Tao, K.: 1969, *J. Rad. Res. Lab. of Japan* **16**, 215.
Hartz, T. R.: 1964, *Ann. Astrophys.* **27**, 831.
Hartz, T. R.: 1969, *Space Sci.* **267**.
Huguenin, G. R. and Papagiannis, M. D.: 1965, *Ann. Astrophys.* **28**, 239.
Kucker, A. F. and Sudan, R. N.: 1969, *Nature* **223**, 1049.
Lin, R. P. and Anderson, K. A.: 1967, *Solar Phys.* **1**, 446.
Newkirk, G. Jr: 1967, *Ann. Rev. Astron. Astrophys.* **5**, 213.
Papagiannis, M. D. and Huguenin, G. R.: 1964, *J. Geophys. Res.* **67**, 1307.
Slysh, V. I.: 1967a, *Soviet Astron.* **11**, 389.
Slysh, V. I.: 1967b, *Cosmic Res.* **5**, 759.
Stone, R. G., Weber, R. R., and Alexander, J. K.: 1966, *Planetary Space Sci.* **14**, 631.
Takakura, T.: 1967, *Solar Phys.* **1**, 304.

23. SUMMARY

LEO GOLDBERG

Harvard College Obs., Cambridge, Mass., U.S.A.

1. Introduction

Many courses have final examinations but never before has a single member of a class been selected to take the examination for all of his fellow students. At my own university, there is usually an interval of two weeks, known as the reading period, between the end of lectures and the beginning of final examinations. I warmly recommend that this custom be followed when future NATO Summer Schools are held in Greece.

In his opening lecture, Dr. Evans reviewed the many outstanding observational and theoretical developments that have transformed the investigation of the corona into a quantitative science. Until about 1930, coronal research was almost entirely limited to the observation and interpretation of brightness and polarization measurements of the white light corona and to vain attempts to solve the mystery of the identification of the coronal emission lines. Then, in the short space of about a decade, came one exciting advance after another, the most important of which were: (1) the invention of the coronagraph by Lyot, (2) the identification of the coronal lines by Edlén, (3) the recognition of the granules as the source of mechanical energy by Biermann and Schwarzschild, (4) Alfvén's classic papers on the behavior of solar ionized matter in a magnetic field, which represented the founding of magnetohydrodynamics, (5) the discovery of solar radio emission by Hey and by Southworth, and (6) the initiation of rocket spectroscopy by Tousey in 1946. Progress during the last two decades has been enormous, particularly in the five years since the last NATO Advanced Study Institute on Solar Physics was held not far from here at Lagonissi. In comparing the two NATO meetings, three major developments are worth noting: (1) the greatly improved understanding of the nature and role of magnetic fields in the corona, both in the inner regions and in the interplanetary medium at 1 AU; (2) the discovery that the temperature gradient in the transition zone between the chromosphere and corona is regulated by conductive heat flow from the corona; and (3) the success of new diagnostic techniques for the determination of electron temperatures and densities from XUV data obtained with rockets and satellites.

2. The Transition Zone

I shall try in this review to summarize what we have learned about the corona from the previous speakers, beginning with the transition zone and progressing outwards to the interplanetary medium. The transition zone has been under investigation ever since the high temperature of the corona was discovered, but its form and location have been terribly elusive until rather recently. Since the temperature in the photosphere decreases outwards, the transition zone must be preceded by a temperature minimum. A model characterized by a minimum temperature of 4300 K at $h \sim 500$ km appears to be consistent both with balloon measurements in the spectral region 100–300 μ and with rocket measurements in the far ultraviolet spectrum near 1500 Å.

The temperature increases gradually from 4300 K at 500 km to 8000 K at approximately 2000 km where, as Dr Noyes explained, the Lyman continuum data demand a steep rise in the temperature. Dr Athay showed how the temperature and density distribution in the transition zone could be determined from a combination of eclipse data and rocket and satellite measurements of solar fluxes in XUV emission lines. It was first proposed by Giovanelli and later demonstrated by Athay that the temperature gradient in at least the upper part of the transition zone is controlled by thermal conduction from the corona. More recently, Withbroe has shown that a very large body of data in the form of spatially-resolved UV intensities obtained with the Harvard OSO-IV experiment can be very well fitted by a model in which the corona is assigned a constant temperature of 2×10^6 K and the temperature gradient in the range $10^5 \leqslant T \leqslant 10^6$ is given by the equation

$$T^{5/2} \, dT/dh = \frac{1}{C}$$

where C is a constant equal to $10^{-12.0}$, corresponding to a constant conductive flux $F_C = 6 \times 10^5$ erg/cm^2/sec. However, the observational uncertainties are such that a slow variation of F_C with height is not unlikely. Below $T = 10^5$ K the derived temperature gradient is no longer consistent with constant conductive flux. Athay and Noyes remind us, however, that when dT/dh is determined by $F_C = $ constant $= 6 \times 10^5$, the temperature increases from virtually zero to 10^5 K in an interval of less than 2 km and point out that the usual ionization equilibrium can hardly apply. For example, an O v ion would on the average travel about 10 km by random walk before it could be ionized by electron impact. Hence, the static ionization theory used in the analysis is incorrect and the conductive flux may retain its constancy well below $T = 10^5$ K.

The structure of the transition zone has also been derived from rocket measurements of the ratio of limb and disk intensities, as reported by Dr Wilson, and from radio measurements of the quiet sun which were described by Dr Noci. In the past, the radio data have given discordant results but the newer work just reported here leads to a structure of the transition zone which is in remarkably close agreement with that derived from XUV data. The limb-to-disk ratios measured by the Culham group for optically thin lines are also consistent with the same model of the transition zone but optically thick lines of transition zone ions seem to be formed several thousand kilometers higher up, as would be expected if such optically thick lines are formed in spicules. Withbroe also shows that the spicules must be optically thick in the Lyman continuum and consequently that they attenuate emission lines seen at the limb when the lines are formed in the lower part of the transition zone and have wavelengths smaller than 912 Å. As Wilson emphasized, the three methods of analysis, based respectively on integrated fluxes from the whole disk (global fluxes, as Athay calls them), on resolved disk intensities and on limb-to-disk ratios, are complementary in revealing the height, shape, and irregularities of the transition zone in quiet regions. The radio data of course refer principally to active regions.

The coronal structure above active regions may now also be inferred from OSO-IV and OSO-VI data. Noyes and his collaborators have found that on the average the coronal temperature above an active region is about 25% higher than in the quiet equatorial zones and that the average electron pressure and conductive flux are greater than in quiet regions by a factor of about 5. Moreover, the higher electron density in active regions decreases the disparity between the electron temperature and the brightness temperature of the Lyman continuum by increasing the rate of collisional ionization and thereby diminishing the overpopulation of the ground level of neutral hydrogen. We have also seen the preliminary results of a much more detailed analysis based on OSO-VI data, which shows remarkable changes in the structure across individual active regions.

3. Thermal Balance

The information now available on the temperature structure of the chromosphere and transition zone provides a good basis for testing theories of the heat balance of the chromosphere-corona and of the relevant heating and cooling mechanisms. Our brief discussion of this question, which was led by Dr Schmidt, shows that the situation is far from satisfactory. It seems reasonably certain that there is a mechanical flux of at least 2×10^6 erg/cm^2/sec associated with acoustic noise from the granulation. This flux passes through the atmosphere unimpeded until a distance of a few scale heights is traversed, when shock waves are formed. About half of the flux, 1×10^6 erg/cm^2/sec, is dissipated into heat and radiated away by the low chromosphere. Another 3×10^5 erg/cm^2/sec is radiated into space by the upper chromosphere, chiefly in Ly-α, and about the same amount is lost by the corona. One of the real puzzles is the fate of the conductive flux F_C from the corona after it reaches the upper chromosphere. Athay argues that the conductive flux cannot be radiated away by the chromosphere, because (a) F_C cannot penetrate into the low chromosphere where most of the energy loss occurs in the form of Balmer radiation and (b) the existence of the transition zone is proof that the upper chromosphere is unable to radiate away the mechanical energy deposited in the transition zone and corona and is therefore also incapable of radiating away the inward conductive flux. Consequently, he and Kuperus have made the intriguing proposal that F_C is converted into spicule motions by heating and compressing the thin layer in which it is deposited. The spicules then return their energy to the corona. One also asks how the mechanical flux manages to reach the corona without being refracted back or absorbed by the transition zone? What is the role of magnetic fields and spicules in channelling the energy? Schmidt suggests that these and other questions about the heating of the corona might well be answered by an experimental determination of how the mechanical flux varies as a function of height.

4. The Density and Temperature of the Corona

We still have much to learn about the transition zone but a great deal of progress has been made since 1953 when Van de Hulst in his review article referred to it as a kind of no man's land. Turning now to the corona itself, we find a variety of excellent intensity measurements from which electron densities and temperatures may be derived. Noyes summarized the results obtained by Withbroe from an analysis of ultraviolet emission intensities measured at a constant height of 2 arc minutes above the limb on OSO-IV spectroheliograms. At this distance above the limb, the electron temperature is found to vary between 1.5–2.0×10^6 K, the higher values corresponding to active regions. The electron density shows considerably more variations, between 4×10^8 cm^{-3} and 3×10^9 cm^{-3} and individual values are in remarkably close agreement with those obtained from K-coronameter measurements on the same day. Since the former is an *rms* value and the latter an *average* value, the close agreement suggests that the corona is homogeneous in density, at least along a tangential line of sight at the given height.

Dollfus has also investigated temperature fluctuations in the corona, by comparing coronal structures in a series of forbidden lines formed at progressively higher temperatures, e.g., Fe x, Fe xiv, Ni xv, and Ca xv. He finds local differences in structure which imply large temperature gradients amounting to about 1% per 1000 km. In a given region, the temperature may also vary with time by as much as $\pm 1\%$ in 10 minutes. The overall average temperature found by Dollfus is 2×10^6 K, which agrees very well with Withbroe's results.

Dollfus and his collaborators have also made regular and prolonged measurements of the white light corona with a K-coronameter on the Pic-du-Midi. By this technique, coronal streamers can be detected as far away as 2 solar radii from the limb. From these synoptic measurements, it can be deduced that the total number of electrons in the corona varies by a factor of 5 between solar minimum and maximum. In recent years, the corona often has been absent in the vicinity of the South Pole for reasons that are not understood. Dr Meyer suggested that the height of the transition zone may be greater at the poles, but in that case the emission from the chromosphere should be much more intense than is observed. It seems more likely that the absence of the corona is simply a consequence of low density.

In the quiet corona, the observations can be represented by rather simple models of the temperature and electron density. In active regions, however, the situation is much more complicated as we have heard from Zirker, Monsignori-Fossi and Neupert. Zirker finds that the intensities of the forbidden lines of several ions of Fe, Ni, and Ca in active regions can be represented by a two-component model in which temperatures in the range 1–1.5×10^6 K are associated with a density of 1.1×10^9 cm^{-3} and material in the temperature range 2.5–5×10^6 K has a density of 3.6×10^9 cm^{-3}.

Monsignori-Fossi has analyzed observations of X-ray fluxes taken with Solrad satellites by calculating the theoretical spectral energy distribution as a function of temperature, electron density and emission measure. She derives a relation between temperature and electron density such that, as the temperature increases from 1.5 to 5 million degrees, the electron density also increases from 2×10^9 cm^{-3} to 10×10^9 cm^{-3}. These values are not in violent disagreement with those found by Zirker.

Most determinations of electron density in the corona are based upon measured intensities which represent averages along the line of sight and there is no sure way of judging the influence of inhomogeneities upon these measurements. In Neupert's lecture, we heard about a new method due to Gabriel and Jordan which makes use of the ratio of the intensities of the forbidden line, $2\,^3S-1\,^1S$, and the intersystem line, $2\,^3P-1\,^1S$, in helium-like ions observed in X-ray spectra. We saw some beautiful examples of such transitions observed from the OSO-V satellite as far up the isoelectronic sequence as Fe xxv. Gabriel and Jordan showed that, for sufficiently high densities, the ratio is sensitive to electron density and calculated the relationship for a series of ions from C v to S xv. Using observations from many sources, they find, for example, that N_e increases from 7×10^{10} cm^{-3} to 5×10^{12} cm^{-3} as T increases from 2×10^6 K to 6×10^6 K. They point out that the magnetic field required to contain a plasma at a temperature of 5×10^6 K at a density of 10^{13} cm^{-3} is about 400 G, which is consistent with fields measured in active regions. From the emission measure derived for one active region, they calculate a volume $V = 5 \times 10^{23}$ cm^3 which corresponds to a filament of length 30000 km and diameter 150 km. X-ray images of the sun show that the sizes of X-ray emitting regions decrease with increasing hardness but the observed condensations are much larger than a few hundred kilometers and therefore the derived volumes can only be explained if there are small unresolved condensations within the observed regions.

Neupert has applied the Gabriel-Jordan method to the X-ray spectra of solar flares for which both continuum measurements and the observation of high stages of ionization of iron seem to require temperatures in the range $20-40 \times 10^6$ K. The electron densities derived are extremely high and lead to rather interesting conclusions with respect to the magnetic fields and energy content of the flare-emitting regions.

5. Coronal Magnetic Fields

Dr Newkirk dealt with the role of magnetic fields in determining the structure and rotation of the corona, in influencing the solar wind and also in guiding the paths of transient events. The fields are so small that they cannot be measured directly by the Zeeman effect. Newkirk and Altschuler express the surface distribution of the field in the photosphere in terms of Legendre polynomials, whose coefficients are determined from observation. This distribution is used to get the distribution of the field above the surface. The inner corona is assumed to be current-free out to some radius $R = R_w$ where the solar wind forces the field lines to become radial and the flow begins. They set $R_w = 2.5\,R_\odot$ by comparing the calculated coronal structures with those observed. In the course of somewhat similar calculations, Schatten obtains $R_w = 1.6\,R_\odot$ by making the field lines at one AU agree with observation. The smaller value is also the maximum height at which U bursts and the tops of coronal loops are observed, but the disagreement with Newkirk is not at all serious considering the various simplifications in the theory itself, notably the neglect of flow at $R<R_w$.

The decrease of the field with height calculated by Newkirk may be compared with values inferred from observations of radio bursts, prominences, filaments and the interplanetary plasma. The empirical values vary from 10^3 G at the photosphere to 10^{-4} G at a distance of 100 $R_\odot$, but the scatter amounts to two orders of magnitude. One difficulty is that the field associated with radio bursts is not typical of the quiet corona. Nevertheless, it does seem that the field drops off more slowly with distance than the calculated value.

The calculated distribution of coronal magnetic fields has two prominent features: (1) there are many active regions with diverging field lines above them, which implies a net flux from active regions, and (2) field lines seem to connect active regions in the two hemispheres. Comparison of the calculated field lines with the corona of November 1966 suggests that streamers do not coincide with active regions as such but with magnetic arcades, extensive regions of magnetic field which tend to be aligned in the north-south direction.

Newkirk's approach is to make approximate calculations based on detailed observations of photospheric fields. In an alternative approach, Pneuman and Kopp assume a simplified model of magnetic field on the disk and then solve simultaneously the magnetic, dynamic and heat-transfer equations to obtain the volume current as a function of height. In this way, they calculate the shape of a coronal streamer resulting from a pair of active regions. These very promising calculations may help us to understand proton flares and the propagation of radio bursts through the corona. For example, a comparison of the calculated field with radioheliograph observations of a burst yields evidence that the shock front is being guided by the magnetic field.

Calculations of coronal magnetic fields make possible the understanding of a wide range of coronal phenomena and therefore it seems very important both to improve the theory further and to provide better coverage of photospheric magnetic fields by magnetographs and synoptic observations of coronal structure.

6. Non-Thermal Radio Radiation

In no context is knowledge of coronal magnetic fields more valuable than in the interpretation of the spectra of radio bursts, a subject that was well illuminated in the lectures by Kundu and Papagiannis. Certainly, one of the most pressing problems in all of solar physics is to discover the generating mechanism for solar bursts, of which five main types have been distinguished:

Type I bursts occur randomly with time over a very wide range of frequencies.

Type II are slow-drift bursts, associated with flares, and consisting of two frequency bands of emission that drift from high frequencies (about 600 MHz) toward lower frequencies in times varying from about 3 to 30 minutes. A type II burst is caused by a disturbance moving outward at about 1000 km sec^{-1}, which is generally believed to be a collisionless MHD shock wave covering a very wide solid angle and exciting plasma oscillations of the frequency appropriate to the local density as it moves out. The two bands are the first and second harmonics of the plasma frequency. The optical counterparts of type II bursts may be the flare-associated waves seen on the beautiful Hα films taken at the Lockheed Observatory and shown here by Sara Smith.

Type III are fast-drift bursts often excited by type II bursts but also occurring very frequently by themselves, over active regions. Tens of thousands of such bursts have been observed at hectometer wavelengths from space vehicles; many of the bursts are not associated with Hα flares. The radiation bursts come from streams of electrons moving out at relativistic speeds and exciting plasma oscillations of progressively lower frequency. They reach maximum brightness temperatures as high as 10^{11} K in times of about 1 second and decay exponentially with about the same time constants. Some bursts (the so-called U bursts) reach a frequency minimum and then turn upward. Evidently, the electrons are guided back to the sun along magnetic lines of force.

Type IV bursts consist of continuum radiation at meter and decameter wavelengths during and following large flares and are usually preceded by type II bursts. The radiation at meter waves seems to be due to synchrotron emission and at decameter wavelengths to plasma waves.

Type V bursts, which follow some type III bursts, are emitted in times of 1 minute or less as low frequency continuum radiation. They are believed to be emission from electrons trapped between the mirror points of arched magnetic field lines.

Both the complexity of these radio events and observational limitations make them very difficult to interpret. They are best studied now, at meter wavelengths with the Culgoora radioheliograph (single frequency, high spatial and time resolution), at decameter wavelengths nearly simultaneously at all wavelengths between 20–60 MHz (one-dimensional position and angular size of emission regions) and at hectometer wavelengths (0.2–5 MHz) with the radio-astronomy Explorer satellite. Kundu anticipates that when simultaneous measurements can be made at all frequencies with both high time and spatial resolution, the following information about the outer corona will be obtained: (1) velocities of the exciting disturbances, (2) the electron density profile of coronal streamers at distances out to 50 solar radii and (3) the temperature profile of the outer corona.

Preliminary results obtained so far from the analysis of type III bursts indicate that the average velocity of the exciting disturbance is about 0.37 c but with a large spread in values that may either be real or due to density inhomogeneities. If the value 0.37c is used, the average electron density in streamers is found to be 15 times the average density derived by Newkirk for quiet solar minimum conditions. The average electron temperature derived from the theory of collisional damping of bursts is in the range $1-4 \times 10^6$ K but at large distances from the sun the temperature falls faster than expected.

Before leaving the corona in the vicinity of the sun, I want to refer to the very detailed and thorough account by Dr. Dunn of green line coronal events seen on his beautiful motion picture films, although he could not possibly achieve, in the time available, his announced intention of teaching us all that he knows about coronal motions. His description of the properties of coronal arches and loops will be very important in keeping the theorists honest, as Zirker has already made clear in his comments on the theory of loop prominences. One interesting dilemma with respect to coronal loops was pointed out by Dr. Meyer. Along an open field line the density distribution will be in hydrostatic equilibrium but along a closed loop, if the pressure difference between the two foot points is greater than 0.5%, the flow along the loop will become supersonic and may be as high as 200 km per second. No such velocities have been observed but it also seems unlikely that the pressure at the two ends of a coronal loop can be so closely equal.

7. The Outer Corona (Interplanetary Medium)

Two problems in connection with the outer corona were discussed by Dr Schmidt: (1) the composition of the solar wind and especially an explanation of the differences, if any, between its composition and that of the inner corona and photosphere; and (2) the two anisotropies in the velocity distribution of ions and electrons as observed from satellites and space probes. The velocity distribution is non-Maxwellian in two respects. First, velocities parallel to the magnetic field are greater than those perpendicular to the field and second, a long tail is found in the distribution in the direction opposite to that of the field. The second anisotropy is greater for proton temperatures than for electron temperatures.

Following the original work by Chapman and Parker which proved the existence of the solar wind on purely hydrodynamic grounds, several new approaches have been made to its further theoretical understanding. These involve (1) adding a viscous transport term to the momentum equation, (2) dividing the energy flow into two parts: electrons and protons, (3) use of the Boltzmann equation, which is valid for both large and small mean free paths, and (4) introduction of the magnetic field in the radial direction. The last approach, by Jockers, neglects collisions, but assumes that electrons and ions are coupled by electrostatic forces which are induced by differences between the densities of ions and electrons of the same velocity. Clearly, the cooling will be more rapid in the direction perpendicular to the field than parallel to the field. Furthermore, the particles move in a combined electrostatic and gravitational potential field. The potential field for protons show a maximum at about 6 solar radii and hence only those with sufficient energy get into the solar wind. Therefore the Maxwellian distribution is deformed as required by observation. The question of what happens to the heavier ions needs to be studied before we can interpret the composition of the solar wind.

The corona at 1 AU has also been studied in novel fashion by Ness and Wilcox, who discovered the sector structure of the interplanetary field. Wilcox and Schatten have shown how the sector structure can be traced back to the distribution of magnetic fields in the solar photosphere. The photospheric fields determine the structure of the coronal fields and the latter are carried away by the solar wind. The boundary between sectors in the corona is always aligned in the north-south direction and extends across the equator into both hemispheres. As Wilcox described it, the sector structure rotates rigidly with a period of 27.03 days, which is equal to the period found in the photosphere at a latitude of 6–8 degrees. There is no good explanation of how the sector boundary can resist shearing by differential rotation over a period of 2–3 months. The latest work is establishing that solar activity is closely related to the sector pattern, e.g., flares are found to occur most often near sector boundaries between active and quiet regions and geomagnetic activity shows a sharp rise at a sector boundary followed by a decline to the next boundary.

The introductory reviews of physical processes in the corona by Drs. Noci and Meyer were invaluable in preparing us for the later lectures and I can only wish that I had the time to say more about them and about the interesting lectures by Drs. Anastassiadis, Macris, Tofani and Xanthakis. Fortunately, they will be published in the Proceedings. I must not fail to express to Dr. Macris and his associates our pleasure and gratitude at having been able to attend such a brilliantly conceived and well organized Summer School. I would like to close by jotting down a list of questions about the corona, most of which were asked but not answered by the lecturers. You may think of them as a problem set and needless to say, anyone who solves one or more of them has an excellent chance of being invited to the next solar institute in this beautiful country.

1. What is the origin of the corona and the solar wind? How does the flux of mechanical energy vary with height in the chromosphere-corona?
2. Does the chromosphere radiate with maximum efficiency and if so, what happens to the conductive flux from the corona? Does it drive the motions of spicules?
3. How can we map the magnetic field in the corona at different heights up to 1.5–2.0 solar radii?
4. How do the magnetic field and the velocity field affect the density profile along a streamer?
5. Is there an appreciable net magnetic flux from active regions? Do field lines connect active regions of opposite polarity in the two hemispheres?
6. How does the magnetic field affect the structure and thermal balance of the corona?
7. Does a significant fraction of the corona above active regions exist in superdense clumps with densities and gas pressures 10^4 times the ambient corona? If so, what physical mechanisms cause and maintain such a distribution?
8. How are particles accelerated and how do they make their way through the tangled web of magnetic lines of force and escape into the interplanetary medium? How are shock waves formed?
9. Are there differences between the composition of the solar wind and that of the inner corona and photosphere, and how may they be explained?
10. What is the source of the two anisotropies in the velocity distribution of ions and electrons in the interplanetary medium as observed from satellites and probes?
11. Why is the sector structure of the interplanetary magnetic field related to solar activity?

INDEX OF NAMES

Aarons, J. 268, 270, 271
Alexander, J. K. 297, 320, 322, 325, 326, 330
Alfvén, H. 35, 37, 333
Alissandrakis, C. VI, 169
Allen, C. W. 37, 38, 40, 220, 221, 227, 271, 311, 312, 314, 315
Allen, J. W. 17, 147
Alon, I. 279
Altschuler, M. O. 35, 66, 71, 76, 78, 123, 337
Alvarez, H. 326, 330
Aly, M. K. 146, 147
Athay, R. G. *36–65*, 142, 156, 195, 196, 200, 201, 205, 226–228, 278, 311, 313–315, 334, 335
Amenitskii, N. A. 272
Anastassiadis, M. 341
Anderson, G. 159, 166
Anderson, K. A. 297, 328
Angelopoulos, O. V
Avrett, E. H. 201

Babcock, H. W. 89
Balasubrahmanyan, V. K. 185
Balmer, J. J. 15, 335
Barrow, W. R. 272
Bartels, J. 68
Bastin, J. A. 270
Bates, D. R. 21
Batstone, R. M. 144
Becker, U. 156
Beckers, J. N. 58, 62, 202
Behring, W. E. 241
Beigmann, I. L. 258
Bely, O. 26, 147
Bessey, R. J. 64
Biermann, L. 5, 8, 219, 220, 333
Billings, D. E. 9, 11, 142–145, 149, 151, 237, 240, 263–265
Black, M. 229
Blackwell, D. E. 319, 326
Blaha, M. 147
Blake, R. L. 239, 246
Boardman, W. J. 142, 143, 149, 151, 240
Bohlin, J. D. 70, 74
Böhm-Vitense, E. 312
Boischot, A. 70, 175, 267
Boltzmann, L. 340
Bourgeois, P. V, *VII–VIII*
Bracewell, R. N. 268

Brandt, J. C. 75
Britt, C. O. 272
Broten, N. W. 271
Brown, L. W. 320
Bruzek, A. 116, 122, 123, 127, 129, 151, 154
Buhl, D. 271, 276
Bumba, V. 91, 92
Burgess, A. 6, 22, 23, 25, 104, 147, 221
Burton, W. M. 52, 53, 229, 231, 316

Canfield, R. C. 42
Castelli, J. 268, 272
Chandrasekhar, S. 64
Chapman, S. 5, 29, 60, 340
Charvin, P. 86
Chevalier, R. A. 147
Chiuderi, C. 50, 309, 311
Chiuderi-Drago, F. 279, 309, 310
Christiansen, W. N. 142, 267, 271, 279, 309
Cillié, G. G. 312
Clark, T. A. 320
Clavelier, B. 70, 175
Coates, R. J. 270, 272, 274, 277, 278
Colburn, D. S. 90, 93
Copeland, J. 272
Couterier, P. 93, 95
Covington, A. E. 271, 279
Cowling, T. G. 29
Culhane, J. L. 247–249, 251, 254, 255, 258

D'Angelo, N. 21
Dara, H. VI
Davidson, A. W. 270
Davis, L., Jr. 75
Delache, P. 14, 45, 61
De Mastus, H. L. 114, 116, 127, 129
Denisse, J. F. 267
Dennison, P. A. 76
Dietz, R. D. 39, 315
Dimitrakos, S. V
Dodson, H. W. 156, 157, 159, 161, 164, 167
Dollfus, A. *97–113*, 115, 118, 123, 140, 141, 145, 242, 336
Doyle, R. O. 320, 321
Dryagin, Yu. A. 272
Dubov, E. E. 38
Dulk, G. A. 79–81
Dumont, S. 42
Dunckel, N. 326–330

INDEX OF NAMES

Dungey, J. W. 35
Dunn, R. B. 40, *114–129*, 140, 141, 143, 145, 339
Duprée, A. K. 25, 26, 40, 45, 47, 48, 141, 147, 217, 223, 224, 228, 311, 313, 315

Eddy, John A. 86
Edlén, B. 1, 237, 333
Elwert, G. 23, 258
Evans, J. W. V, *1–12*, 66, 114–116, 118, 122, 125, 127, 128, 147, 168, 242, 246, 333

Fainberg, J. 295–297, 322, 326–329, 331
Fälthammer, C.-G. 35
Fan, C. Y. 76
Fedoseev, L. I. 272
Felli, M. *267–286*
Ferraro, V. C. A. 35
Firor, J. W. 279
Fisher, R. R. 81, 83, 152
Freeman, F. F. 249
Friedman, H. 3
Frost, K. J. 244, 247, 251

Gabriel, A. H. 144, 212, 219, 239, 242, 243, 249, 337
Garz, T. 223
Gingerich, O. J. 311
Ginzburg, V. L. 267, 325
Giovanelli, R. G. 37–40, 176, 334
Glencross, W. N. 266
Goldberg, L. V, 19, 26, 40, 45, 47, 48, 123, 141, 147, 192, 221, 223, 224, 228, 262, 263, 311, 313, 315, *333–341*
Goody, R. M. 270
Gorokhov, N. A. 270, 272
Graedel, T. E. 295, 297, 300, 322, 325, 326, 330, 331
Grant, N. 272
Griem, H. R. 14, 21

Haddock, F. T. 279, 295, 297, 300, 322, 325, 326, 330, 331
Hagen, J. P. 270, 272, 278
Hakura, Y. 331
Hall, L. A. 15, 223, 224
Hansen, Richard T. 76
Hartz, T. R. 295, 297, 321, 322, 325–327, 329–331
Harvey, J. W. 70, 71, 153
Harvey, Karen L. *156–167*
Hebb, M. H. 20
Hedeman, E. R. 156, 157, 159, 161, 167
Helliwell, R. A. 326, 330
Hey, J. S. 271, 333
Higgs, A. A. 271
Hinteregger, H. E. 46, 58

Hirayama, T. 147
Holweger, H. 15
House, L. L. 51, 238
Howard, R. 76, 89
Hudson, H. 262, 263
Hughes, V. A. 271
Huguenin, G. R. 319, 320
Hulst, H. C. van de 273, 277, 312, 326
Hundhausen, A. J. 75
Hyder, C. L. 86, 153

Ivanchuk, V. I. 71
Ivanov-Kholodnyi, G. S. 21, 39, 221, 224, 313–315

Jaeger, J. C. 268
Jager, C. de 38, 39, 219, 311
Jefferies, J. T. 10, 43, 146, 147, 153, 154
Jordan, C. 22, 23, 40, 104, 112, 144, 147, 212, *219–236*, 237–239, 242, 243, 248, 249, 255, 258, 337

Kai, K. 56, 70, 81, 82, 84, 167
Kakinuma, T. 279
Kalkofen, W. 203, 205, 206
Karimov, M. G. 122, 124
Kingston, A. E. 21
Kirschner, R. P. 141
Kislyakov, A. G. 270, 272
Kleczek, J. 114, 120–122, 127, 129, 153, 171
Kock, M. 223
Koeckelenbergh, A. 309, 311, 313
Komonovich, E. V. 39, 314
Kopp, R. A. 39, 51, 53–57, 59, 69, 71, 72, 75, 219, 338
Koyama, S. 313–315
Kreplin, R. W. 257
Kucker, A. P. 321
Kulsrud, R. M. 35
Kundu, M. R. 49, 66, 274, 277, *287–307*, 338, 339
Kuperus, M. 38, 39, 51, 53–57, 59, 62, 64, 71, 219

Labrun, N. R. 271, 279
Labs, D. 16
Lambert, D. L. 147
Landini, M. *257–266*
Leblanc, Y. 93, 95, 182, 183, 273, 274, 282
Leighton, R. B. 8
Leroy, J. L. 99, 115
Le Sequeren, A. M. 282
Levy, G. S. 76
Lin, R. P. 166, 297, 328
Lingenfelter, R. E. 70, 305
Livingston, W. C. 76
Low, F. J. 270

Lüst, R. 35, 153
Lyot, B. 107, 108, 114–117, 120, 122, 123, 128, 146, 333

Macris, C. J. *V–VI*, 35, 129, *168–178*, 341
Malitson, H. H. 322, 325, 326, 330
Malville, J. McKin 86
Mandel'stam, S. L. 258
Martijn, D. F. 48, 267
Maurice, E. 99
McCabe, M. K. 81, 83, 303
McWirter, R. W. P. 21
Meekins, J. K. 244
Meekins, J. F. 254
Menzel, D. H. 20, 42, 168, 312
Meyer, F. *29–35*, 123, 159, 336, 340, 341
Minnett, H. C. 271
Mitchell, F. H. 270, 272
Modisette, J. L. 75
Molchanov, A. P. 271, 278, 280
Monsignori Fossi, B. C. *257–266*, 336
Moreton, G. E. 156, 164
Munro, R. H. 212, 213
Murcray, D. G. 270, 285
Murcray, F. H. 270, 285
Musman, S. 8

Nagane, K. 271
Nakagomi, J. 263
Neckel, H. 16
Ness, N. F. 89, 90, 91, 93, 340
Neupert, W. M. 140, *237–253*, 254, 336, 337
Newkirk Jr., G. A. 4, 12, 35, *66–87*, 122, 123, 140, 141, 283, 328, 329, 337–339
Newstead, R. 274, 277
Nikol'skii, G. M. 21, 39, 221, 224, 313–315
Nishi, K. 263
Nishizaki, R. 331
Noci, G. *13–28*, *308–316*, 334, 341
Noyes, R. W. 140–142, *192–218*, 274, 277, 311, 334, 336

Obridko, V. N. 91, 92
O'Brien, P. A. 279, 281, 282
Orrall, F. Q. 10, 56, 115, 147, 153, 154
Oster, L. 221, 311–313
Osterbrock, D. E. 9, 38, 62, 219
Owaki, N. 144, 145

Papagiannis, M. D. *317–332*, 338
Parhtasarathy, R. 279
Parker, E. N. 5, 6, 153, 340
Pawsey, J. L. 267
Petford, A. D. 319, 326
Petschek, H. E. 31
Phillips, K. J. H. 248, 249, 251, *254–256*
Piddington, J. H. 39, 313, 314

Plumpton, C. 35
Pneumann, G. W. 69, 71, 72, 75, 76, 338
Pottasch, S. R. 39, 40, 44, 45, 48, 221–223, 225, 227, 255, 258, 262, 263, 313–315
Pounds, K. 242, 246
Preston, G. W. 268
Purcell, E. M. 14

Ramaty, R. 70, 305
Ramsey, H. E. 114, 156, 161, 165
Reber, E. E. 270, 273
Regemorter, H. van 15, 16, 222, 250
Reidy, W. P. 266
Richardson, R. S. 156
Riddle, A. C. 81, 83, 271, 279, 280
Roberts, W. O. 10, 40
Rösch, J. 99
Roy, J. R. 70, 71
Rozelot, J. P. 147
Rugge, H. R. 144, 151, 240, 246
Rusch, W. V. 270
Rust, D. 70, 71

Saiedy, F. 270
Saito, K. 71, 144, 145, 263–265
Salmanovich, A. E. 270
Schatten, K. H. 12, 66, 69, 71, 76, 79, 95, 96, 337, 340
Schatzman, E. 37, 38, 40, 219
Scheffler, H. 273, 281
Schmidt, H. U. 32, 35, 66, 335, 340
Schmidt, K.-H. 5
Schirmer, H. 38, 219
Schnable, G. 114
Schwarzschild, N. 8, 219, 333
Seaton, M. J. 20, 22, 23, 104, 221
Severny, A. 70
Sheeley, N. R. 208
Sheridan, K. V. 282, 283
Shimabukuro, F. I. 269, 271, 274–278
Shlouskii, I. S. 37, 39, 267, 314
Short, J. A. 270, 271
Simon, G. 41
Simon, M. 70, 270, 275, 283, 284
Skumanich, A. 42
Slijsh, V. I. 321, 322, 324–326, 330
Slobin, S. D. 270
Smerd, S. F. 79, 81, 279, 280, 301
Smith, S. F. 156–167, 339
Southworth, R. B. 333
Speer, R. J. 19, 234
Speybroeck, L. van 121, 122, 243, 244
Spitzer, L, Jr. 29, 51
Stacey, J. M. 269, 271, 274, 275
Staelin, D. H. 270
Stanier, H. M. 279
Stelzried, C. T. 94, 96, 270

Stewart, R. T. 306
Stone, R. G. 295–297, 319, 320, 322, 325–331
Straiton, A. W. 270, 272
Strezhneva, K. M. 271
Sturrock, P. A. 35
Sudan, R. N. 321
Suzuki, T. 147
Swartz, M. 240, 246, 254, 255
Swarup, G. 267, 271, 279

Takakura, T. 70, 321
Tandberg-Hanssen, E. 70, 71, 279
Tao, K. 331
Thomas, R. N. 38, 52, 59, 61, 63, 201
Tlamicha, A. 271, 274–276
Tofani, G. *267–286*, 341
Tolbert, C. W. 270, 272
Tonsey, R. 3, 4, 15, 240, 333
Tsobanoglou, A. VI
Tsubaki, T. 145
Tsuchiya, A. 271

Uchida, Y. 38, 159
Ulmschneider, P. 39, 43

Vaiana, G. S. 74, 143, 249–251
Vainstein, L. A. 258
Valdeze, J. 76–78
Valnicek, B. 156
Veisig, G. S. 271
Vernazza, J. E. 201

Waldmeier, M. 114, 128, *130–139*, 263

Walker, A. B. C. 144, 151, 240, 246
Walker, T. A. 270
Warburton, J. A. 267, 271, 279
Warwick, J. W. 295
Weaver, H. F. 272
Weber, E. J. 75
Weber, R. R. 320
Weiss, A. A. 303, 306
Weymann, R. 38
Whitehurst, R. N. 270, 272
Wilcox, J. M. 12, 76, *88–96*, 340, 341
Wild, J. P. 71, 79–81, 85, 166, 168, 280, 294, 299, 300, 306
Wildt, R. 313
Williams, W. J. 270
Wilson, R. 21–23, 140, 219–236, 334
Withbroe, G. L. 41, 52, 53, 141, 195–202, 215, 334, 336
Woolley, Sir R. 17, 37, 38, 40, 220, 227, 311, 312, 315
Wort, D. J. H. 270, 272
Wulfsberg, K. N. 270, 271

Xanthakis, J. V, 179–191, 242, 341

Yabsley, D. E. 267

Zaitsev, V. V. 306
Zheleznyakov, V. V. 271–273, 281, 306
Zirin, H. 39, 51, 70, 153, 168, 237, 251, 275, 315
Zirker, J. B. 20, 56, *140–155*, 336, 337, 340

ASTROPHYSICS AND SPACE SCIENCE LIBRARY

Edited by

J. E. Blamont, R. L. F. Boyd, L. Goldberg, C. de Jager, Z. Kopal, G. H. Ludwig, R. Lüst,
B. M. McCormac, H. E. Newell, L. I. Sedov, Z. Švestka, and W. de Graaff

1. C. de Jager (ed.), *The Solar Spectrum. Proceedings of the Symposium held at the University of Utrecht, 26–31 August, 1963.* 1965, XIV + 417 pp.

2. J. Ortner and H. Maseland (eds.), *Introduction to Solar Terrestrial Relations. Proceedings of the Summer School in Space Physics held in Alpbach, Austria, July 15–August 10, 1963 and Organized by the European Preparatory Commission for Space Research.* 1965, IX + 506 pp.

3. C. C. Chang and S. S. Huang (eds.), *Proceedings of the Plasma Space Science Symposium, held at the Catholic University of America, Washington, D.C., June 11–14, 1963.* 1965, IX + 377 pp.

4. Zdeněk Kopal, *An Introduction to the Study of the Moon.* 1966, XII + 464 pp.

5. Billy M. McCormac (ed.), *Radiation Trapped in the Earth's Magnetic Field. Proceedings of the Advanced Study Institute, held at the Chr. Michelsen Institute, Bergen, Norway, August 16–September 3, 1965.* 1966, XII + 901 pp.

6. A. B. Underhill, *The Early Type Stars.* 1966, XIII + 282 pp.

7. Jean Kovalevsky, *Introduction to Celestial Mechanics.* 1967, VIII + 427 pp.

8. Zdeněk Kopal and Constantine L. Goudas (eds.), *Measure of the Moon. Proceedings of the Second International Conference on Selenodesy and Lunar Topography held in the University of Manchester, England, May 30–June 4, 1966.* 1967, XVIII + 479 pp.

9. J. G. Emming (ed.), *Electromagnetic Radiation in Space. Proceedings of the Third ESRO Summer School in Space Physics, held in Alpbach, Austria, from 19 July to 13 August, 1965.* 1968, VII + 307 pp.

10. R. L. Carovillano, John F. McClay, and Henry R. Radoski (eds.), *Physics of the Magnetosphere. Based upon the Proceedings of the Conference held at Boston College, June 19–28, 1967.* 1968 X + 686 pp.

11. Syun-Ichi Akasofu, *Polar and Magnetospheric Substorms.* 1968, XVIII + 280 pp.

12. Peter M. Millman (ed.), *Meteorite Research. Proceedings of a Symposium on Meteorite Research held in Vienna, Austria, 7–13 August, 1968.* 1969, XV + 941 pp.

13. Margherita Hack (ed.), *Mass Loss from Stars. Proceedings of the Second Trieste Colloquium on Astrophysics, 12–17 September, 1968.* 1969, XII + 345 pp.

14. N. D'Angelo (ed.), *Low-Frequency Waves and Irregularities in the Ionosphere. Proceedings of the 2nd ESRIN-ESLAB Symposium, held in Frascati, Italy, 23–27 September, 1968.* 1969, VII + 218 pp.

15. G. A. Partel (ed.), *Space Engineering. Proceedings of the Second International Conference on Space Engineering, held at the Fondazione Giorgio Cini Isola di San Giorgio, Venice, Italy, May 7–10, 1969.* 1970, XI + 728 pp.

16. S. Fred Singer (ed.), *Manned Laboratories in Space. Second International Orbital Laboratory Symposium.* 1969, XIII + 133 pp.

17. B. M. McCormac (ed.), *Particles and Fields in the Magnetosphere. Symposium Organized by the Summer Advanced Study Institute, held at the University of California, Santa Barbara, Calif., August 4–15, 1969.* 1970, XI + 450 pp.

18. Jean-Claude Pecker, *Experimental Astronomy.* 1970, X + 105 pp.

19. V. Manno and D. E. Page (eds.), *Intercorrelated Satellite Observations related to Solar Events. Proceedings of the Third ESLAB/ESRIN Symposium held in Noordwijk, The Netherlands, September 16–19, 1969.* 1970, XVI + 627 pp.

20. L. Mansinha, D. E. Smylie and A. E. Beck, *Earthquake Displacement Fields and the Rotation of the Earth. A NATO Advanced Study Institute. Conference Organized by the Department of University of Western Ontario, London, Canada, 22 June–28 June, 1969.* 1970, XI + 308 pp.

21. Jean-Claude Pecker, *Space Observatories.* 1970, XI + 120 pp.

22. L. N. Mavridis (ed.), *Structure and Evolution of the Galaxy. Proceedings of the NATO Advanced Study Institute, held in Athens, September 8–19, 1969.* 1971, VII + 312 pp.

23. A. Muller (ed.), *The Magellanic Clouds. A European Southern Observatory Presentation: Principal Prospects, Current Observations and Theoretical Approaches, and Prospects for Future Research. Based on the Symposium on the Magellanic Clouds held in Santiago de Chile, March 1969, on the Occasion of the Dedication of the European Southern Observatory.* 1971, XII + 189 pp.

24. B. M. McCormac (ed.), *The Radiating Atmosphere. Proceedings of a Symposium Organized by the Summer Advanced Study Institute, held at Queen's University, Kingston, Ontario, August 3–14, 1970.* 1971, XI + 455 pp.

In preparation:

25. G. Fiocco (ed.), *Mesospheric Models and Related Experiments. Proceedings of the 4th ESRIN-ESLAB Symposium, held at Frascati, Italy, July 6–10, 1970.*

26. I. Atanasijević, *Selected Exercises in Galactic Astronomy.* 1971, XII + 140 pp.